PLANT PATTERNING

Structural and Molecular Genetic Aspects

Previous Books by Esra Galun

Frankel, R. and Galun, E. (1977) *Pollination Mechanisms, Reproduction and Plant Breeding*. Springer Verlag, 281 pp.

Galun, E. and Breiman, A. (1997) *Transgenic Plants*. Imperial College Press/ World Scientific, 376 pp.

Galun, E. and Galun, E. (2001) *The Manufacture of Medical and Health Products by Transgenic Plants*. Imperial College/World Scientific, 332 pp.

Galun, E. (2003) *Transposable Elements — A Guide to the Perplexed and the Novice*. Kluwer Academic Publishers (Springer Verlag), 335 pp.

Galun, E. (2005) *RNA Silencing*. World Scientific, 456 pp.

PLANT PATTERNING

Structural and Molecular Genetic Aspects

ESRA GALUN

The Weizmann Institute of Science, Israel

World Scientific

NEW JERSEY • LONDON • SINGAPORE • BEIJING • SHANGHAI • HONG KONG • TAIPEI • CHENNAI

Published by

World Scientific Publishing Co. Pte. Ltd.

5 Toh Tuck Link, Singapore 596224

USA office: 27 Warren Street, Suite 401-402, Hackensack, NJ 07601

UK office: 57 Shelton Street, Covent Garden, London WC2H 9HE

Library of Congress Cataloging-in-Publication Data

Galun, Esra, 1927-

Plant patterning : structural and molecular genetic aspects / Esra Galun.

p. cm.

Includes bibliographical references and index.

ISBN-13: 978-981-270-408-5 (hardcover : alk. paper)

ISBN-10: 981-270-408-6 (hardcover : alk. paper)

1. Plant pattern formation. I. Title.

QK731.G32 2007

567.918--dc22

2007016773

British Library Cataloguing-in-Publication Data

A catalogue record for this book is available from the British Library.

Typeset by Stallion Press

Email: enquiries@stallionpress.com

Printed in Singapore by Mainland Press Pte Ltd

This book is dedicated to my grandchildren,

Michal, Reuth, Guy, Nir, Ran, Shai, Chen and Dor

PREFACE

大學

物有本末、事有終始、知所先後則近道矣

Every entity has Roots and Tips of branches
And every entity has a Beginning and an End
The one who is putting his mind to the Former
and to the Latter
He is Approaching the WAY!

(From the DA XUE – The Great Learning, one of the Four Books of the Chinese cultural heritage)

"Alle Theile der Pflanze – Stengel und Wurzel ausgenommen – können auf die Form des Blattes zurückgeführt werden; sie sind nichts anderes als Modifikationen derselben."

All parts of the plant – except the shoot and the root – can be attributed to the structure of the leaf; they are nothing but modifications of leaves.

(From the Doctoral Thesis of Caspar Friedrich Wolff, submitted in 1759 to the University of Halle, Germany, in Latin; the German version is of 1812).
Places and times may have strange associations: shortly after this German version was published in Leipzig there was the great defeat of Napoleon at the "Battle of the Nations" ... in Leipzig).

These two mottos represent two extremely different lines of thought. In the Chinese motto, the tips of the branches and roots were used as metaphors. It was probably intended to bring up two issues. One is that the entities should be studied thoroughly, from the roots to branch tips. The other issue is that a thorough knowledge can lead a person to the WAY. The Way has several meanings. Kung fu-Tze (Confucius) meant by Way morality and value. The Way can also mean all the events happening between the first step and the final goal.

In the German motto, from Caspar Friedrich Wolff, the plant is not a metaphor. It serves as an example of an approach to understand the development of organisms. Wolff was a pioneer in this approach, although he was ignored in the recent literature on plant development. This literature cites the claim of Goethe that floral members are leaves that underwent metamorphosis. Goethe's claim was made several years *after* the statement of Wolff.

This book aims to serve to bridge the gap between the metaphoric approach of Da Xue and the "down-to-earth" approach of Wolff. I shall attempt to update a wide range of readers on what we know about the processes that regulate the formation of patterns in plants. I shall also indicate clearly what is not yet known. By "plants" I mean angiosperm plants. Other, lower plants, will be mentioned only rarely. The deliberations will be on different levels of complexity, such as single cells, meristems, tissues and plant organs (e.g. roots, shoots, floral members). Wherever applicable the use of methodologies such as genetics, molecular biology and structural analyses will be described, and I shall attempt to show how the application of these methodologies led to additional knowledge on the control of plant patterning.

As with my previous books (Galun and Breiman, 1997; Galun and Galun, 2001; Galun, 2003; Galun, 2005a) the present book is aimed at a wide range of readers. Basic knowledge of plant biology, genetics and molecular biology should be helpful for the readers. However, where required, the various studies on patterning and the methodologies that were used in these studies, will be introduced and explained. Thus, the book should be of interest and useful for the novice in molecular developmental biology as well.

Because the book is comprehensive with respect to its topics (dealing with different plant components), scholars who are engaged in the study of specific aspects of plant differentiation, may benefit from the overall view of the present book. On the other hand, this book should also be useful as a companion-text for courses on plant development.

I am grateful to my wife, Professor Margalith Galun, and to my son, Professor Eithan Galun, who read parts of the manuscript and provided useful remarks. I am also grateful to the staff members of my Plant Sciences Department for useful discussion, among them Professor Dan Atsmon and Professor Moshe Feldman, and especially Dr. Yuval Eshed, who provided very useful advice throughout the whole process of writing this book. I thank professor Dr. Rudolf Hagemann, of the University of Halle (Germany), for the German motto from thesis of C. F. Wolff. Thanks are due to Mrs. Suzanne Trauffer Ramakrishna for her professional typing. I very much appreciate the efforts of my publisher, World Scientific Publishing Co., especially of Ms. Sook Cheng Lim, scientific editor at WSPC, for the editing and production of this book. I acknowledge the professional work of the Graphic Arts Department of the Weizmann Institute of Science for performing very professional work on the figures and on the cover. I further acknowledge the permissions to use figures and tables that were granted by the respective publishers and authors. I provided the sources of these figures and tables below each figure or table and the full citations are listed in the list of references. These citations should mean that I am grateful for the consents of the publishers to use their copyright material in this book. My special gratitude goes to Dr. Nillie Weinstein. Dr. Weinstein showed great interest in my writing throughout my last four books and since then has become an inseparable component of my writings. She encouraged me to go on writing this book. I and the book benefitted immensely from her wisdom and knowledge, especially, but not only, in philosophy.

Esra Galun
Rehovot, Israel
7th April 2007

CONTENTS

Chapter 1

INTRODUCTION

Biological Patterning: From Awareness to Understanding

The attitude of humans towards patterns of plants and animals has gone through several transitions during the cultural evolution of man. First, there was primarily a bewilderment of nature, and this bewilderment concerned primarily the physical world and less the biological nature. Later came an epoch during which the developmental stages that result in mature animals and plants were followed and described. Only subsequently were experimental procedures employed in order to understand the processes and mechanisms that control development and lead to the formation of patterns in tissues, organs and whole organisms.

The following sentences from an ancient poem exemplify the phases of awareness and bewilderment of nature. The poem, in total, reports the fascination with the physical world and the greatness and omnipotence of God. Less emphasis is put on animals and only minor attention is directed towards plants:

> "... He causeth the grass to grow for the cattle and the herb for the service of man ... The trees of the Lord are full of sap; the cedars of Lebanon, which he hath planted; where the birds make their nests; as for the stork, the fig trees are her house. The high hills are a refuge for the wild goats; and the rocks for the conies ... O Lord, how manifold are thy works! ... So this great and wide sea, wherein are things creeping innumerable, both small and great beasts."
>
> *(The Book of Psalms, Chapter 104, Hebrew Bible, King James version)*

This Psalms chapter is attributed to King David, and was composed in about 1000 BC. Only several hundred years later did the early Greek philosophers (e.g. Thales of Miletus, ca. 636–546 BC, and his Ionian school) divert from pure mythology to questions regarding the evolution of nature. About 200 years later another Greek philosopher, Hippocrates of Cos (ca. 460–370 BC), known mainly as a physician, handled the development of organisms by rationalization. He based his claims about the formation of organisms on the then accepted three entities of nature: heat, fluidity and solidity. About 60 years later emerged a "giant": Aristotle (384–322 BC). Aristotle had a holistic approach that took into consideration the physical as well as the biological world. According to him there are levels of ascending entities: non-moving entities (e.g. rocks), plants, non-speaking animals and a speaking animal (man). Aristotle and his Liceum school introduced the concept of causality. For him, man is at the "top" of the hierarchy. The other entities of nature exist to serve man, who is the ultimate goal of the universe; hence, Aristotle was what we call now a teleologist, attributing a strive or even an intention to entities trying to attain a goal. Moreover, the teleological explanation was for him the key for the proper study of organisms. If we take an example from the plant world, the development of an orange tree is meant to attain an orange fruit.

Attributing purpose to created entities did not start with Aristotle. It can be deduced already from the Psalms poem: "... grass to grow for the cattle ... cedars of Lebanon ... where birds make their nests ... high hills are refuge for the wild goats ... the rocks for the conies." For the development of animals, Aristotle considered two possibilities. One was that mature animals are fully represented, structurally, in the young embryo. Then there is only growth and expansion. The other possibility, favored by Aristotle, was that new structures emerge during differentiation from embryo to adult. Aristotle did not deal only with principles but also dealt in detail with organisms, including the description of numerous plants. But, he was limited by the lack of "tools", thus his descriptions were based merely on what the naked eye can see.

After Aristotle, the study of the development of organisms, including plants, went into "hibernation" for about 1900 years. Novel tools are the key to progress in biological investigations. This shall be exemplified in several chapters of this book. Hence, a new era in biological investigation started in the early 17th century. In 1603 Federico Cesi founded in Italy the Licean academy, the first post-Renaissance academy that intended to promote scientific knowledge. The name of this Academy was derived from the lynx, which was considered to have very sharp vision. The new tool that provided the Academy with a much sharper sight was the magnification lens. Lenses could be used to look at celestial bodies (by the telescope) and this was performed by a prominent member of the Licean Academy, Galileo Galilei. Other members of the Academy utilized lenses (i.e. microscopes) to look into the details of organisms, including developing plants and their reproductive organs. The endeavors of the members of the Licean Academy in the early 17th century heralded the beginning of the modern era of plant development.

Microscopic studies were not only performed by the Liceans in Italy. Microscopic observations were also conducted by others, such as Robert Hooke (1635–1703) of England and Antony van Leeuwenhoek (1632–1723) of the Netherlands. Here I would like to retrieve a wrong assumption that I made in a previous book (Galun, 2003). My assumption concerned the philosopher Baruch (Benedict) Spinoza (1632–1677) who was also born in the Netherlands but made his living from the polishing of magnification lenses. Born in the very same year as Spinoza, van Leeuwenhoek lived twice as long as Spinoza. My (humoristic) assumption was that the life of Spinoza was shortened from the dust of polishing lenses for van Leeuwenhoek's microscope lenses. I was wrong! For van Leeuwenhoek, who was a businessman, microscopic observations were a hobby. But, he polished his lenses by himself!

I shall not describe the progress in this field of science during the recent 400 years, and refer the interested readers to relevant text books, should they want to learn more about this. But, it should be noted that during the last 100 years additional tools have become available. I shall mention only some of them. About 100 years ago genetics became a useful tool. This was followed by biochemical genetics and the discovery of the structure of DNA and its role in storage, memory and replication of information. Subsequently a plethora of molecular biology tools were developed.

Moreover, microscopic and ultramicroscopic techniques were vastly improved and served as very useful tools in the investigation of development. Finally, computers were recruited for these investigations.

Using these tools in various combination contributed towards an increase of our understanding of development in general and to a fast progress in plant patterning studies. Notably, there was always an interaction between studies on plants and studies on animals. Tools and methodologies developed for plants (e.g. genetics, awareness of transposable elements) were quickly utilized in studies of animals and *vice versa*. Moreover, in some specific cases, investigators of animals (e.g. E. Coen, E. Meyerowitz) shifted from studying animal development to studying plant development and took their tools with them. As we shall see below, the development of patterns in plants differs from that in animals, but there are also common aspects and even homologous genes. Consequently, studies on pattern formation in plants can benefit from knowledge obtained from animal studies.

Recommended Previous Books and the Outline of this Book

There are several outstanding texts that deal with development and pattern formation in animals and plants. Of these, I would like to mention three. The first is a book that focuses on plants, by T. Sachs (1991). Much of this book is based on the author's own study of tissue and organ formation in plants. Albeit that this book was written over 15 years ago before very important tools, especially in molecular biology, were developed, it is still useful. As noted above, the tools are of prime importance to biological investigation, and in the past 15 years there has been a "jump" in our understanding of pattern formation. For example, Sachs' book was written before the rather elaborate process of floral member patterning became an intensive subject of investigation. Nevertheless, this book provides a good background to plant patterning and illuminates important aspects of this issue.

There are two other, rather different books, that deal with development and pattern formation. While being very different, the two books were published almost simultaneously and by the same publisher (one by Oxford University Press and the other by both Oxford University Press and Current Biology). The book of E. Coen (1999), entitled *The Art of the Genes*, is rather unique: it recruits art, science and philosophy to explain pattern formation in plants and animals. The book provides information from genetic and molecular-genetic studies as well as uses the artwork of several masters (e.g. da Vinci, Dürer, Hals, Rembrandt, Velazquez, van Gogh, Picasso, Escher, Magritte and Robinson) to explain how a final, rather complicated pattern can evolve. Unfortunately, the figures, which are an important component of Coen's book, are without color, while color, as a metaphor, is a central issue for the deliberations of this book. The book is well written and has the typical flair of its gifted author. It stops short of providing detailed information on the development and pattern formation of the cells, tissue and organs of plants, but constitutes a very good introduction to these patternings.

The third book (L. Wolpert, 1998) has a unique authorship. The whole text was written by Wolpert, a world-renowned expert on development, but he was assisted by first-rate experts; moreover, each of the 15 chapters was reviewed by several specialists before being finally written by Wolpert. The many clear and colored illustrations in this book render even complicated issues understandable to the novice. The great majority of the book deals with animal development but there is also one chapter (out of 15) that is devoted to plant development. This chapter is based on publications until 1996 (with one exception, published in 1997). Since numerous and very relevant studies were conducted in plant patterning during recent years, an update and expansion on what is known in plant patterning is now desirable.

This book will start with an overview on patterning and indicate which are the main solved and unsolved questions in plant patterning. It will also point towards common and uncommon developmental regulatory mechanisms in animals and plants. The book is then divided into two main parts. Part A shall deal with basic issues: factors and regulatory mechanisms that are common to several types of cell-, tissue- and organ patternings. Plant hormones, plant cell division and cell cycles, and the plant's cytoskeleton will be briefly discussed. There shall also be a summary of plant phylogeny that led to flowering plants (angiosperms), as well as a brief discussion on the emergence of "novel" genes that affect patterning (i.e. gene duplication followed by mutation and selection). Finally, the novel approach of *systems biology* and its relevance to plant patterning will be discussed briefly.

Part B deals with specific kinds of patterning, and the patterning of tissues and organs are described (e.g. embryos, roots, shoots, leaves). The regulated transition from vegetative shoots to reproductive shoots (inflorescences) that is unique to angiosperms will be elaborated. Finally, the patterning of the typical organ for sexual reproduction in angiosperms, the flower, is handled in detail.

An Overview of Biological Patterning

By "biological patterning" it is meant the patterning of specific individual cells, groups of cells (e.g. tissues), organs, and whole organisms. In this book these patternings will be related to cells, tissues and organs of flowering plants (i.e. angiosperms). The awareness that all organisms are composed of cells emerged from the studies and rationalizations of the German botanist, Mathias Jakob Schleiden (1804–1881), and the German physiologist, Theodor Schwann (1810–1882). Although stationed in different universities, they arrived (in 1838/39) at the same conclusion: that all plants and animals are composed of cells. It was subsequently stated that all cells are derived from the division of previous cells. The conclusion of Schleiden and Schwann and the subsequent statement became known as the "Cell Theory". This theory was then the cornerstone of studies on the development of organisms.

In angiosperms, where cell movement is extremely rare (it occurs during pollination) and can be neglected, patterns are the results of one or more of the following cell activities:

> Cell division: Cells may stop dividing or divide at various frequencies and with various durations of the cell cycle; the division can either be into two similar cells, or asymmetric, resulting in cells of different sizes and forms; the plane of the division may be at different angles.

Changes of cell form: After division, cells may either become approximately isodiametric or attain specific forms, such as very elongated (e.g. the pollen tubes and cotton-seed hairs that can reach a length of several centimeters); their periphery may be smooth or highly indented (as in leaf-epidermal cells that appear as pieces of a jigsaw puzzle).

Changes in the composition of cells: The cells' contribution to pattern is dependent on the cytoskeleton (this will be elaborated in Part A of this book); the cell walls may attain different compositions and shapes, as is typical of tracheal cells.

Termination: Cells may attain their fully differentiated state as leaf-mesophyll cells that carry out the photosynthesis. Under regular conditions these are "terminal" cells that will cease further division, although after their isolation (as protoplasts) and culture, they can resume cell division (see Galun, 1981). Similarly, leaf cells, covering the mesophyll layers, and which under natural conditions are terminal, may be induced to rejuvenate in tissue culture, divide, and form flower-like structures. Terminal cells can be "eliminated" as root-cup cells that are shed to the soil.

Plant cells have three characteristics by which they differ from animal cells. In animals there is a clear separation of cell lines at a very early stage between cells that will form the sexual organs and produce the gametes and the rest of the cells that will form the vegetative organs of the animal. There is no such early "division of labor" in angiosperms. Another feature that separates angiosperms from animals is the omnipotence (or at least pluripotence) of plant cells. This is exemplified by the stem apical meristems that can yield, after due cell division, either vegetative or reproductive organs. The third characteristic feature of plants is that their cells are (almost always) surrounded by a rigid cell wall. Such cells walls do not exist in animal cells.

The Wolpert book, which deals primarily with animal development, introduced the various model systems by providing a "conceptual tool kit". The book focuses on animal embryos to indicate that development is the emergence of organized structures from an initial, very simple group of cells. The development is thus divided into five processes. These developmental processes are neither independent of each other, nor strictly sequential. The processes are classified in the animal embryo but human embryos served also as a reference. The first process that follows fertilization is a period of rapid cell division without increase in the total mass of the embryo. The second process of development consists of a spatial and temporal pattern of cellular activity in the embryo. The second developmental process also involves the laying down of the overall body-plan and the main axes of the embryo (e.g. anterior and posterior ends and dorsal and vertical sides). The axes can be seen as a coordination system to define locations in the embryo. This patterning process includes also the allocation of cells to the different germ layers (ectoderm, mesoderm and endoderm). The third process of embryo development consists of further morphogenesis and also involves cell migration. In the fourth developmental process differentiation of cells occurs, in which the cells become structurally and functionally different from each other. At least 250 clearly distinct types of cells were categorized in the human body. The fifth process is growth; meaning increase in size of tissue and organ initials that is produced by cell multiplication, increase in cell size and/or depositions of extracellular material. The above summary of embryogenetic phases can be recognized in birds and mammals; thereafter emerges the chicken or the newborn. This emergence signals, for animal biologists, the end of their main interest in development and pattern formation. Not so in angiosperms. It is the fully formed embryo (inside the mature seed) that is the starting point of the main interest of plant biologists in the patterning of angiosperms.

There is an additional feature that is of major importance in animal (embryo) development. This is the folding of sheets of cells. These foldings change the contacts between neighboring cells as in the formation of furrows during the differentiation of the neural system. Although recent studies in plants hint to communication between neighboring cells, the communications by which cells influence each other play a much more important role in animal differentiation. Similarly, programmed cell death (apoptosis) was revealed in plants, but in animals apoptosis has a major role in the differentiation of organs (for example in the development of hands and feet it leads to the formation of fingers and toes, respectively). This brings us back to the importance of the Cell Theory. We can now pose developmental questions in a more focused manner: How are genes controlling cell behavior? Is it possible to relate cellular behavior to gene expression? While this question may be regarded as simple, the answers are very intricate. We shall learn, during further discussions, that cells of the very same animal or plant react differently to the same gene products! Cells in different tissues and organs are already programmed in different manners, and this different programming renders them to react differently to the same additional gene product.

Wolpert (1998) stressed the distinction between genes that encode proteins that have "housekeeping" activities (as enzymes involved in the cells' metabolism) and genes that encode proteins that are tissue-specific or cell-specific and thus make cells different from one another. The latter proteins are the major players in differentiation and patterning while "housekeeping" proteins have no such roles. In both cases the levels of the respective proteins in a given cell is determined by what is generally termed "gene expression". But, behind this term are a plethora of mechanisms that control this expression. Gene expression shall be elaborated in Part A, below. How many genes are involved in the development of animal embryos? Wolpert (1998) gave rough estimates. His estimate was that for fly and vertebrate embryo-development there are between 1000 and 50,000 such genes; and he mentioned that the total number of genes estimated to be encoded in the human genome is about 80,000. These estimates clearly show that important genetic information was added since Wolpert's book was published, because after the human and the mouse genomes were sequenced, the total number of genes in the genomes was found to be less than half of Wolpert's estimate. Hence, the complexity of an organism is not only manifested by the number of its genes. Some plants, for which the genomic sequence is already established, have not much less genes than humans. Surely the number of genes is not the only measure of the ...complexity. This is an emerging issue, and outside the scope of this book.

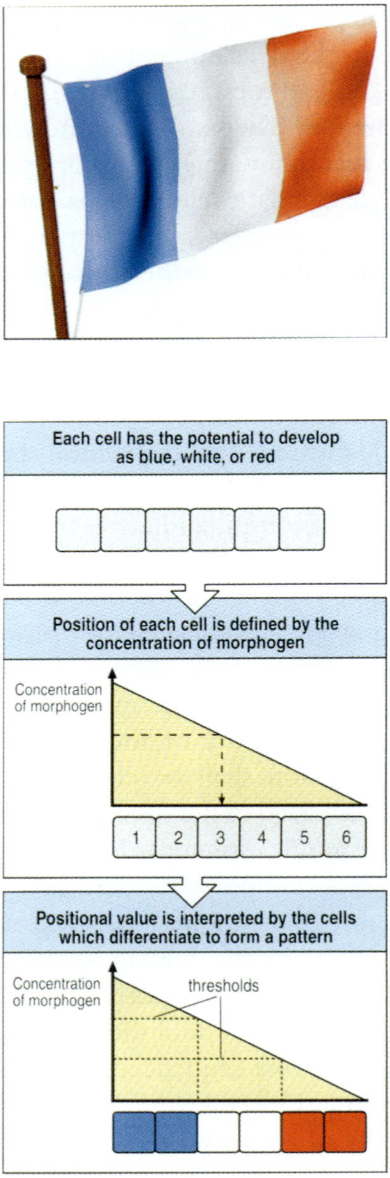

Fig. 1. The French flag model of pattern formation. Each cell in a line of cells has the potential to develop a blue, white, or red color. The line of cells is exposed to a concentration gradient of some substance and each cell acquires a positional value

The "conceptual tool kit" of Wolpert deals also with a model for pattern formation that is based on the French flag (see Fig. 1). It shows a color-based solution for position information that leads to acquisition of an *identity* and a *positional value* by cells. To demonstrate this model, a line of six cells is exposed to a gradient of a morphogen so that cell number one in this line is exposed to the highest level of the morphogen; the last cell in the line is exposed to the lowest level of the morphogen. The level of the morphogen is thus interpreted by the cell as its position. The two cells that are exposed to the highest morphogen level now "know" that they are at the anterior end of the line and differentiate into blue, according to the predetermined genetic program. The two cells exposed to the lowest level "know" that they are at the posterior end and differentiate into red.

Hence the model involves two requisites: the cells perceive their position by the morphogen level, and the cells have the capability to differentiate a color according to their position. This model also requires a threshold, meaning that as the level of morphogen is reduced gradually, there is an abrupt change in response (from blue to white and from white to red). This model seems abstract and not relevant to real patterning. But it is readily applicable for understanding patterning. We shall see, in the chapter of Part B that deals with the patterning of angiosperm flowers, that Coen and Meyerowitz (1991) applied a basically similar model to explain the differentiation of the four kinds of floral members in the model plant *Arabidopsis thaliana* (Arabidopsis). The patterning of Arabidopsis' floral members will be discussed at some length in Part B, Chapter 12.

---

defined by the concentration at that point. Each cell then interprets the positional value it has acquired and differentiates into blue, white, or red, according to a predetermined genetic program, thus forming the French flag pattern. Substances that can direct the development of cells in this way are known as morphogens. The basic requirements of such a system are that the concentration of substance at either end of the gradient must remain different from each other but constant, thus fixing boundaries to the system. Each cell must also contain the necessary information to interpret the positional values. Interpretation of the positional value is based upon different threshold responses to different concentrations of morphogen. (From Wolpert, 1998.)

The book of Enrico Coen (1999) intends to lead the reader to a basic understanding of "how organisms make themselves". We are fortunate that in Coen we have a combination of a talented and prolific scientist who has a keen interest and knowledge of art and philosophy, as well as the capability to write clearly for a wide range of readers. Before handling specific cases of development in animals and plants, Coen elaborates on his central metaphor. He leads the readers to the recognition that pattern formation in organisms is akin to an elaborate drawing executed by an artist on a canvas. The artist starts with a sketchy image and in a course of interactions between the existing drawing and the artist, the drawing passes through a process of additional details until the final, rather elaborate, picture is created. Hence, an organism is not the result of meticulously following a preexisting recipe or blueprint, as happens in the manufacture of a car, but rather a *creative* process of interactions. Only, in the development of an organism there are not two separate entities: the artist and the picture; in development, the organism is capable of *creating itself*.

To clarify another aspect of development in animals and plants, Coen returns to the metaphor of car manufacture. A car can be produced on the basis of a detailed manual of instructions. But for that a person is required who can read, understand and interpret those instructions. Obviously, the instructions themselves cannot interpret the instructions; the latter shall therefore not be able to manufacture the car. But an egg can manufacture its mature image. As for cars, manufacturing instructions and interpretation of the instructions by production engineers, I had a lesson in 1961. I was a post-doctoral fellow at the California Institute of Technology in Pasadena. Together with other post-doctoral fellows, I was invited to visit the assembly line of Ford cars in Southern California. We saw the huge hangar in which different car components were moved, overhead, to the main assembly line. It looked like many streams flowing into a main river. It was clear that each stream carried the same type of component (e.g. doors, wings, bodies) but the components were not identical in shape and color. When we reached the exit of the assembly line, the sequence of cars coming out was surprising. The cars were of different models and of different colors. We expected that, for some hours, exactly the same model and colour would exit the assembly line until another model would be assembled. The guide of the Ford company told us that the assembly line is marketing-oriented: each agent of Ford cars submits his order — an order may contain 20 different cars — and the assembly line is producing the cars order by order. We asked: "Why are there no mistakes, and does the 'stream' not occasionally bring a yellow door to a red car in the 'river'?" The guide had a short answer: "We don't have engineers who make such mistakes"; meaning that an engineer who makes a mistake in interpretation of the production instructions is fired right away. Well, coming back to biology, a mistake in the instruction or in their interpretation of patterning that concerns a vital tissue will kill the respective organism or prevent its ability to propagate. Nature, just as a manufacturing company, acts without mercy!

If an organism is likened to a computer, we have a case that the program (the software) is responsible for producing the hardware. In computers the software is completely separate from the hardware. And the hardware (the machine) has to be there *before* one can run the software, which has the instructions for building the hardware. We reach an impossible circular situation. In the development of an organism the program is recorded in the genome and the hardware is the organism itself. In the case of computers the solution is easy: the basic hardware is man-made. But organisms have to perform the development by themselves. One way out of this is to imagine a computer that can build itself, meaning that its software and hardware are interdependent. At least by some complicated mechanism such an imaginary computer would be able to modify itself. But this idea does not appeal to Coen, and he came up with an original idea of how development really proceeds; he contemplated how humans make things, and by "things" he primarily thinks of the production of art. Hence, the production of a work of art by a human artist became the favored metaphor for the process of the development of organisms. The continuous interaction between the artist (or the composer or the poet) with what is gradually being produced, does not serve Coen to *explain* development, but rather is intended to furnish a *viewpoint* that shall guide the reader through the intricate process of development, in order to better come to grips with the elaborate scenario of plant patterning.

An important facet for understanding development and patterning is a knowledge of what is happening inside the cells that undergo this process, meaning primarily the ability to follow in detail the changes in the activity of genes (and changes in the levels of gene products) that are causally related to stages of pattern formation. The molecular-genetic tools that enable such a follow-up started to become available only in the past three decades. These tools are still being improved by computer-assisted methodologies. Some of these tools are mentioned below (Part A), but it is recommended that readers, who are novices in this field and wish to acquire detailed knowledge on molecular-genetic methodologies, study relevant texts.

We shall witness throughout this book that two different cells may respond in very different ways to the very same outside stimulant. This means that a cell may be predisposed in such a way that it reacts to a given affector in a unique manner, or even does not react at all. For example, a mesophyll cell will not enter into a new cycle of cell division when exposed to a morphogen (e.g. a plant hormone, such as an auxin or a cytokenin) while the same morphogen will induce division in some other cells. Likewise a specific cell in the shoot apex of a stem may respond to a given protein (e.g. a transcription factor) that is furnished internally, while another cell at a slightly different location will show a different response to the same protein, or may not respond at all. We shall meet such specific predispositions of cells in examples of plant patterning that shall be detailed in Part B of this book.

The term *morphogen* will appear during several discussions of patterning. The term has a functional meaning: these are substances that flow and can affect the behavior of cells, or even specific regions of a cell. But, chemically they may be either defined or undefined because not all of them were characterized. The characterized morphogens are of very different kinds. They may be plant hormones (as noted above), messenger RNAs (mRNAs as in fertilized fly eggs), or proteins and protein complexes. The term was actually coined by a mathematician/computer expert, Alan M. Turing. Turing was a first-rate mathematician and a pioneer in computer science. He became famous because of his role in the Enigma project, during World War II, that deciphered the German code. Later he returned to mathematics to study *Hydra* development, and used this model for the development of algorithms of flow mechanics. Indeed a 35-page paper suggested that during *Hydra* development there is a flow of a morphogen. But neither the morphogen nor the channel of flow were characterized at that time (1952). Soon after this publication Turing committed suicide.

Some morphogens, such as the growth regulators, auxin and gibberellins, flow unidirectionally from a considerably distant source; others, such as brassinosteroids, are produced nearer their target cells. It is plausible that the same cell is affected by more than one morphogen, arriving from different directions. It is easy to conceive that a cell can sense the level of a morphogen, but does the cell sense the *direction* of the flow of a morphogen? Let us assume that the cell is capable of sensing a *gradient*. If this is correct, then in a cell that is exposed to a gradient of a morphogen, one end of the cell's envelope will sense a higher level of the morphogen than the envelope on the opposite side. An imaginary line connecting the two ends will then indicate the direction of the flow. Precise, specific and accurate analyses of morphogens, at the cellular and subcellular levels, which are not yet available, should furnish answers to the above-mentioned questions. We shall recall that morphogens such as auxin do not "pierce" the plant cell as an arrow: entering at one end and coming out at the other end. Rather, there are receptors at the cell's membrane that are affected by the auxin and transfer (by a rather elaborate process, as shall be discussed in Part A) the signal into the internal volume of the cell. In general, the details of the reactions of plant cells to a flow of morphogens are still enigmatic.

Solved and Unsolved Major Issues of Plant Patterning

The details of the major issues in plant patterning that have already been solved will be presented in Part B of this book. Here I shall provide only an overview of them. First, the structural issue of plant patterning — at the levels of the naked eye, the microscopic observation and the electron-microscopic observation — was solved; at least for model plants such as Arabidopsis and snapdragon (*Antirrhinum majus*). The intracellular structures, such as the cytoskeleton, were successfully analyzed; in many cases with the aid of specific labelling and confocal microscopy.

Gene expression, at the level of mRNAs and proteins was readily evaluated in cells and tissues. Major progress was made in the understanding of plant patterning by using genetic methods. By *forward genetics*, a plethora of mutants having specific phenotypes, concerning patterning, was isolated. Such mutants are appropriate material for further molecular and structural studies. In *reverse genetics* the investigator focuses on a certain sequence of nucleotides and can ask what role this sequence plays in such cellular activities as cell differentiation and plant patterning. This is done by mutating the specific sequence. There are various methods to achieve this. One method that emerged in recent years is RNA silencing (Galun, 2005a). In this method the sequence of nucleotides in the DNA is left intact but a specific transcript is destroyed (or very much reduced), or the translation of the transcript is strongly reduced. Also, the chromatin that encompasses a given nucleotide sequence may be modified to reduce the transcription of the respective nucleotide sequence.

The advantage of the RNA silencing procedures is that the silencing of a given DNA sequence can be regulated spatially (i.e. in specific cells and tissues) and temporarily (i.e. at different developmental stages). Thus the establishment of refined tools to investigate plant patterning is demonstrated in the combination of molecular-genetic methods, computer-based analyses and instrumentation that now enable us to screen changes in the expression of a vast number of genes in cells and tissues that undergo differentiation or have defined mutations. Such throughput analyses will be presented in Part B of this book, when dealing with specific cases of investigating patterning.

A final "solved problem" to be indicated in this overview is the availability of the full sequence of the DNA in the nuclear genome of several plants that serve as models for the study of plant patterning. Such genomic sequences are now available for Arabidopsis and rice (*Orysa sativa*); partial sequences are available for additional plants and the full sequences of additional plants are expected. Moreover, the respective sequence data is readily accessible to all investigators. The functionality of these genomes is progressively being understood and is very useful for pattern investigations. In parallel, genomic sequences and their functionality are available already in fungal organisms and several animals, such as worms (*Caenorhabditis elegans*), insects (*Aedes* sp., *Drosophila melanogaster*), fish (*Denio rerio*) and mammals (*Mus musculus, Homo sapiens*). The respective data are readily available by computer search and are of great help to investigators engaged in the study of plant patterning.

As for unsolved questions of plant patterning, we have a logical problem: it is difficult to specify all the mysteries of systems with which we are not fully familiar. In Part B several systems of patterning in cells and organs of plants shall be handled, and it will become evident that a lot of knowledge on these systems has been accumulated in recent decades, but in no system do we possess a complete understanding. A central enigma was already alluded to above when the basic ideas on development, put forward in the books of Wolpert (1998) and Coen (1999), were briefly reviewed.

It became evident that for a proper participation of a cell in patterning, the cell has to be *predispositioned*. Only predispositioned cells will respond correctly to internal and external affectors. Predisposition can take place only after the cells "know" their position in space (relative to other cells in their immediate and/or less immediate surrounding) and time (the developmental phase). How does a cell get to "know" its position? Wolpert (1996), in a "Perspective" review, addressed the issue of positional information in animals that was later elaborated on in his above-mentioned book (Wolpert, 1998). But clearly this question is still far from a complete answer.

I would like to mention one exception. The brown alga (*Fucus*) releases egg cells from the fronds of female plants and sperm cells from fronds of its male plants. Fertilization takes place in the seawater of the ocean (and can be performed in a Petridish). The fertilized egg cell settles on the substrate (rock, or the bottom of the dish). When polarized light is illuminating this egg, the predisposition takes place. Consequently, apical hairs will develop at the side of the egg that faces the source of light and rhizomes will develop at the opposite end of the fertilized egg. Interestingly, the fertilized egg "remembers" its orientation during several cell division cycles. This is evident because the development of apical hairs and rhizomes can be inhibited by high osmotic pressure in the medium. The high osmotic pressure does not inhibit cell division. When later the osmotic pressure of the medium is reduced, the apical hairs and the rhizomes develop at the sites that are expected from the light exposure (Galun and Torrey, 1969; Torrey and Galun, 1970). This is apparently a simple system of subcellular orientation, because a simple physical stimulation is involved; but in reality it requires an elaborate mechanism of light perception by cell membranes and further transduction of the light signal. Light and proper osmotic pressure are probably required for causing the predisposition that initiates the molecular-genetic process that leads to the development of apical hairs and rhizomes. After the fertilized egg cell "knows" that it has a dark substrate below it and sunlight from above, it is able to develop the rhizomes and the apical hairs at the proper sites.

When we are concerned with a group of cells, as in the apical meristem of a plant shoot, we are faced with another yet unsolved question: how does a cell know that it is located at a certain place in this meristem? It can be at the very upper tip or at the flanks of the apical dome, and consequently will be predisposed differently. We do have clues, such as the flow of auxin, but not detailed knowledge. One reason for the lack of precise knowledge, with respect to auxin flow, is that present analytical methods do not furnish a quantitative evaluation of auxin levels in individual cells. Notably, such an evaluation is possible for many defined proteins (as transcription factors) or RNAs (as mRNAs).

We should remember that for understanding development, it is not only important to know when and how the expression of specific genes is activated but also when and how the expression is stopped and the gene products are eliminated.

The importance of the correct sensing by a cell — *where* its position is — is central in patterning. A few examples will help to demonstrate this importance. For a cell to start differentiating into a trichome, it has to sense that it is a leaf-epidermis cell on either the upper (adaxial) or lower (abaxial) side of the leaf. Also, a cell has to sense that it is a leaf-epidermal cell before it will initiate division towards the formation of stomata. For other kinds of differentiations, such as the formation of tracheal cells, a cell must sense that it is located in a file of cells projected from the root or the shoot apical meristem that is destined to become vascular tissue. Likewise, in order to become predisposed for the participation in the formation of a floral member (as sepal, petal, stamen or carpel), the cell has to sense *where* in the dome of the reproductive shoot apex it is located.

How do cells achieve such a sensing of their location? In most cases this question does not yet have a full answer. We do know that cells communicate by sending signals to each other and in some cases (as shall be detailed in Part B), protein, RNA or hormonal signals were identified. In specific cases signals can move a long way, as in the defense against viral infection, when the movement of RNAs complexes were found to travel along the plant. In other cases, as photoinduction of the change from vegetative shoot apices to reproductive apices (i.e. inflorescence), there are indications of a movement of a signal from a photo-induced leaf to the shoot apex. But the signal is not fully characterized. Long- and short-distance movements of RNA and proteins in plants can now be traced by elegant procedures (see Kim, 2005 and Aoki *et al.*, 2005).

In summary, while vast progress has been made in recent years with respect to plant patterning, there is a long way to go to full understanding. The central question — how a three-dimensional plant pattern emerges from linear information, stored in the chromosomal DNA — still awaits clarification. An almost full answer to this question does exist in a subcellular entity: the T_4 bacteriophage. The host of this phage is the *Escherichia coli* bacterium. Like other T-even phages, it has an icosahedral head (with a diameter of about 700 Å). The head consists primarily of three proteins. The phage contains also a long tail, a connecting neck with a collar and long "whiskers" (tail fibers) and a complex base plate (see Fig. 2). The head contains dsDNA, in which all the genetic information of T_4 is stored. Figure 3 demonstrates the genetic map of T_4 (since the publication of this figure, additional details were provided).

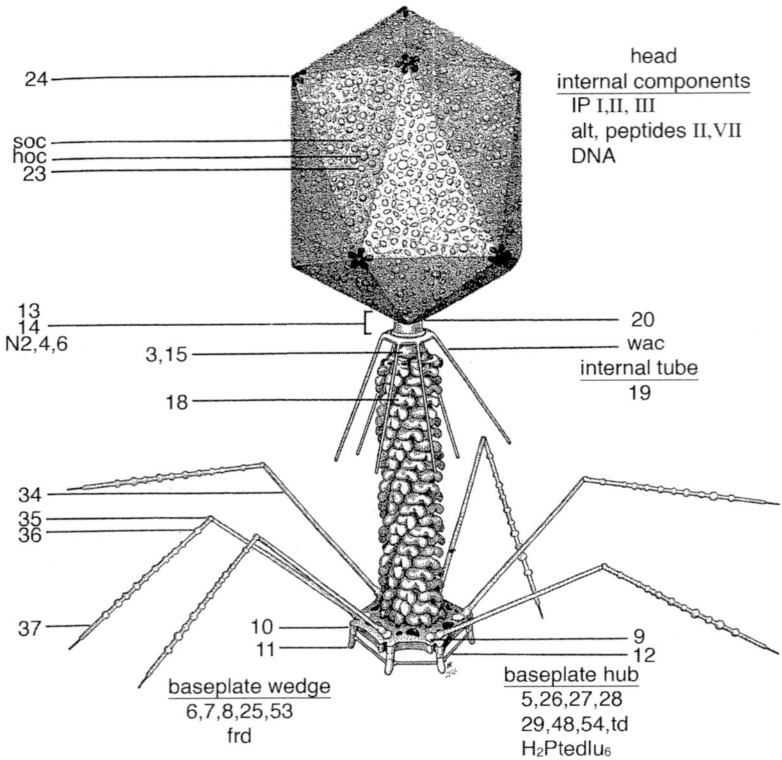

Fig. 2. Structure of the T4 virion based on electron microscopy at 2- to 3-nm resolution. The locations of protein components are indicated by gene number except for several unknown connector proteins, called N 2, 4, and 6, and the baseplate component dihydropteroyl hexaglutamate, called $H_2Pteglu_6$. The portal vertex composed of gp20 is attached to the upper ring of the neck structure, inside the head itself. The internal tail tube is inside the sheath and itself contains a structural component in its central channel. The baseplate contains short tail fibers made of gp12; these are shown in a stored or folded conformation. (From Mosig and Eiserling, 1988.)

During infection of the host bacterium, one or more of the tail fibers contact the lipoproteins of the host's cell wall and the phage is positioned perpendicularly to the surface of the host. The base plate approaches the cell wall, the tail contracts and causes the tube, which is inside the tail, to pierce the bacterial cell wall. The content of the head (mostly dsDNA) is then injected into the bacterial cell. Initially

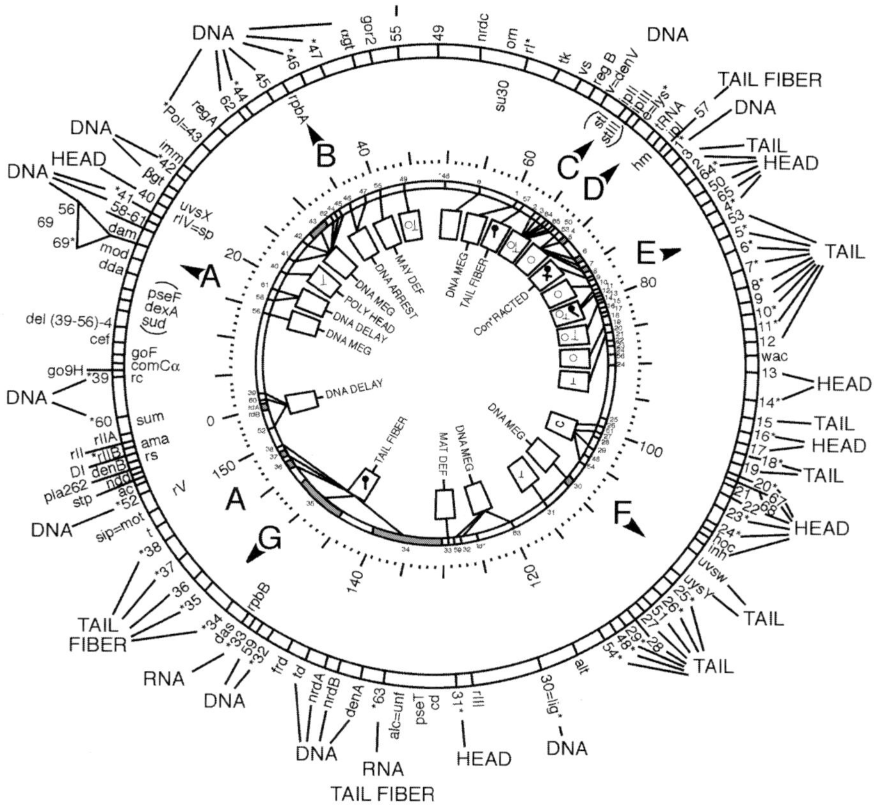

Fig. 3. A map of the known T4 genes. Distances are based on recombination frequencies (*inner circle*), on electronmicroscopy and restriction enzyme analysis (*middle circle*), and on the probability of packaging cuts (*outer circle*). The arrowheads labeled with large capital letters point to origins of DNA replication. (From Mosig and Eiserling, 1988.)

polymerases of the host are utilized by the phage, but with respect to all later metabolism and the construction of the phage progeny, the phage is autonomous and based on the genetic information of the phage.

The construction of new phage particles is a very elaborate process. In a way it is akin to the production of cars in the Ford assembly plant mentioned above. Different components, such as heads and baseplates with tails, are assembled separately and then put together until the mature phages are constructed. Again, one can visualize the process as streams coming together into a main river. The process is exemplified by the assembly of the head (see Fig. 4) and the tail (see Fig. 5). The numerous proteins required for the construction of the phage components interact in an orderly manner and then the components are combined. All this is orchestered by a set of several hundred phage genes, (almost) all of which were characterized and sequenced. But this orchestra is without a known conductor! Some additional details on the assembly, protein components and three-dimensional structure of T_4 are still being furnished by novel methods (such as cryoelectron microscopy). Despite this ongoing study, we already have a fair picture of how the linear information in the DNA

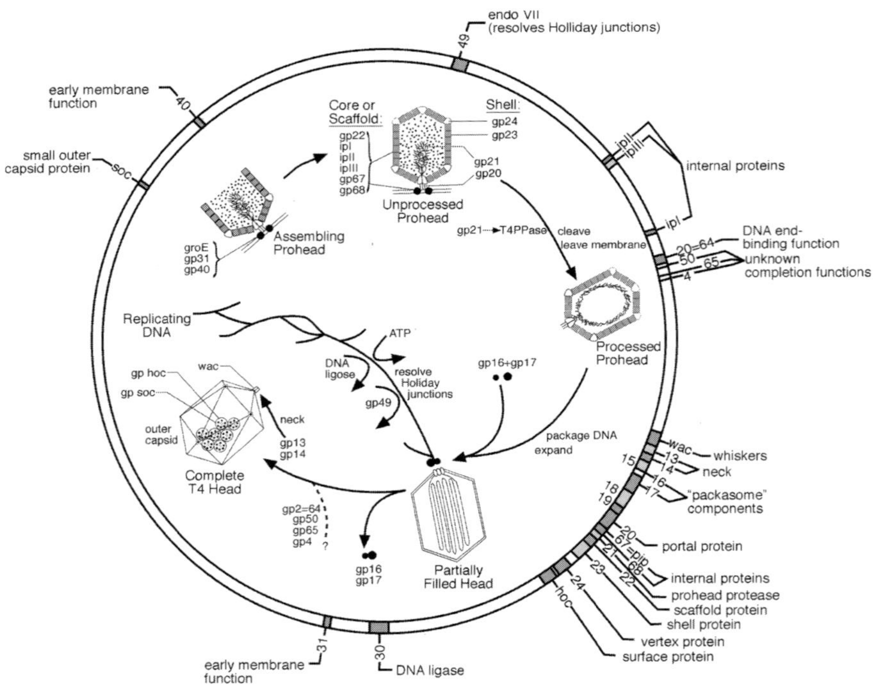

Fig. 4. Assembly and maturation of the T4 prohead and location of genes that control these processes. Interspersed tail genes are indicated on the inside of the circle. (From Mosig and Eiserling, 1988.)

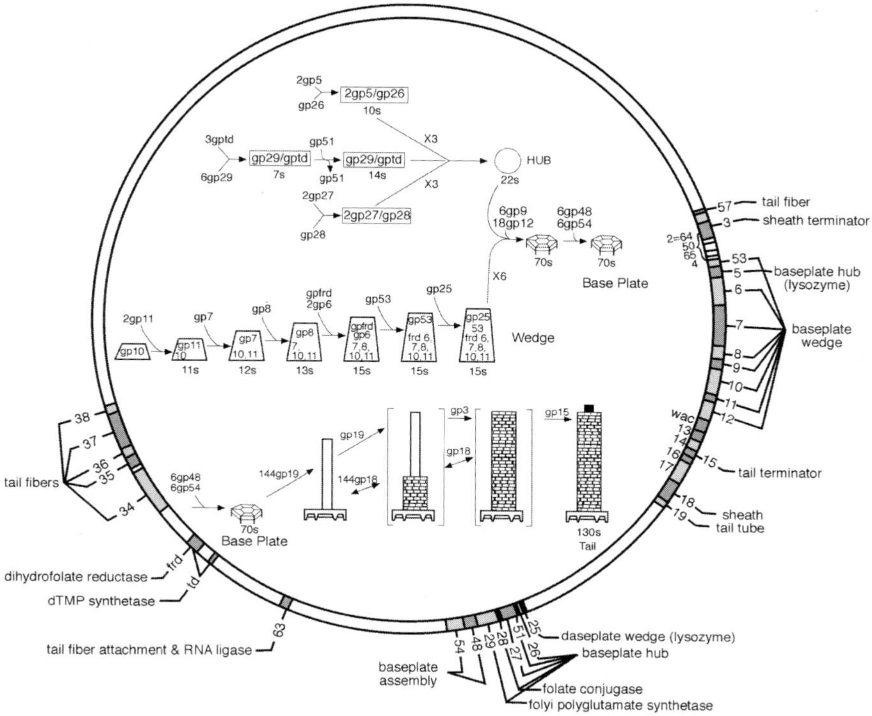

Fig. 5. Assembly pathways for both the hub and wedge parts of the baseplate, and the steps in the tail assembly, showing the positions of genes for tail and tail fiber synthesis and assembly. Head genes interspersed among tail genes are shown inside the circle. (From Mosig and Eiserling, 1988.)

of T_4 leads to a three-dimensional structure. Moreover, the thus constructed phage has several functional capabilities: it can recognize its host (*E. coli* bacteria), inject its DNA into this host, and is capable of replication.

Patterning in Angiosperms versus Patterning in Animals

Meyerowitz (2002) discussed the comparison between plant and animal development. He briefly traced the evolution of these two kingdoms of organisms as the basis of his consideration, and noted that the earliest signs of life on Earth are from about 3.8 billion years ago;

eukaryotic cells appeared about 2.7 billion year ago (as indicated by hydrocarbon biomarkers); the last common ancestor of animals and plants, which already went through the endosymbiotic inclusion of mitochondria from an alpha proteobacterium, is estimated to have evolved about 1.6 billion years ago. Thereafter was the phylogenetic split: the second endosymbiotic event was the uptake of the chloroplast from a cyanobacterium-like single-cell organism. Evidence for multicellular plants and animals indicates that these organisms emerged only after several hundred million years, about 600 million years ago. Therefore, the evolutions of multicellular plants and multicellular animals occurred separately.

By patterning, we mean development of multicellular structures. Hence, it is evident that the mechanisms that regulate patterning in plants and patterning in animals evolved separately. It is thus no wonder that plants and animals use different regulators for patterning. Nevertheless, animals and plants have some surprising similarities in the overall logic of development. A metaphor comes to mind. In my previous book I mentioned the metaphor of the mosque in Acre that was built by Achmad (Gàzar) Pecha, the Turkish ruler of Lebanon and Syria, in 1781. For building his mosque Achmad Pecha used different pillars that he collected from ruins of old Roman temples along the Eastern coast of the Mediterranean Sea. The old pillars were modified to fit their new purpose. This metaphor shall be coined as "The Pillars of the Mosque of Acre". Here I shall modify the metaphor. Let us assume that a church was also built, north of Acre, and it was revealed that the builder of this church also used modified pillars from the very same ruined Roman temples. Now we have two rather different entities, but both are based on very similar components. Like other metaphors, this one does not fit exactly to reality.

As for patterning, an important component of plant patterning that we shall meet below are MADS box genes: a family of genes encoding transcription factors. These are master-regulatory genes that are active in several plant patternings as the specification of the radial pattern of flowers (in Arabidopsis and other plants). While very few MADS box genes were also detected in animals, the latter are not the common master-regulators of pattern in animals. Animals use family members of the HOX homeobox genes for patterning. The latter genes have no sequence similarity to MADS box genes. Plants also have a few HOX-like genes, but again, these genes are not the master-regulators of plant patterning. It thus appears that both HOX-like genes and MADS-like genes existed in the last common ancestor of plants and animals but then they attained different roles in these two groups of organisms.

The metaphor on the mosque and the church and the ancient Roman pillars is thus relevant to the patterning of multicellular organisms. Meyerowitz (2002) listed several additional animal and plant proteins that have specific roles in the patterning of these organisms. The respective amino acid sequences in these proteins from the two groups of organisms can be similar, but the role of these proteins are different in these two groups. There are also proteins in plants that have a role in patterning but have no equivalents (in sequences or in roles) in animals. For example, there are no genes for phytochrome in animals. It is assumed that phytochrome-encoding genes were introduced into the plant genome by "horizontal transfer" from cyanobacteria about one billion or more years ago. These genes were then utilized in plants for the perception of red and far-red light, in a cascade of mechanisms that control growth and differentiation in response to red/far-red light.

Plants also contain genes that encode ethylene receptor proteins and no equivalents to these genes were found in animals. In plants, ethylene is a growth/differentiation hormone. Ethylene probably also has a role in pathogen response in plants. It was suggested that the ethylene receptor-encoding genes, also derived from cyanobacteria and by "horizontal transfer", were introduced into the plant genome. As with the phytochrome genes, it is plausible that such a transfer could have happened after plants diverted from animals, but before they became multicellular organisms, about one billion or more years ago.

Part A

BASIC ISSUES RELEVANT TO PLANT PATTERNING

Issues such as plant hormones, the cytoskeleton of plant cells, cell division, angiosperm phylogeny and systems biology as well as the means of regulation of gene expression in plants will appear repeatedly when the patterning of specific plant cells, tissues and organs is handled in Part B. Therefore, Part A shall provide the main known characteristics of these issues. The scope of this book does not allow a comprehensive description of each of these issues. Hence the features of these issues that are related to patterning will be presented only briefly. For additional information the readers shall be referred to specific literature, such as books and reviews.

Chapter 2

PLANT HORMONES

Plant hormones are also termed plant growth regulators or phytohormones. Evidence of their existence was provided about 100 years ago. These hormones are relatively simple chemical components and all of them, but one group (brassinosteroids), are very different from animal hormones. Consequently, their modes of regulating plant cell activities are very different from the modes of action of animal hormones. The first recorded clue for the existence of a plant hormone was provided by Charles Darwin and his son, Francis Darwin. They found that when dark-grown oat seedlings were exposed to lateral light, the seedlings bent towards the light. A signal from the seedling apex seemed to be involved in this bending. Later, this signal was found to be indoleacetic acid (IAA). We shall see below that most of a plant's hormones are involved in the regulation of more than one process of growth and development. Also, in many cases, several types of plant hormones interact in one process of growth and development.

Ethylene

As narrated by Crozier *et al.* (2000), the first clue that ethylene can regulate plant growth (etiolated pea seedlings) came from greenhouse experiments conducted in St. Petersburg in 1886 (!). For many years it was not clear whether ethylene was a natural and endogenous growth regulator or only the application of ethylene caused morphological alterations. The identity of ethylene in plant tissue and its release from plants followed appropriate analytical methods (e.g. gas chromatography) that became available in the early 1960s. Meanwhile, it became evident that the application of ethylene caused many physiological effects and changes in morphogenesis. Among these were the facilitation of senescence in flowers and fruits, abscission of leaves, and changes of sex expression. The latter phenomenon, the suppression of stamen development during the very early stage of cucumber floral bud differentiation, was a subject of my own research in the late 1950s and early 1960s (summarized in Frankel and Galun, 1977).

The first clue for such an ethylene effect came (again) from greenhouse experiments, but this time by Minina (1938) and conducted in Kiev. She found that flue gas (which was later found to contain ethylene as the main effector) caused femalization of cucumber flowers. Minina substantiated her results in further studies, but the verification of her studies by investigators outside the Soviet Union was recorded only many years later by myself and others (e.g. McMurray and Miller, 1968), when, rather than treating plants with gaseous ethylene, plants were treated with 2-chloroethylphosphonic acid ("ethephon"). The latter compound decomposed inside the plant to ethylene.

While the application of the ethylene-releasing compound suppressed stamens and caused femalization of the floral buds of cucumber, antidotes that inhibited the impact of ethylene, such as aminoethoxyvinyl glycine (AVG), caused enhanced maleness: stamens appeared in flowers of genetical-female cucumber plants. Why this happened was not revealed at the time, but, in practice, it was a very significant phenomenon because changing the sex expression was very useful in the production of hybrid cucumber seeds.

Minina was a pioneer in the study of sex expression in plants. However, in 1948, during the All Union Congress on Agricultural Sciences, organized by T.D. Lysenko in Moscow, her career was curtailed. Minina's husband was one of the few Soviet geneticists who resisted the demand of Lysenko to declare that Mendelian genetics was nonsense. In response, Lysenko sent the couple to northern Siberia. Minina was assigned to study the sex expression of northern forest trees. Minina fulfilled her assignment, and in 1979 she published a book on her studies. Ironically, after Lysenko did not succeed to "train" wheat to resist cold climates, he himself was removed from office in 1965 and probably died in cold Siberia.

These events remind me of the very old passage of Hillel, a Jewish scholar in Jerusalem (ca. 60 BC to ca. 10 AD). Hillel saw a skull floating on the water and spoke to the skull: "Because you killed people, others killed you and those who killed you will also be killed" (Mishna, Avot, Chapter 2/7). As an exception, Minina was allowed to attend the Botanical Congress in Leningrad in 1975, where I met her. I wondered why she was still confined to Siberia while Lysenko was deposed 10 years ago. I received a convincing answer: Minina and her husband were dissidents, and in the Soviet Union dissidents should stay in Siberia forever.

Here is a final remark on my favorite scholar Hillel. While generations of scholars searched for happiness of the individual, Hillel had the whole of humanity in mind. To attain harmony among humans he had one rather modest demand: do not do to others what you do not like to be done to yourself! If we could abide by Hillel's wisdom, the world would be completely different. For human harmony, love to all is not essential, but avoidance of bad deeds to others is crucial!

Synthesis of ethylene and its elimination

Plants contain a "methionine cycle" in which methionine is generated. In the presence of ATP, methionine, adenosine and the enzyme S-adenosyl-L-methionine (SAM) synthase, the compound SAM is synthesized. In a further step, 1-aminocyclopropane-1-carboxylic acid (ACC) is produced by the enzyme ACC synthase. Finally, ACC is oxidized by ACC oxidase to ethylene (see Fig. 6). By changes in the metabolic pathway that leads to ethylene, plants can also deplete the ACC pool and reduce the rate of ethylene production. The enzyme ACC synthase is probably the limiting step in the production of ethylene. It was found that the level of ACC synthase was elevated by external factors, such as wounding and lithium chloride, as well as during climacteric fruit ripening. There are several isoforms of ACC synthase. These differ among plants, but the same plant may contain several isoforms of this enzyme, and the level of each of these may be increased by different effectors (e.g. wounding, fruit ripening and stresses).

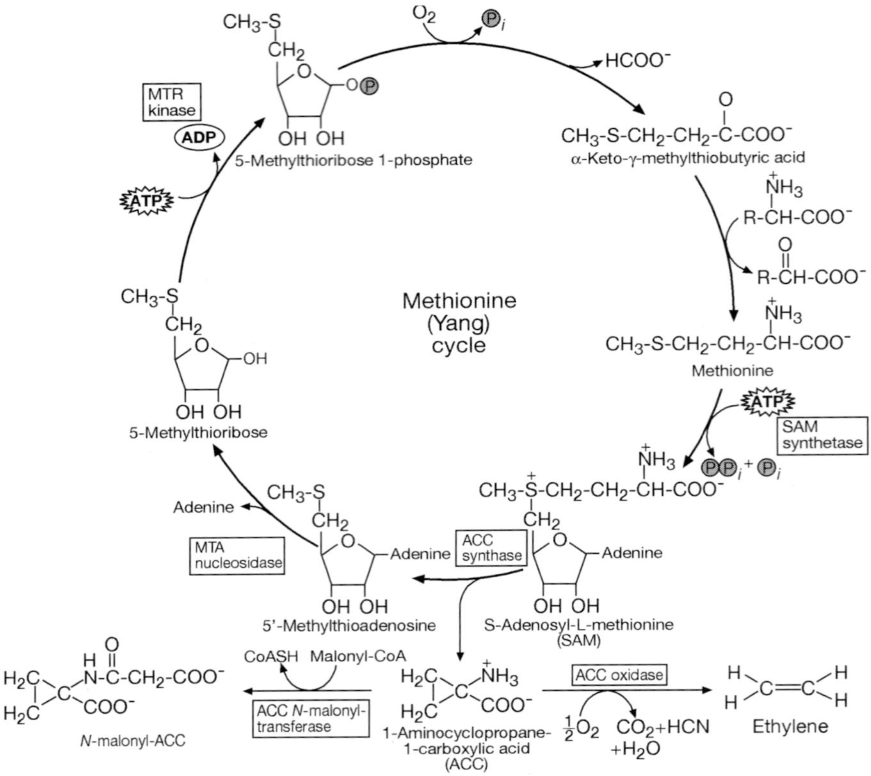

Fig. 6. The methionine cycle and ethylene biosynthesis. Ethylene is synthesized from methionine by way of SAM and ACC. The enzymes that catalyze these three steps are ATP:methionine S-adenosyl-transferase (SAM synthase), S-adenosyl-L-methionine methyl-thioadenosine-lyase (ACC synthase), and ACC oxidase. 5′-Methyl-thioadenosine, a product of the ACC oxidase reaction, is salvaged for the resynthesis of methionine through the methionine cycle. If the methylthio-group from SAM were not recycled, methionine availability and ethylene biosynthesis would probably be restricted by sulfur availability. By converting ACC to N-malonyl-ACC instead of to ethylene, plants can deplete the ACC pool and thereby reduce the rate of ethylene production. (From Crozier *et al.*, 2000.)

The activity of ACC synthase can be reduced by several inhibitors, furnishing a tool for investigators to study the physiology of ethylene. The production of ethylene in plant tissue may have an autocatalytic effect, generating more ethylene. However, in some cases, ethylene may constrain further ethylene production. The enzyme ACC synthase is probably the main target of these autocatalytic and autoinhibitory processes in ethylene production. The conversion of ACC to ethylene is processed by ACC oxidase in a process by which ACC — in the presence of oxygen, Fe^{2+} and ascorbate — is converted into ethylene, dehydroascorbate, CO_2, and cyanide. The CO_2 activates ACC oxidase. The cyanide is further detoxified by conversion to nontoxic compounds. Due to the instability of ACC oxidase, the isolation and purification of this enzyme was problematic and it was then identified by molecular cloning. Messenger RNAs (mRNAs) encoding specific ACC oxidases were thus isolated and the transcription of respective ACC oxidases could be inhibited by methods such as antisense suppression in transgenic plants (e.g. see tomato ripening in Galun and Breiman, 1997). When the conversion of ACC into ethylene was investigated in plant tissue, the results indicated that the transcription of the mRNA for ACC oxidase is a main step in the regulation of ethylene production.

There is evidence that ethylene can be oxidized (with the generation of CO_2) or converted to ethylene oxide and ethylene glycol. However, the levels of ethylene are reduced, mainly by the release of ethylene from the tissues and organs of plants. This release is very different from the release of other plant hormones; this is due to the fact that ethylene is a very volatile gas that can swiftly pass aqueous and lipid phases. Hence ethylene can easily diffuse from the plant into the environment (and when the outside air is filled with ethylene, this hormone will enter by diffusion into plant tissues).

Ethylene signal transduction

The reproducible and simple assay for the "triple response" (inhibition of epicotyl and root cells, swelling of the hypocotyl, and horizontal growth habit) in dark-grown seedlings that are exposed to ethylene furnished an efficient means to isolate Arabidopsis mutants that are defective in the biosynthesis of ethylene, in the perception of ethylene or in the propagation of the ethylene stimulus. Further, mutants that are overproducers of ethylene were isolated. This arsenal of mutants that became available, mainly in Arabidopsis, served in further experiments to identify mutants that are defective in ethylene signal transduction. For example, Kieber *et al.* (1993) isolated, by the triple response, an Arabidopsis mutant, termed *ctr1*, that constitutively exhibits seedlings and adult phenotypes as in wild-type plants that are treated by ethylene. The investigators also found in *ctr1* plants a constitutive expression of other genes that in wild-type plants are only expressed after ethylene treatment.

The *CTR1* gene was cloned and four different mutant alleles were revealed. The gene *CTR1* was found to encode a protein that is a putative serine/threonine protein kinase. Kieber *et al.* (1993) suggested that the protein encoded by the wild-type gene *CTR1* is active as a negative regulator in the ethylene pathway, and thus the elimination of this protein causes the phenotype of a constitutive ethylene response. Ecker (1995), who updated the ethylene signal transduction pathways known at the time, also stressed the relevance of the accumulated information for the horticultural practice. Ecker (1995) listed 15 ethylene mutants in Arabidopsis.

I would like to note that while Arabidopsis is a very useful model plant, there are great differences with respect to ethylene signal transduction among dicot species, even between closely related species such as tomato and potato. Due to the volatility of ethylene, seedlings grown in a closed space can "speak" to each other by the release of ethylene. The isolation of numerous ethylene mutants in Arabidopsis led to a suggestion of the order of action of the ethylene signal transduction genes. Hence, two genes, *EIN1* and *EIN4*, act before the negative regulator *CTR1*, and several *EIN* genes (e.g. *EIN2*, *EIN3*, *EIN5*, *EIN6*, *EIN7*) are operative after *CTR1*. Several of these, and additional ethylene mutants, have a pleiotropic effect in Arabidopsis as well as in other plants (e.g. tomato). For example, an enhancement of root hairs in seedlings that is caused by stresses and by ethylene application is the regular phenotype in seedlings that are not treated with ethylene but carry the *ctr1* mutation. This is actually an example of the connection between ethylene and root epidermis patterning.

It is noteworthy that while by 1995 a plethora of ethylene genes, including those revealed by the elimination of hypocotyl hook formation, was accumulated, the receptor of ethylene was still enigmatic; nor was it clear where in the plant cell this receptor was located. A later study by Hua and Meyerowitz (1998) implicated the ETR1 protein (and its complexes with other proteins) as the putative receptor of ethylene. However, there was a surprise. Contrary to most known receptors of hormones, the putative receptor(s) of ethylene had a negative effect: in the absence of the receptor, the plant behaves as though it is constantly supplied with ethylene. Additional information (e.g. Gamble *et al.*, 1998) revealed that the putative ethylene receptor protein (ETR1) requires Mn^{2+} for its phosphorylation (later studies indicated that the cofactor is Cu^{2+}). Indeed, ethylene has unique features among hormones. Rather than facilitating the cycle of cell division in synchronized plant cell suspension, it inhibited the G2-to-M transition and caused cell death when applied at a certain phase of this cycle (Herbert *et al.*, 2001).

An important finding about the ethylene receptor (ETR1) was made by the team of Eric Schaller in Durham, New Hampshire (Rodriguez *et al.*, 1999; Chen *et al.*, 2002; Gao *et al.*, 2003). These investigators located ETR1 in the endoplasmic reticulum (ER) of Arabidopsis cells rather than on the plasma membrane. The "anchoring" of ETR1 to the ER was attributed to the amino-terminal half of ETR1. It thus became evident that the scene of the signal transduction of ethylene in plant cells takes place in the ER. Furthermore, there is a five-member family of receptors (ETR1, ERS1, ETR2, ERS2, and EIN4) that is involved in ethylene perception. CTR1, as mentioned above, functions as a negative regulator downstream of the ethylene receptors, but it is considered to be a component of the receptors' complex.

In a further study (Guo and Ecker, 2004), the Ecker team focused on the EIN3 protein of Arabidopsis. This protein was previously assigned to act downstream of CTR1 in the pathway of ethylene response (e.g. Kieber *et al.*, 1993). EIN3 was revealed as a key transcription factor in ethylene response, mediating ethylene-regulated gene expression and morphological responses. The levels of EIN3 increase rapidly in response to ethylene. However, in the absence of ethylene, EIN3 is quickly degraded by an ubiquitin/proteosome pathway. When this degradation is prevented by mutations in genes that activate this degradation, these mutants lead to the retention of EIN3 and, consequently, to a phenotype of constitutive ethylene. Conversely, when the wild-type proteins of these genes are over-expressed, the EIN3 is destabilized and the respective transgenic plants are insensitive to ethylene.

Clearly, the ubiquitin/proteosome pathway negatively regulates ethylene response by targeting EIN3 for quick degradation. In fact, in the absence of ethylene, the half-life of EIN3 in Arabidopsis is only 30 minutes. The quick lowering of a signal transduction protein is not unique to the ethylene response system; similar ubiquitin degradations were also found with respect to transcription factors involved in the responses of other plant hormones, such as IAA and gibberellins (GA).

Exactly how ethylene prevented the degradation of EIN3 was not fully understood. However, Achard *et al.* (2003) provided evidence that a family of nuclear growth repressors, termed DELLA proteins, are involved in ethylene-imposed growth regulation. DELLA proteins were previously found to be involved in GA-induced growth regulation of wheat, maize, barley and rice. In these crops, DELLA proteins appear to restrain growth, whereas GA relieves this restraint. Achard *et al.* (2003) provided evidence showing that, in Arabidopsis, DELLA proteins have a role in the perception of ethylene and auxin. In the presence of these hormones, DELLA is altered and causes an inhibition of growth. In the presence of GA, DELLA is degraded and consequently this inhibition is relieved.

By the end of 2003, about 10 years after intensive molecular genetic studies, sufficient information was accumulated for a review article by Guo and Ecker (2004). The main links of the chain — which starts from perception of ethylene by an ER-anchored protein (ETR) through chain links such as CTR1 and EIN2 to EIN3, and from there into the nucleus, where transcription factors regulate the transcription of specific ethylene-related genes — were summarized in this review (see Figs. 7 and 8). Several of these links were characterized with respect to their structure and function, and assigned in Arabidopsis to specific genes. For example, the five proteins that serve as receptors of ethylene on the ER have transmembrane polypeptides, and their anchorage to the ER requires Cu^{2+}. The five receptors differ,

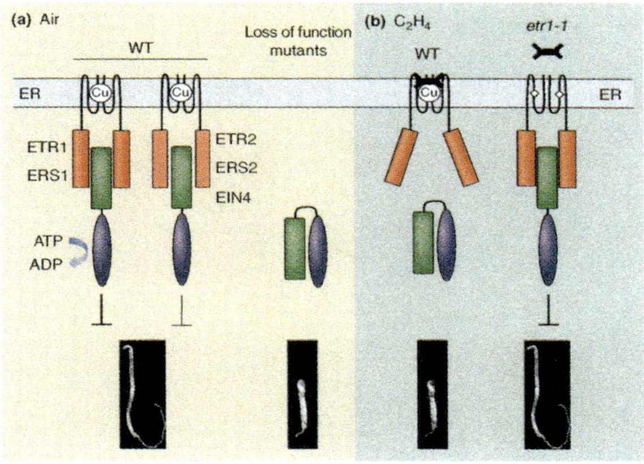

Fig. 7. Proposed signaling mechanisms of the ethylene receptors and CTR1 at the ER. There are two types of ethylene receptors in Arabidopsis. Type-1 receptors (ETR1 and ERS1) contain a conserved histidine kinase domain, while type-II receptors (ETR2, ERS2 and EIN4) contain a degenerate histidine kinase domain that is predicted to lack catalytic activity. The histidine kinase activity observed in type-I receptors is not required for receptor signaling. The amino-terminal sensor domains of the receptors contain a copper cofactor (Cu) that is needed for ethylene binding and are associated with the ER membrane. The amino-terminal domain of CTR1 interacts with the histidine kinase domain of the receptors, at a higher affinity with type-I members than with type-II members. This interaction allows CTR1 to be localized to the ER. **(a)** In the absence of ethylene, the receptors are predicted to remain in a functionally active state, and able to interact with CTR1. CTR1 is activated by association with the ER-bound receptors and represses the downstream ethylene responses by a mechanism that requires its carboxyl terminal Ser/Thr kinase domain. Loss of the function of multiple receptors (in LOF mutants) leads to the dissociation of CTR1 from the ER. The soluble CTR1 is inactivated, possibly through self-association with a negative regulatory domain as has been shown for some Raf kinases. As a result, the repression of downstream ethylene responses is relieved. The loss of two type-I receptors results in ethylene response phenotypes that are more severe than those that result from the mutation of any other combinations of two LOF receptors. This greater response may be unrelated to the fact that type-I receptors possess histidine kinase activity. Instead, it could be due to the fact that the affinity of type-I receptors for the amino-terminal domain of CTR1 is stronger than that of type-II receptors. **(b)** When ethylene is present, it binds to the sensor domain of the receptor and presumably causes a conformational change, resulting in an inactive receptor. CTR1 is then released from the ER and also becomes inactivated. In contrast, dominant mutations in the receptor that disrupt ethylene binding may lead to constitute receptor-CTR1 interaction and the repression of downstream components in the ethylene signaling pathway. (From Guo and Ecker, 2004.)

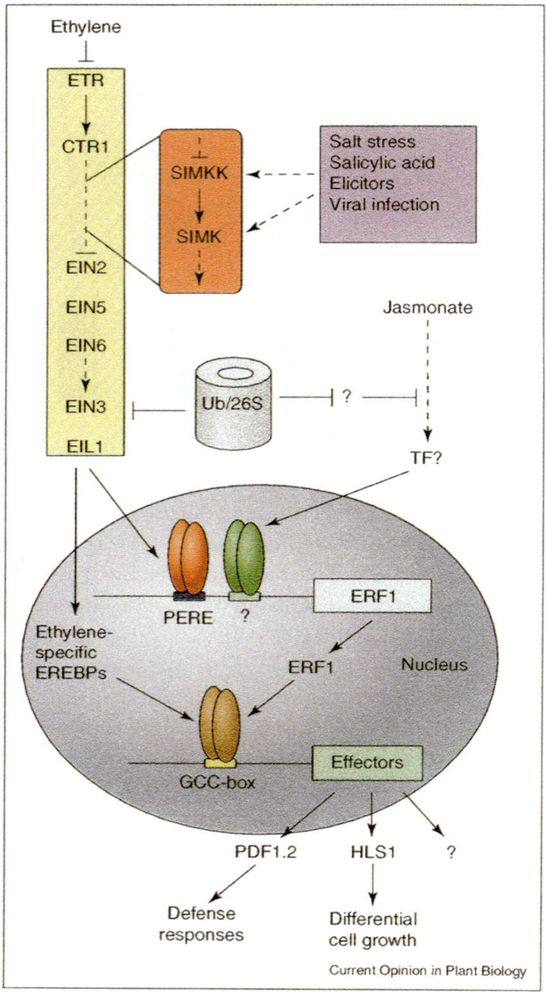

Fig. 8. A model for the ethylene response pathway in the regulation of gene expression. Ethylene gas is perceived by a family of ER-associated receptors (ETR). Ethylene binding is proposed to inhibit receptor function. CTR1 is proposed to be activated by the unoccupied receptors via physical interaction with them, and is inhibited upon binding of ethylene by the receptor. A MAPK module, consisting of SIMKK and SIMK, is proposed to act downstream of CTR1, although the biochemical consequence of this MAPK pathway is not evident. Because many biotic and abiotic stimuli activate the SIMKK/SIMK pathway, it remains to be determined whether their activation is dependent upon the functions of the ethylene receptors and CTR1.

and were divided into two types. All have a histidine kinase domain, but this domain is probably functional only in type 1 receptors (ETR1, ESR1). The role of this kinase (if there is such a role) was not clarified.

A noteworthy point is that the binding of ethylene (to the amino-terminal end) *inactivates* the receptors' function, contrary to the *activation* of receptors by an outside signal in many signal transduction systems. The serine/threonine kinase of the next link (CTR1) is thus active only in the absence of ethylene binding to the receptor. CTR1 itself is also a negative regulator of the downstream links of the chain (e.g. EIN2). The downstream links themselves (EIN2, EIN5 and EIN6) are positive regulators and activate EIN3, which is a nuclear transcription factor. EIN3 can thus regulate the transcription of ethylene-responsive element-binding proteins (EREBPs) that have

---

Downstream components in the ethylene pathway include several positive regulators (EIN2, EIN5, EIN6 and the transcription factors EIN3 and EIL1). The level of EIN3 protein is controlled by ethylene, possibly via the proteasome (Ub/26S). The primary ethylene signaling pathway components (indicated in yellow) are required for all known ethylene responses and, to date, none have been found to respond to signals other than ethylene. Branch points in the ethylene response pathway may lie downstream of EIN3/EIL1. Several EREBP transcription factors are known to be immediate targets of EIN3/EIL1, which can bind to a primary ethylene response element (PERE) in the promoters of *EREBP* genes. One EREBP, called ERF1, is also involved in JA-mediated gene regulation. It is likely that an as yet unidentified JA-regulated transcription factor (TF) may also bind to the promoter of *ERF1* to activate its expression. Therefore, the promoter of ERF1 might function to integrate signals from both the ethylene and JA signaling pathways. Other EREBPs may act in a similar manner to integrate the actions of ethylene with developmental signals and/or other hormone signals. Many EREBP proteins are known to regulate gene expression through interaction with a *cis*-element called the GCC-box, which is found in several ethylene-responsive genes, including *PDF1.2* and *HOOKLESS1* (*HLS1*). These genes encode effector proteins that are needed to execute a wide variety of ethylene responses, from disease resistance to differential cell growth. "?" represents an unknown factor or element. Arrows and t-bars represent positive and negative effects, respectively. Solid lines indicate effects that occur through direct interaction while dotted lines indicate effects that have not yet been shown to occur through direct interaction. (From Guo and Ecker, 2004.)

domains that bind to a "GCC-box" (actually a AGCCGCC box), termed ERF domains (Hao *et al.*, 1998).

Why are there five different ethylene regulators in the ER of Arabidopsis? Mutation in only one or two of them does not result in a clear phenotype; but when three or four of the receptors are defective, there appears the constitutive ethylene response (in the absence of ethylene). Also, each of the two types of genes, when defective, can be compensated by transgenes of the same type of receptor gene, but not by transgenes of the other type. It also appears that the receptors that bind to the ER membrane are instrumental in joining the next link, CTR1, to the ER. CTR1 is functional only when receptors are bound to the ER. This binding of CTR1 is stronger to the carboxy-terminal kinase domain of type I than to type II. Was the kinase domain of type I receptors retained for its binding to CTR1, and did the kinase activity then become irrelevant? This question was not pursued.

While the proteins encoded by *EIN2*, *EIN5* and *EIN6* may affect the downstream cascade of ethylene induction, their exact role requires further clarification. On the other hand, two further links, the EIN3 and EIL1 proteins, are essential factors in this cascade. In Arabidopsis, there are six members of the EIN3 family to which EIN3 and EIL1 belong. However, only the latter two proteins are clearly involved in the constitutive triple response of dark-grown Arabidopsis seedlings. These two proteins are therefore regarded as primary transcription factors of the ethylene system. Other plants also have several EIN3-like proteins (e.g. at least five in tobacco, three in tomato and two in mung beans). While EIN3 and EIL1 are themselves transcriptional factors, their own expression is not controlled by changing their transcription level. It seems that their activity is regulated posttranscriptionally, such as by the lowering of the degradation of the protein in a ubiquitin/proteosome pathway.

What is the next link, after EIN3/EIL1, in the ethylene response cascade? A good candidate for this link is the ethylene response factor 1 (ERF1). The ERF1 has a protein motif that can bind to a GCC-box in the promoter of several genes; among them, genes involved in pathogen defense and jasmonic acid. We should consider that for the activation of a given gene (e.g. with a GCC-box in its promoter), more than one transcription factor protein is required for the activation of transcription. Hence, a complex of several proteins may be needed for full activation of a gene. Such complex situations will probably be resolved when *systems biology* will be applied to plant biology. A glimpse of where such combination studies can lead, with respect to the ethylene cascade, was provided by Stepanova and Alonso (2005), who constructed a respective "pathway image" map that starts with ethylene and ends with ethylene genes and defense genes. Also, Stepanova *et al.* (2005) found a connection between ethylene and auxin (see below) after characterization of root-specific ethylene-insensitive mutants of Arabidopsis.

Interaction of ethylene with other plant hormones

The interaction of the cascade of ethylene signal transduction with jasmonic acid was noted above. Stepanova *et al.* (2005) showed that the ethylene-triggered inhibition of root growth in dark-grown Arabidopsis seedlings is mediated by the action of the genes: *WEAK ETHYLENE INSENSITIVE2/ANTHRANILATE SYNTHASE* $\alpha 1$ (*WEI2/ASA1*) and *WEI7/ANTHRANILATE SYNTHASE* $\beta 1$ (*ASB1*). These are genes that encode α- and β-subunits of a rate-limiting enzyme of tryptophan biosynthesis (anthranilate synthase) and, hence, are involved in the synthesis of auxin. By upregulation of these genes (*WEI2/ASA1* and *WEI7/ASB1*) that can be imposed by ethylene, accumulation of auxin in the tips of Arabidopsis primary roots is caused. Loss of function of these genes prevents the ethylene-mediated increase of auxin. Hence, ASA1 and ASB1 are a node of interaction between ethylene responses and auxin biosynthesis in Arabidopsis. This node is probably not unique to Arabidopsis, and additional nodes between the pathways of other plant hormones may become evident in the future (see Fig. 9).

Auxin

Historical background with a personal note

The history of indoleacetic acid (IAA) as a major plant hormone can be traced back to the 19th century, when Theophili Ciesieski studied

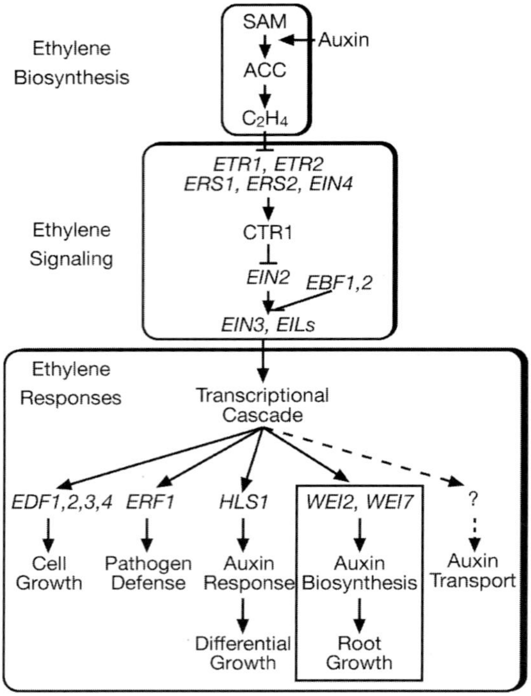

Fig. 9. Schemes of the ethylene biosynthesis, signaling, and response pathways. Key elements of the pathways are displayed in a linear manner and are connected by arrows. A previously established link between auxin and ethylene biosynthesis is indicated. The new branch in the ethylene response pathway that connects ethylene with auxin biosynthesis is highlighted by a box. SAM, *S*-adenosyl-Met. (From Stepanova *et al.*, 2005.)

geotropism in plants and Charles Darwin with his son Francis investigated geotropism and phototropism. Both phenomena were suggested to be driven by the flow of a morphogen that later was identified as IAA. Later, in 1926, Fritz Went found that auxin diffused basipetally from the tip of oat coleoptiles. Went then went to the USA and became a professor at the Division of Biology of the California Institute of Technology (Caltech). In the plant section of this division was also Professor James Bonner. At Caltech, Went built the first fully acclimatized set of greenhouses and called them Phytotron.

The relations between the two plant biology professors during the 1950s were far from idyllic, and Fritz Went went to St. Louis (Missouri) to build the gigantic fully acclimatized greenhouse in that city's Botanical Garden.

Bonner had a low opinion of auxin research and plant physiologists. I attended his lectures (while being a postdoctoral fellow at Caltech) in 1961 and saw Bonner drawing two lines on the blackboard. One line represented the increase in population in the US, with projection to future years. The other line represented the increase in the number of plant physiologists in the US, with projection of future years. The lines for the future years crossed each other, meaning that in the future there shall be in the US more plant physiologists than people. The Caltech students were smart enough to deduce Bonner's character from these lines. While not excelling in modesty, Bonner was an outstanding scholar, investigator and author. Already in the early 1960s, he started the study of chromatin remodeling as a means for gene regulation in plants — a subject which became fashionable 40 (!) years later (see Appendix in Galun, 2003).

Much before the Darwins and Went revealed the roles of auxins in plants, actually many millions of years earlier, bacteria were utilizing the auxin effect for their parasitic nourishment. The parasitic bacterium *Agrobacterium tumefaciens* introduced genes (*iaaM* and *iaaH*) for the synthesis of IAA into its plasmids. A fragment of this plasmid (T-DNA) containing these genes is inserted, after infection, into the nuclear genome of the host plant, and there (together with a gene for cytokinin synthesis) these genes cause the formation of crown galls (tumors). In these tumors, the bacteria multiply and feed on metabolites that are produced by the host (and induced by the parasite). However, these metabolites cannot serve the host plant (see Galun and Breiman, 1997).

The involvement of auxin with plant physiology and patterning includes a component of degradation. This degradation of proteins is not confined to auxin-related proteins, nor is it confined to plants; but due to its role in auxin effects, it will open the discussion of this plant hormone.

The ubiquitin–proteasome system for the selective degradation of proteins

As noted above, the ubiquitin system is shared by all eukaryotic organisms in which it was searched. The main components of the system are (1) the 76-amino-acid ubiquitin; (2) ubiquitin-activating enzyme, E1; (3) ubiquitin-conjugating enzyme, E2; (4) ubiquitin protein ligase, E3 (which provides specificity to the substrate protein that is destined for degradation); and (5) the 26S proteosome (where the degradation takes place). For a detailed description, refer to reviews such as those of Hershko and Ciechonover (1998), the investigators who (together with Irwing Rose) received the 2004 Nobel Prize in Chemistry for the discovery of the ubiquitin system.

The role of the ubiquitin–proteasome pathway in plants was reviewed by Moon *et al.* (2004) and Smalle and Vierstra (2004). Here I shall provide only a brief description of the ubiquitination, which shall serve as a background for the further dealings with auxin effects on plant patterning. Let us start from the end, i.e. with the evolutionary conserved 26S proteosome. This is a very long (ca. 1700 amino acids) protein that is shaped like a barrel (with a lid and a "base"), into which the proteins that are destined for degradation are inserted and degraded into amino acids. In many ubiquitin systems, including those of plants, the entrance into this degradative barrel is restricted to proteins that were complexed with a recognition protein that selectively binds (due to specific recognition signals) to candidate target proteins by specific motifs. This recognition protein, termed E3 ubiquitin ligase, is instrumental in binding several hundred E3 proteins, so that many target proteins can be selectively destined for degradation in the 26S proteosome. The E3 ubiquitin ligase will not bind to the target protein unless it is undergoing a specific process.

Now we start from the beginning. The ubiquitin itself, a short protein (76 amino acids), is activated by phosphorylation of a Gly residue at its C-terminal end. This activation is processed, in the presence of ATP, by the E1 ubiquitin-activating enzyme. Subsequently, the ubiquitin is bound to a Cys residue of E1 in a thioester linkage. In a further step, the activated ubiquitin is transferred to an active site,

Cys residue, of an E2 ubiquitin carrier protein. It then enters the aforementioned E3 ubiquitin ligase. The activated ubiquitin may be transferred to E3 by E2, and then the ubiquitin is transferred to the target protein, or the ubiquitin is transferred directly from E2 to the target after the target protein is complexed with E3. The process of transferring the activated ubiquitin to the target protein is commonly repeated several times, so that several units of activated ubiquitins "mark" the target protein: it is now ready for degradation in the 26S proteosome.

This is a vast simplification of the ubiquination process. There are additional steps and components that should be mentioned. According to the transfer of ubiquitin from E2 to the target protein, either directly or via E3, the E3 ligase complex is termed RING/U-box and HECT, respectively. In the system with RING domain E3, this domain can be further divided into single-subunit RING/U-box E3s and multi-subunit RING E3s. The latter contain complexes such as SCF, CUL3-BTB and APC; we shall meet some of these complexes below because mutations in them can abolish ubiquination and result in defects in normal patterning. In plants (e.g. Arabidopsis), there are many more RING/U-box ubiquination systems than HECT systems, meaning that the activated ubiquitin is commonly transferred directly from E2 to the target protein which is bound to the RING/U-box E3. Further details are provided by the reviews of Smalle and Vierstra (2004) and Moon *et al.* (2004).

While in plants the RING/U-box type of E3 is more prevalent than the HECT type, in animals the HECT type is rather abundant. We shall see below that much information has been accumulated in plants on the SCF complex of the RING/U-box type. SCF has four subunits: SKP1, CDC53 (or Cullin), the F-box and a RING finger protein, RBX1. Each of these subunits has different forms. Hence the combination of subunits with the various forms causes a vast number of combinatory possibilities of binding to various specific target proteins.

The involvement of the ubiquitin system with auxin effects became apparent in several studies. One such study was that of Xie *et al.* (2002) in N. H. Chua's laboratories at the Rockefeller University (New York) and at the National University in Singapore. These investigators focused on NAC1, a member of a family of transcription factors that is involved in the transduction of the auxin (IAA) signal for lateral root development. This NAC1 can be selectively degraded by the ubiquitin system if its RING-box includes a wild-type SINAT5. Overexpression of SINAT5 reduces the number of lateral roots in Arabidopsis, while a mutation in SINAT5 increases the number of lateral roots. The expression of NAC1 can also be increased by treatment with a proteasome inhibitor. This suggests that SINAT5 targets NAC1 for ubiquitin-mediated degradation and downregulates the auxin signal.

The rather elaborate selective degradation of target proteins by the ubiquitin system is an obvious source for the bewilderment of nature. Proteins that must be eliminated are being degraded in an environmentally friendly manner: the "waste" (amino acids) can be recycled by the cell.

Auxin synthesis

To avoid confusion, the term "auxin" shall be used for indole-3-acetic acid (IAA) and the related compound, indole-3-butyric acid (IBA). There are two synthetic hormones that cause auxin-like effects: 2,4-dichlorophenoxyacetic acid (2,4-D) and 1-naphthalacetic acid (NAA); as well as modifications of the two latter compounds. I shall briefly describe the synthesis of IAA and IBA and their conjugates. IAA, in its free form, is the major auxin in plants and is synthesized by all plants (not only in angiosperms), as well as in some microorganisms (as noted above, by genes in the plasmid of the bacterium *A. tumefaciens*, which is integrated into the host genome). Although the synthesis of auxin in plants has been studied for about 50 years, the exact processes is not yet fully revealed.

Several factors hamper progress in this field. One of them is that the same plant may have different pathways for this synthesis and there are several pathways in different organs of the same plant. Moreover, during the development of a plant, the pathway may change, at least quantitatively. Because of the redundancy of pathways, single mutants will usually not completely prevent the synthesis of auxin, rendering the genetic approach a difficult task. In spite of these difficulties, fairly clear pathways of IAA and IBA syntheses have emerged. The reviews of Crozier *et al.* (2000) and Woodward and Bartel (2005) provided details on these pathways and listed the respective references. Below, only a short summary of these syntheses, degradations and conjugations of auxins will be presented.

As described schematically in Fig. 10, chorismate is the common starting point of several pathways. From indole-3-glycerol phosphate, the pathway may lead to either the tryptophan (Trp) pathway or to the

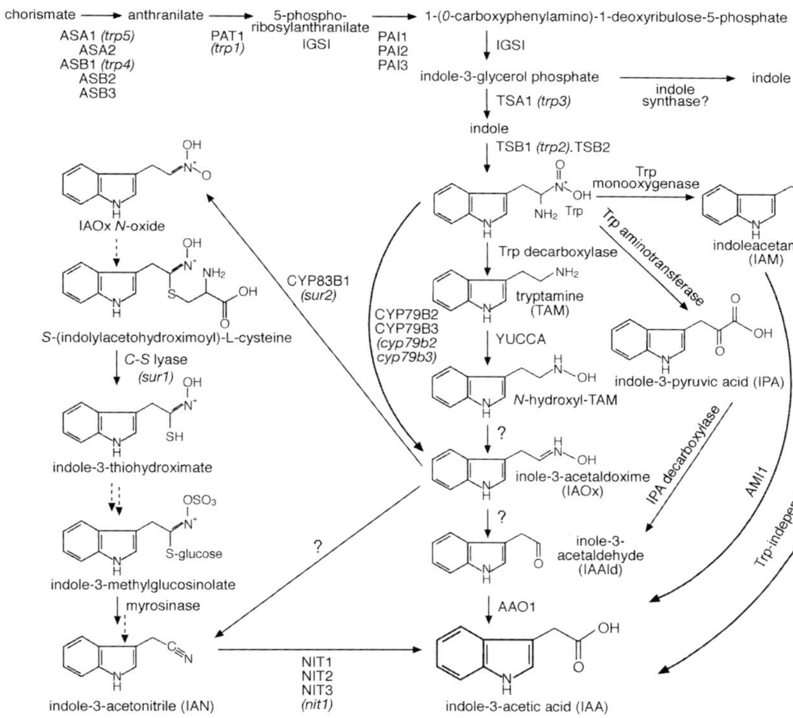

Fig. 10. Potential pathways of IAA biosynthesis in Arabidopsis. *De novo* IAA biosynthetic pathways initiate from Trp or Trp precursors. Arabidopsis mutants are in lowercase italics. Suggested conversions for which genes are not identified are indicated with question marks. Trp biosynthesis and the P450-catalyzed conversion of Trp to IAOx are chloroplastic, while many Trp-dependent IAA biosynthetic enzymes are apparently cytoplasmic. (From Woodward and Bartel, 2005.)

Trp-independent pathway (e.g. through indole). The former (Trp) pathway may proceed either through indole-3-pyruvic acid (IPA) or through indole-3-acetaldoxime (IAOx). There may be another shunt of the conversion of Trp to indoleacetamide (IAM) by a monooxygenase and from IAM to IAA. Also, IAOx may be converted to indole-3-acetonitrile (IAN) and IAN may then be converted to IAA. The enzyme indole-3-butyric acid synthase can reversibly change IAA to IBA.

The conjugated compounds of IAA and IBA are schematically shown in Fig. 11. The conjugates of IAA with alanine (IAA-Ala) and with leucine (IAA-Leu) are reversible, and the conjugates may serve as a storage for free IAA. Also, the conjugate of IAA with glucose (IAA-glucose) may serve as storage. On the other hand, other conjugates, such as IAA-Asp and IAA-Glu, are destined for the destruction of IAA.

Auxin transport

The two plant auxins, IAA and IBA, can clearly be defined as hormones. They are synthesized away from the location in which they cause a response. However, are they also morphogens in the sense that auxin gradients are perceived by plant cells? If the perception of a gradient does take place, it can be instrumental for cells to sense their location. This would fit the "French flag" concept of morphogen gradients advocated by Wolpert (1996), as mentioned above in the Introduction.

There are theoretical considerations about the role of auxins as positional information determinants. First, the cell may be capable of measuring a gradient when the level of auxin at one location of the cell's surface (i.e. on the plasma membrane) is higher than the level at the opposite side of the cell. As the difference in level is expected to be very small, this requires great sensitivity to auxin levels. Another consideration is that auxin does not flow freely through the cell. There are elaborated influx and efflux mechanisms in which auxin is complexed with specific proteins. Hence, the specific rates of influx and efflux will actually determine the level of auxin in a given cell.

We should also take into account that auxin responses are of two main types. In one type, the rate of *cell division* is regulated; while in

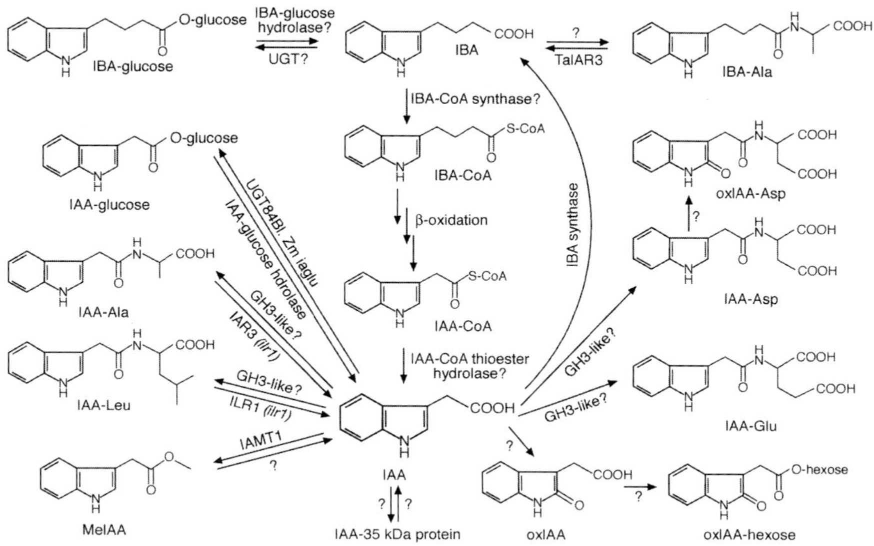

Fig. 11. Potential pathways of IAA metabolism. Suggested conversions for which plant genes are not identified are indicated with question marks. A family of amidohydrolases that apparently resides in the ER lumen can release IAA from IAA conjugates. ILR1 has specificity for IAA-Leu, while IAR3 prefers IAA-Ala. Maize (Zm) iaglu and Arabidopsis UGT84B1 esterify IAA to glucose; the enzymes that form and hydrolyze IAA-peptides have not been identified. IBA is likely to be converted to IAA-CoA in a peroxisomal process that parallels fatty acid β-oxidation to acetyl-CoA. IAA can be inactivated by oxidation (oxIAA) or by formation of non-hydrolyzable conjugates (IAA-Asp and IAA-Glu). IAA-amino acid conjugates can be formed by members of the G3/JAR1 family. OxIAA can be conjugated to hexose, and IAA-Asp can be further oxidized. IAMT1 can methylate IAA, but whether this activates or inactivates ISS is not known. IBA and hydrolyzable IAA conjugates are presumably derived from IAA; biosynthesis of these compounds may contribute to IAA inactivation. Formation and hydrolysis of IBA conjugates may also contribute to IAA homeostasis; the wheat (Ta) enzyme TaIAR3 hydrolyzes IBA-Ala. (From Woodward and Bartel, 2005.)

the other type, the *shape* of the cell may be regulated, as by asymmetric expansion of cell walls. The latter type requires responses at specific intracellular sites outside the nucleus; while for the control of cell division, the perception probably involves the nucleus. These and additional considerations of auxin flow and its role in *positional information* were provided by several reviews (e.g. Friml, 2003; Woodward and Bartel, 2005; Kepinski and Leyser, 2005a, 2005b; Callis, 2005; Blilou *et al.*, 2005).

Numerous studies clearly indicated that auxins are synthesized in plant meristems as well as in young leaves. In the root meristem, this synthesis is confined to cells between the quiescent center and the root cap (the columella initials). There is currently no good and direct method to quantify auxin at the cellular or intracellular levels. Such quantitative analyses can be done only in tissues or groups of cells. The transport of auxin in Arabidopsis roots is now fairly well established. There are two opposite flows in young roots that are separated spatially. In the center of the *meristematic zone* of the root, the flow is towards the root cap; while at the periphery of this zone, the flow is away from the root cap. It should be noted that the meristematic zone as well as the *elongation zone* of the root (that is located proximal to the former zone) are not vascular cells. The vascular cells are differentiated only in the more distal zone, the *differentiation zone*. Hence, the two-directional flow of the auxin in the root tip is confined to the cell-to-cell polar flow. This latter flow is slower than the flow of auxin in the phloem, which typically transports auxin from apical tissue to the roots.

The cell-to-cell flow of IAA requires influx and efflux protein complexes that belong to the AUX1/LAX (AUXIN1/LIKE-AUXIN1) family and to the PIN (PIN-FORMED) family, respectively. The PIN proteins were found to be located asymmetrically in the membrane of different plant cells. The AUX1 proteins were also found to be located asymmetrically, but in opposite sides of these cells. While AUX1 is a component of the influx in the cell-to-cell transport of auxin, it probably also plays a role in the phloem transport, because it is instrumental in loading auxins from leaves into the phloem and later in unloading auxins from the phloem to the root, from where the polar, cell-to-cell transport will carry auxin further basipetally (in the stele, the endodermis and cortex; but not in the root epidermis, where the transport is in the opposite direction).

More specifically, the auxin efflux regulator in the meristematic zone of Arabidopsis roots is probably PIN4. In the columella initial cells, the PIN3 was identified; while in the elongating zone of the root, PIN1 was located. When the root is placed horizontally, it seems that the PIN3 at the columella initial cells moves into an asymmetric position and causes differential elongation of one side of the cell wall, thus causing the gravitational bending of the root tip. Likewise, PIN3 is probably also involved in phototrophic bending of hypocotyls. Indeed, the lateral transfer of auxin in shoots is impaired in *pin3* mutants. As the gravity-induced root bending is causally related to starch-containing particles (statoliths), it is probable that the statoliths are implicated with PIN3-protein activity.

Interestingly, the efflux proteins (e.g. PIN1 and PIN3) are capable of undergoing rapid change in intracellular location. They may move along the actin cytoskeleton and cycle between the plasma membrane (where they participate in the efflux of auxin) and the endosomes. Hence, the activity of the PIN proteins can be changed by relocation. Clearly, relocation is faster than synthesis/degradation of these proteins. The auxin transport inhibitor TIBA (triiodobenzoic acid) prevents the cycling of PIN proteins. Also, the vesicular-trafficking inhibitor, BFA (brefeldin A) inhibits the PIN1 movement from endosomal compartments to the plasma membrane (e.g. Geldner *et al.*, 2001, 2003), adding support to the claim that PIN cycling is essential for polar transport of IAA. There are also indications that in light-grown Arabidopsis, flavonoids negatively regulate auxin transport. On the other hand, members of another group of proteins, the MULTIDRUG RESISTANCE-like proteins (MDR), are necessary for the polar auxin transport in some dicot and monocot plants. Moreover, based on genetic studies with mutants which have phenotypes that suggest impairment of auxin transport, several additional proteins are probably involved in auxin transport. Is the IAA transport system the same as the transport system for IBA or for the synthetic auxin 2.4-D? Studies with specific mutants suggest that the transport systems are different.

Auxin signaling

During recent years, there has been impressive progress in understanding the signal transduction of IAA. An update by the review of

Leyser (2002) stressed the gap that existed between the vast information on auxin-related responses of growth and development phenomena and the understanding, at the molecular level, of how auxins are causing these phenomena. This gap has been considerably closed in more recent years. Auxin is probably a prime player in plant patterning. Hence, a brief review of the present knowledge on auxin perception and transduction shall be presented.

A guiding concept for auxin perception → transduction → effect is the following: auxin is complexed with a receptor. The complex then regulates the expression (of a cascade) of intermediary transcription factors. These factors trigger specific responses that are temporarily and spatially defined (e.g. asymmetric expansion of specific cells, cell division of specific cells). This general concept can have several variations. For example, in some steps, there could be a positive or a negative regulation. Moreover, some intermediates may cause ubiquination of specific proteins. The removal of the latter proteins can release the inhibitory effects of these proteins. The first step of this concept is the perception of IAA. Such a perception is commonly mediated by a receptor (or several receptor) protein(s). The concept becomes even more complex because the auxin signal could be transduced through several pathways. Consequently, we should search not for *the* IAA receptor, but for *several* receptors (Leyser, 2002).

Historically, the hunt for proteins that bind auxin, and thus are candidates as auxin receptors, yielded many proteins (see Napier *et al.*, 2002). One of these proteins is the auxin-binding protein ABP1, which was described in maize in 1972 by Hertel *et al.* (1972). ABP1 has a strong affinity to the synthetic auxin NAA, and mutations in ABP1 have phenotypes which suggested ABP1 to be a candidate for *the* receptor of IAA. This implied that ABP1 could be the first link in a chain of transductions that lead to specific IAA responses. However, further studies did not unequivocally verify that ABP1 has such a role in auxin signaling.

Two consecutive publications in *Nature* by Dharmasiri *et al.* (2005) and by Kepinski and Leyser (2005b) identified the F-box protein, TIR1 (transport inhibitor response 1) as an auxin receptor. The F-box is a short sequence of amino acids that is found in proteins which form a complex termed SCF (named for its components: SKP1, Cullin, F-box protein), and also includes RING-box proteins. We mentioned this SCF while dealing with the ubiquitin–proteasome system, since this complex is an essential component of the selective degradation (ubiquination) of proteins. The SCF that includes the TIR1 as its F-box component was termed SCF^{TIR1}.

The identification of TIR1 as the receptor of auxin was based on the following considerations and experimental results. Plants contain two families of transcription factors that are involved in auxin response: the auxin response factor (ARF) and the AUX/IAA proteins. The ARF proteins bind directly to the DNA of genes, and can either activate or repress the genes' transcription, depending on the ARF protein. In Arabidopsis, there are 29 members of the transcriptional repressors AUX/IAA. The AUX/IAA can dimerize with and repress the activity of ARFs. Thus, when AUX/IAA proteins are eliminated (i.e. by ubiquitination), the activity of ARFs will be derepressed. The AUX/IAAs are targeted for degradation by the SCF complex. The target specificity of the SCF is determined by its F-box protein. In the case of the AUX/IAA target, the relevant F-box protein is a leucine-rich repeat TIR1 and possibly a small number of structurally-close homologs.

Previous information indicated that correct interaction between AUX/IAA and SCF^{TIR1} is essential for auxin response; mutations in either of these will abolish auxin effects. That means that the auxin-induced degradation of AUX/IAA requires an effective interaction between AUX/IAA and SCF^{TIR1}. Various experimental approaches clearly showed that auxin is bound directly and reversibly with TIR1. For example, the H^3-IAA that is bound to SCF^{TIR1} can be washed out by unlabeled IAA (as well as with NAA), but not with tryptophan. When a mutated TIR is used for complexing with SCF, the IAA will not bind. In summary, there is good evidence for concluding that TIR1 is an auxin receptor that mediates (probably by causing conformational changes) rapid degradation of AUX/IAA by ubiquitination.

For rapid responses, such as the auxin-induced plasma membrane H^+-ATPase activity (associated with cell elongation), it was proposed that no transcriptional alterations take place. Rather, the TIR1–auxin association could cause degradations of other proteins (by ubiquitination). However, the verification of such an assumption requires further experimental work. *In vitro* experiments also showed that whatever impact IAA has on SCF^{TIR1}, this impact is temporary: it is kept only in the presence of IAA and is reversed when IAA is removed. Such a reversible effect is compatible with the responses of plant tissues to IAA. The studies of Dharmasiri *et al.* (2005) and of Kepinski and Leyser (2005a), together with previous binding studies, led to the following possible scenario for IAA signal transduction:

- Specific cells produce the complexes of SCF^{TIR1};
- There are many different kinds of SCF^{TIR1} because hundreds of types of F-box proteins were found in investigated plants (e.g. in Arabidopsis);
- SCF^{TIR1} could be located in specific subcellular locations in the cells;
- In the absence of IAA, the activity of an auxin-regulated gene is shut off by a repressor;
- When IAA arrives and binds to the TIR1 of SCF^{TIR1}, it causes a decisive modification of SCF^{TIR1};
- Modified SCF^{TIR1} is then able to interact efficiently with a target (e.g. the AUX/IAA transcriptional repressor protein);
- The interacting complex renders the target acceptable by a ubiquitination system;
- The degradation of a repressor will cause derepression of the auxin-regulated gene.

In reality, IAA signaling is vastly more complex. For example, auxin rapidly and transiently induces the accumulation of at least three families of transcripts. There is a vast number of F-box proteins. Hence, there is a great number of different SCFs. Woodward and Bartel (2005) listed twelve SCF regulatory genes from Arabidopsis, and described the phenotypes resulting from their loss-of-function mutations.

There is also a seemingly counterintuitive phenomenon: the expression of the gene for AUX/IAA, which is a repressor of auxin-regulated genes, is elevated by auxin application. Interestingly, the genes that have an auxin-induced expression share a common sequence in their upstream regulatory region: TGTCTC (or a slight variance of this sequence). The sequence was termed auxin-responsive element (AuxRE). Now, after the genomes of plants (e.g. Arabidopsis, rice) have been sequenced, a computer search for AuxRE motifs may reveal additional auxin-responsive genes and test them experimentally for their auxin responsiveness.

There could be additional links in the chain of events between the binding of auxin to a receptor and the derepression of a final auxin-regulated gene.

Temporal and spatial effectors (e.g. developmental cues, environmental factors) also participate in the auxin-regulated gene expression. Also, additional *bona fide* auxin (IAA) receptors may be revealed in the future. In short, IAA signaling, the mode of auxin impact on development and cell division, as well as the interactions of auxin with other plant hormones is merely beginning to be revealed.

Interaction with other hormones

Numerous investigations, especially physiological studies, indicated that there are links between auxin and several other plant hormones: cytokinins, ethylene, gibberellins, jasmonic acid, brassinosteroids and abscisic acid. Describing all these links is beyond the scope of this book, and readers interested in these links are referred to Woodward and Bartel (2005). Such links were revealed by the defective responses to other hormones in mutants defective in auxin response. Interestingly, several of these links are channeled through the SCF-mediated degradation of regulatory proteins. Other links are probably channeled through protein phosphorylations. Still other links may involve transport mechanisms.

Cytokinins

Historical background

The early studies on cytokinins clearly benefited from the envy/rivalry among scholars, as indicated in the motto of a previous book (Galun and Galun, 2001) that was taken from the Talmud (Bavli), which was codexed in the fifth century AD. The motto says: "The envy of scholars will increase wisdom (or knowledge)". These studies involved two talented scientists who came to the US from Europe: F.C. Steward, from England, came to Cornell University; and Folke Skoog came from Sweden to the University of Wisconsin (Madison). What happened with respect to studies on cytokinins during about a dozen years in the middle of the 20th century was reviewed in very vivid detail and with humor by Richard Amasino (2005). Here, I shall provide only a short summary of the main events.

Both Steward and Skoog were pioneers in studying the growth of plant explants in tissue culture. Steward used carrot root sections and cultured them in different media. Finding that coconut components improved the *in vitro* cultures, he started to characterize these components. Skoog tried to influence the type of growth as undifferentiated callus, shoots and/or roots, resulting from tobacco stem sections that were cultured in different media (e.g. with different ratios of added auxin and adenine). Skoog obtained inconsistent results with coconut components that were used to supplement culture media for the growth of plant explants. He wrote to Steward, asking for information on the coconut components that may improve his cultures. Steward responded by telling Skoog to take his hands off the coconut issue. This triggered the fighting spirit of Skoog, who years ago was a competing athlete in the 1932 Olympic games. Skoog thus turned to Professor Frank Strong, a natural products expert in the Department of Biochemistry at the University of Wisconsin. A very efficient collaboration between the laboratories of Skoog and Strong started. Strong's lab handled the chemistry, while Skoog's lab performed the tissue cultures and the bioessays. By 1952, the first report on this collaboration was published, showing that one fraction of the extract from coconut meat was 4000 times more active than the crude extract. Yeast extract and salmon sperm DNA were also tested for activity in the growth of plant explants. The latter studies were mainly conducted by a very talented postdoctoral fellow, Carlos Miller, who joined Skoog's lab in 1951.

Luckily, the results with different batches of yeast and DNA were inconsistent! Indeed, it was found that fresh batches were not active; however, extracts from autoclaved DNA contained a very active fraction.

By wisely combining experimental results, biochemical knowledge, a lot of work, collaborations between the two labs, and intuition, Miller isolated, in December 1954, a crystalline material that was highly active in bioessays. The Strong lab quickly characterized this crystalline material. This was submitted for publication in January 1955, and was published on March 5 in the *Journal of the American Chemical Society* (Miller *et al.*, 1955a and 1955b).

Further collaborations, especially between Miller and the Strong lab, identified this material as 6-furfurylamiopurine, in which the furan (a degradation product from the deoxyribose of DNA) is linked to carbon 6 of adenine. The chemically synthesized compound was identical to the crystalline material, and was coined ***kinetin***. Hence, the first cytokinin was not isolated from plants. It probably (but not surely) does not exist in living plant tissues. Other purine derivatives were then synthesized and tested in bioessays. One of them became popular as an ingredient in culture media: 6-benzyladenine. Meanwhile, the Steward lab (Shantz and Steward, 1955) identified 1,3-diphenylurea in coconuts. But, the diphenylurea was actually contaminated from the equipment used to extract large quantities of coconuts, although it was later found that diphenylurea can bind to a cytokinin receptor. Moreover, diphenylurea was later found as an inhibitor of cytokinin oxidase; hence, diphenylurea may just prevent cytokinin degradation. The first natural cytokinin, zeatin, was isolated only 10 years later (Letham and Miller, 1965) from immature maize seeds (see Fig. 12).

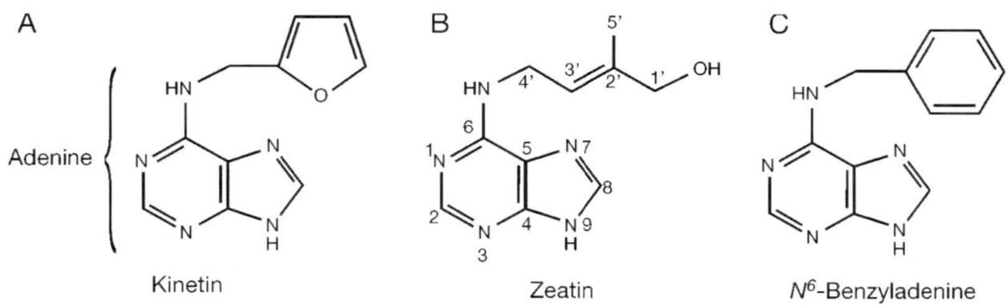

Fig. 12. Cytokinin structures. **(A)** Kinetin is a synthetic cytokinin generated when DNA is autoclaved. **(B)** Zeatin was the first endogenous cytokinin isolated from plants. **(C)** N^6-Benzyladenine (BA) is a synthetic compound with cytokinin activity. (From Crozier *et al.*, 2000.)

An overview on cytokinin's effects

In early studies, cytokinins were considered primarily as factors required for cell division. Later, several other effects of cytokinins on plant development were revealed. Soon after the isolation of kinitin, Amos Richmond and Anton Lang found that kinitin affected the protein content of detached *Xanthium* leaves and delayed the decomposition of chlorophyll. Crozier *et al.* (2000) listed 32 cytokinins in addition to kinetin (6-furfurylaminopurine). Many of these were used in plant physiology studies. There was a claim that natural cytokinins (such as zeatin) are synthesized in the roots and then transported in the xylem to the shoot where they participate in apical dominance and delay in leaf senescence. However, some experimental results question such a transport. It is probable that cytokinins are commonly synthesized near the site where they are required. Infection of plants with *Agrobacterium* transfers into the host plant a gene that encodes an enzyme which is active in the cytokinin biosynthesis pathway. The coding sequence of this gene can be fused with specific promoters that are either active constitutively (e.g. the 355 CaMV promoter) or are activating expression only after due induction (e.g. a tetracycline-dependent promoter). The chimeric gene can then be used in genetic transformation.

In the case of the constitutively active promoter, the transgenic tobacco plants had a phenotype of cytokinin-treated plants, such as abundant growth of lateral branches and retardation of leaf senescence. The cytokinin content was also strongly elevated. By the use of inducible promoters, the morphological change could be localized to the site of the application of the inducing compound. Likewise, fruit maturation in tomato could be drastically modified by genetic transformation with a chimeric gene that has a fruit-specific promoter and a coding region for cytokinin synthesis. Fruits of tomato transformed with such a chimeric gene are specifically changed, resulting in deepred pigmentation and vastly elevated cytokinin level.

It should be noted that there is a rapid interconversion in plants, between nucleotide cytokinins (i.e. with a phosphatized ribose), nucleoside cytokinins (i.e. with a nonphosphatized ribose), and free cytokinin. Hence, when the nucleoside or the nucleotide cytokinins are applied to a plant or a plant tissue, these may be converted quickly to cytokinin. Although the natural and active cytokinin in plants is considered to be the *trans*-isomer of zeatin, it was claimed for some time that the natural *trans*-zeatin cannot be a breakdown derivative of tRNA that would lead to *cis*-zeatin. However, there are indications that isomerization of *cis*- to *trans*-zeatin can occur in plant cells. Hence, the tRNA as a source of active cytokinin cannot be excluded. The exact pathway(s) of *in vivo* cytokinin biosynthesis, as shall be noted below, is not yet fully clarified. Neither were all the genes and encoded enzymes involved in the pathway(s) identified. Moreover, it seems possible that this biosynthesis differs among angiosperms and/or that different pathways exist in the same plant.

There are three major sites for cytokinin biosynthesis: root tips, shoot tips and maturating seeds. The molecular genetic mechanisms that restrict this synthesis to these locations and the regulation of the level of synthesis are subject for further studies.

Summary of cytokinin biosynthesis, degradation and bioconversions

The *in vivo* synthesis of cytokinins in plants is a rather complicated and not yet a fully understood issue. Reviews on this topic were provided by Binns (1994), Crozier *et al.* (2000) and Mok and Mok (2001).

It is plausible that adenosine 5′-monophosphate (5′-AMP) is the common precursor of cytokinin biosynthesis. From 5′-AMP, the N^6-(Δ^2-isopentenyl) adenoside 5′-phosphate ([9R-5′-P] iP) is produced by the binding of dimethylalkyl diphosphate at N^6 of the adenosine. From the latter compound, the pathway may go either through the removal of the phosphate and then the removal of ribose to form N^6-(Δ^2-isopentenyl) adenine (i^6Ade) and by further hydroxylation to zeatin. The [9R-5′-P] iP (i^6AMP) could be hydroxylated to [9R-5′-P] Z, then the latter can be dephosphorylated to [9R] Z (9-ribosylzeatin), and finally the removal of ribose could result in zeatin. Another possibility is that [9R] iP (i^6Ado) is hydroxylated to 9-ribosylzeatin ([9R] Z). There are additional possibilities as noted in the abovementioned reviews (e.g. Mok and Mok, 2001).

The adenine ring can be glycosylated *in vivo*, at the side chain, to result in O-glucoside cytokinin. Such glucosylated cytokinins were found in several plants. These can probably be quickly converted *in vivo* to cytokinins, such as zeatin. Hence, the glucosylated cytokinins could serve as "storage" cytokinins. Glucosylation can also be at the N^7 of the adenine ring, however, such glucosylations are probably not reversible. Several other modifications at the N^3, N^7 and N^9 adenine ring will lead to several additional cytokinin-like compounds that were actually reported in several plants. A considerable number of the genes that are involved in such modifications, and the respective enzymes that are encoded in these genes, were studied rather actively. However, this subject is outside the scope of the present book.

It is noteworthy that in the nopaline-type crown gall system (see Galun and Breiman, 1997), where a bacterial (*Agrobacterium*) gene, *ipt*, is transferred to the host plant, the cytokinin production pathway goes through [9R-5′P] iP (i^6 AMP) which is converted to [9R-5′P] Z by transhydroxylation. This hydroxylation is performed by an enzyme of the host plant. The Ti-plasmid of the octopine-type agrobacteria contains a *tzs* gene (rather than an *ipt* gene), but the two genes share considerable homology. Interestingly, when the coding sequence of the *ipt* gene was put downstream of either its own natural promoter or other promoters, and the construct was used to transform plants or plant tissues, as noted above, the results were rather bizarre, such as inability to differentiate roots, and even inability to produce shoots and roots. However, the resulting unorganized tissue was autonomous with respect to both auxin and cytokinin. Clearly, such genetic transformation did not furnish simple answers to the role of cytokinins in normal plant patterning (see Binns, 1994, for review).

Perception and signal transduction of cytokinins

Several studies suggested that the actual active cytokinins in angiosperms are isopentyladenine (i.e. N^6 (Δ^2 isopentenyl) adenine, abbreviated as iP) and zeatin. The perception of these cytokinins is most probably performed by the "two-component" system. Kakimoto (2003) and Ferriera and Kieber (2005) furnished comprehensive reviews on the signal transduction of cytokinins in plants. In brief, this system usually consists of a *signal-sensing* domain (input) positioned in the extracellular space and a *signal-transducing* domain that is positioned in the cytoplasm. This system is acting as a histidine kinase, and the input signal (e.g. cytokinin) causes the phosphorylation of the His in the transducing (transmitter) domain. From there, the phosphate can be transmitted to Asp in the receiver domain of a *response regulator.* The protein CKII was identified in plants (e.g. Arabidopsis) as clearly involved in the binding of cytokinins. The extracellular, cytokinin-binding domain in Arabidopsis was termed CHASE (cyclase/histidine kinase–associated sensory extracellular).

Further information implicated the protein CRE1/WOL/AHK4 as a histidine kinase and cytokinin receptor. It was suggested that cytokinins activate a phosphor relay in which the CRE1/WOL/AHK4 is operative. Indeed, the latter protein complex has an input signal domain, a transmitter domain in which the His can be phosphorylated, and a receiver domain in which the Asp can be phosphorylated. The signal is then transferred to the nucleus where additional transductions are taking place, as is schematically described in Fig. 13. This scheme was slightly modified by Ferriera and Kieber (2005), and should be regarded as a concept rather than the final process. Figure 13 contains several question marks; hence, how exactly cytokinin-responsive genes are regulated is still enigmatic.

It is plausible that such regulations are not identical in all plants, neither in different tissues of the same plant. One observation will demonstrate the enigma. It seems that cytokinins positively regulate cell division (e.g. facilitating the G1 to S transition in the cell cycle) in the shoot apice, while cytokinins downregulate cell division in roots. As with IAA, the quantitation of cytokinins in individual cells is problematic. Many of the assumptions concerning the role of cytokinins in gene regulation are based on experimental methodologies in which cytokinins were applied to the plant (or to the plant cells), transgenic plants with enhanced cytokinins were used, mutants that caused enhancement or reduction of cytokinins were employed, or plants with an enhanced degradation of cytokinins (e.g. by cytokinin oxidases) were used.

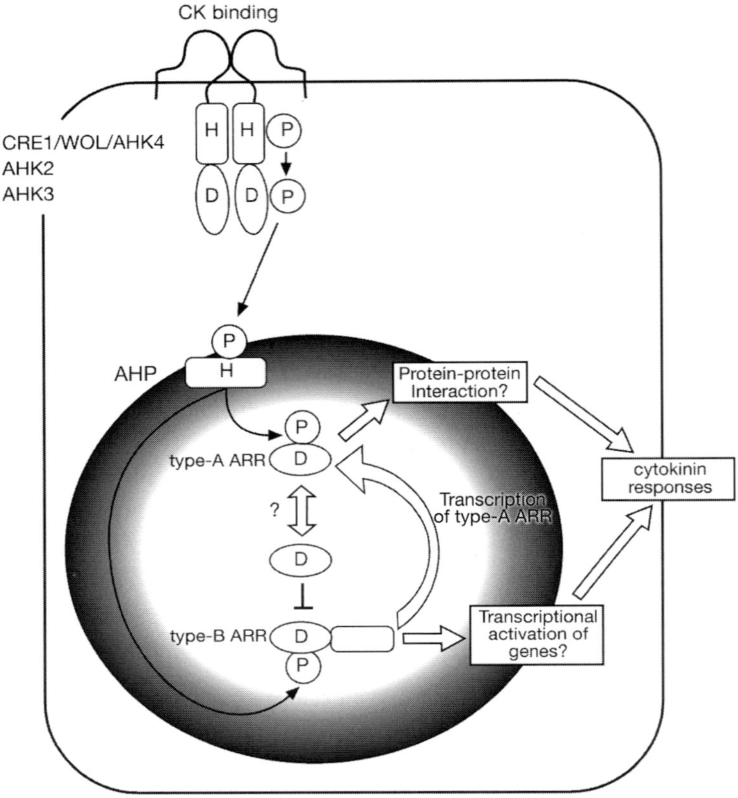

Fig. 13. A model of cytokinin signal transduction in Arabidopsis. The three cytokinin receptors — CRE1/WOL/AHK4, AHK2, and AHK3 – bind to cytokinins and initiate phosphotransfer. The phosphoryl group, indicated by *encircled letter P*, is then transferred to the HPt domain proteins, AHPs. Phosphorylated AHPs are translocated to the nuclei and probably transfer the phosphoryl group to type-A ARRs and/or type-B ARRs. How phosphorylation regulated transcriptional activity of type-B ARRs is an open question. In this model, type-A ARRs function as a molecular switch of type-B ARR activity. Type-A and/or type-B ARRs may have output functions, which is also not well understood. (From Kakimoto, 2003.)

Gibberellins

The awareness of gibberellins emerged in Japan in 1926 due to Eiichi Kurosawa, who studied the *bakanae* disease in rice. This disease caused excessive growth in rice, but the tall plants lacked grains. It was then revealed that the disease was caused by the fungus, *Gibberella fujikuroi,* that produced a diterpenoid compound that was termed gibberellin. Only in the early 1950s were British and US scientists starting to study gibberellins. The language barrier and the Second World War probably caused this time gap. Incidentally, such time gaps between Japan and the West happened also with respect to other plant biology subjects. I witnessed one of them.

In 1969, I made an extensive tour of Japan and visited companies that were engaged in hybrid seed production. When I called on a group of small islands in a bay near Sendai, I saw that each island was growing a different line of *Brassica oleracea* (cabbage) that served as a parent for hybrid seed production in a system based on self-incompatibility. Self-incompatibility in *Brassica* was discovered in the West only in the mid-1950s, while in Japan it was known 20 years earlier and used for the production of hybrid seeds (see Frankel and Galun, 1977). Again, the language barrier and the Second World War were probably the reasons for this time gap.

Some effects of gibberellins on plants

The early observations on the effects of gibberellins (GAs) on plants were based on the application of GAs to plants. First, gibberellic acid (GA_3) was used to treat plants. The GA_3 (see Fig. 14) has a double bond between C1 and C2, hydroxyls at C3 and at C13, and a carboxyl at C7. GA_3 can be obtained in large quantities from cultures of the fungal genus, *Gibberella*. The most common effect of GA_3 was the elongation of shoots, especially in dwarf cultivars of crop plants. However, GA_3 also affected flowering. It could induce flowering in plants that usually require photoperiodic signals for flowering. GA_3 also caused flowering, at normal temperature, of plants that usually require a cold treatment for flowering. Other effects included the retardation of leaf and fruit senescence and promotion of germination. The latter phenomenon was used commercially in the production of beer, because GA treatment of barley grains facilitates α-amylase activity, thus shortening the malting process.

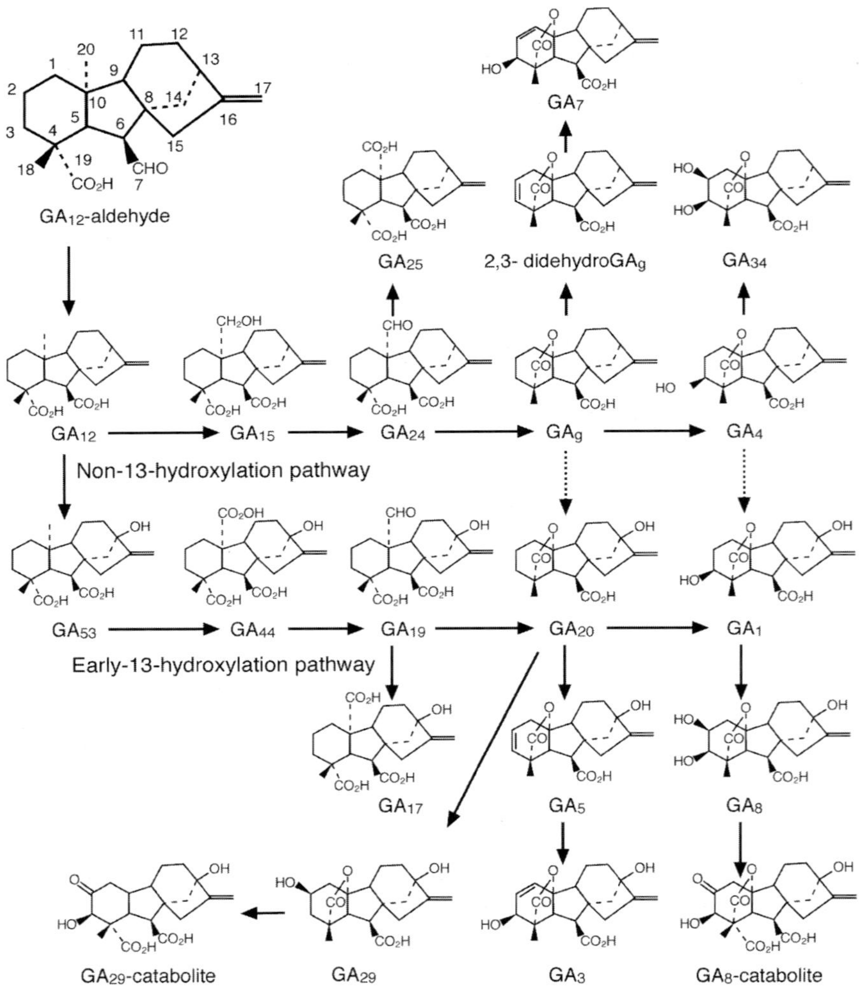

Fig. 14. Gibberellin biosynthesis pathway from GA_{12}-aldehyde. (From Hedden and Kamiya, 1997.)

My personal involvement with GA treatment concerns sex expression in cucumber. GA treatment converted female plants (with only pistillate flowers) into monoecious plants (with both staminate and pistillate flowers; Galun, 1959a). This shift in sex expression actually resulted from the suppression of the first flower in the axils of female plants; the subsequent flowers that do not develop in untreated female plants, but do develop after GA treatment, are staminate (male) flowers. The other GA-responsive organ in cucumber is the tendril. The latter elongates vastly in response to GA (Galun, 1959b).

My GA experiments revealed further phenomena that were reported also by others. First, plant species differ considerably in their response to GA. For example, while in cucumber (*Cucumis sativus*) GA does affect sex expression, no such effect is imposed in the melon (*C. melo*). Moreover, the most effective GAs in cucumber are GA_4 and GA_7, while GA_3 is less effective. The effectiveness of GAs in various plants differs considerably. Also, the endogenous levels of the various kinds of GAs are different in various plant species. The impact of endogenous and applied GAs on plant patterning shall be handled in Part B of this book.

Biosynthesis

A detailed account on the synthesis of GAs in plants is far beyond the scope of this book. There are over 150 GAs and the pathways of syntheses ramify in various manners. The reviews of Hedden and Kamiya (1997) and Crozier *et al.* (2000) are recommended as starting points for GA syntheses. More recent information is reported by Lange *et al.* (2005).

In common with the biosynthesis of terpenoid plant hormones (e.g. brassiosteroids, abscisic acid) and other derivates of sesquiterpenes, diterpenes, triterpenes and tetraterpenes, the biosynthesis of GA goes through isopentenyl diphosphate → dimethylallyl disphosphate → geranyl diphosphate → farnesyl disphosphate → geranyl disphosphate. The first key enzyme of this pathway that leads specifically to GAs is copalyl diphosphate synthase (CPS), which yields copalyl diphosphate. The latter is converted to *ent*-kaurene by *ent*-kaurene synthase (KS), which is apparently a limiting link in GA synthases. Both CPS and KS have transit peptides that traffic these enzymes to plastides, where the respective biosyntheses take place.

Several enzymatic reactions convert *ent*-kaurene to GA_{12}-aldehyde and from the latter compound to various GAs, as suggested in the scheme of Fig. 14. It should be noted that the GA syntheses in plants differ from those of fungi. From the GA precursor, *ent*-kaurene, there is a series of reactions, such as oxidations, hydroxylations and decarboxylations that yield either C19 or C20 types of GAs (meaning, having or lacking a carbon that is attached to C10; see Fig. 14 for the numbering of the carbons in GAs). The same plant species may contain several different GAs and different plant species differ in the assortment of GAs that they contain.

While the KS transcript is abundant in developing seeds and in growing tissues, the CPS transcript is expressed in lower levels. Some dwarf mutants lack enzymes that are crucial for the GA pathway. Hence, they can be rendered to normal growth by the application of GA. Some conversions are crucial to render GAs active, such as the 3β-hydroxylation, although some GAs that lack this 3β-hydroxylation are probably active in some plants. It appears that, at least in some plants, light and phytochromes play a role: in cowpea (*Vigna sinensis*), far-red light increases the production of GA_1 from GA_{20} by enhancement of 3β-hydroxylation. GA levels can also be regulated by catabolisms, as via GA_{20} oxidase enzymes.

The signaling pathway

Although the gibberellin receptor (perhaps there is more than one receptor) in plant cells is still enigmatic, several features of the signaling of these plant hormones were already revealed (see Harberd *et al.*, 1998; Richards *et al.*, 2001; Olszewski *et al.*, 2002; Gomi and Matsuoka, 2003; Fleet and Sun, 2005; and Swain and Singh, 2005).

To summarize the GA signaling that was revealed, we shall start at the "end". Here, there is a key protein, termed GAI. When the GAI (unmutated) protein is bound to GA (or a GA signal), it does not retard normal growth. However, in the absence of GA (or a GA signal), the GAI will repress growth phenomena. The normal GAI protein contains a sequence of 17 amino acids, close to its amino-terminal side. Because the first five amino acids of this sequence are Asp, Glu, Leu, Leu and Ala, this sequence was termed according to the single-letter designation of these five amino acids: DELLA. The protein encoded by the *gai* mutant is devoid of DELLA and, consequently, does not bind GA (or a GA signal). The result is similar to the lack of GA, meaning there is a repression of growth phenomena. But this repression cannot be derepressed by GA. In fact, *gai* mutants have a higher GA level than wild-type (GAI) plants. The wild-type GAI protein has a long C-terminal region, which is assumed to cause the repression (unless the GAI is bound with GA or its signal). This long region is homologous to the protein SCARECROW (SCR) that was found (in Arabidopsis) to regulate cell division during root development. Both GAI and SCR, as well as other proteins of the GRAS family, contain the DELLA regions and are grouped in the DELLA proteins. These proteins also contain nuclear localization signals (NSLs), and hence are confined to the nucleus, where they probably serve as transcription factors.

An additional Arabidopsis gene, *SLEEPY1* (*SLY1*), was revealed that comes into play in GA signaling. *SLY1* encodes a putative F-box of the SCF complex (see section on ubiquitination in the previous chapter). When *SLY1* is mutated (*sly1*), there is an accumulation of the DELLA proteins, probably due to the lack of DELLA protein ubiquitination. Under normal conditions, GA induces the degradation of the DELLA (repressor) proteins by the ubiquitination pathway. This degradation does not occur in *sly1* mutants, nor in the absence of GA.

While *GAI* and *SLY* were identified in Arabidopsis, several similar and additional genes were found in other dicot plants as well as in monocots (such as barley, wheat, rice and maize). Such genes were reviewed by Gomi and Matsuoka (2003) and shall not be detailed here. Some Arabidopsis genes (e.g. *gai-1*) were transferred to monocots (e.g. rice plants) and seem to act there as in Arabidopsis. This does not mean that the whole process of GA signaling is identical in all angiosperms.

Apparently, not all GA signalings pass through the DELLA proteins (SLR1 and REPRESSOR of ga1) by releasing the repressive activity of these proteins — there is a direct route by the PHOTOPERIOD-RESPONSIVE1 (PHOR1) protein that induces the activity of the GA-response gene, although the production of PHOR1 is triggered by the GA signal in a manner that does not involve DELLA proteins. However, PHOR1 has a U-box domain and consequently may also have a role of ubiquitination of DELLA proteins. PHOR1 may thus act as an E3 ubiquitin lyase that specifically degrades the DELLA proteins in the GA signaling pathway.

The crucial role of ubiquitination and hormonal control of growth and development is coming to light through the investigation of DELLA proteins in Arabidopsis. The two proteins, GAI and REPRESSOR, of ga1–3 (RGA) are two closely related putative transcription factors of the DELLA protein family. This family includes RGL1, RGL2 and RGL3 (which are RGA-LIKE1, 2 and 3, respectively). All these are probably negative regulators of GA. For RGA, there is evidence that both GA and auxin promote protein degradation, while ethylene increases the stability of this protein. Hence, all these three plant hormones regulate growth via RGA protein levels. While a great deal of information on the understanding of DELLA function and regulation has accumulated, further research is required for a better understanding of the components of GA signal transduction and the interactions of GA with other plant hormones (e.g. auxin and ethylene).

An overview of gibberellins in plant morphogenesis

Although major issues of GA in growth and differentiation, as well as steps of the GA signaling and the interactions of GA with other plant hormones were revealed in recent studies, many gaps in our understanding of these themes are still awaiting clarification. One such subject is the primary receptor(s) of GA and the verification of the localization of this (or these) receptor(s). Is the localization confined to the plasma membrane?

With respect to the interactions of GA with other plant hormones, here is a list of such functional interactions:

> <u>Germination</u>: GA interacts positively with auxin and ethylene, as well as with brassinolide and negatively with abscisic acid;

Apical hook formation in dark-grown seedlings: GA interacts positively with ethylene and auxin;

Hypocotyl elongation in seedlings: GA interacts with ethylene, auxins and brassinolide;

Stem elongation: GA interacts with auxin;

Root elongation: GA interacts positively with auxin, and negatively with ethylene;

Leaf shaping: GA interacts with auxin and cytokinin;

Floral member differentiation: This is a complicated issue, GA is probably required for the development of petals and stamens, but ethylene seems to suppress staminate flowers in cucumber.

These interactions were reviewed by Fleet and Sun (2005), and it is expected that further information will accumulate on this subject in future years.

In the aforementioned review, Fleet and Sun listed 18 genes involved in GA modulations of plant morphogenesis (see Table 1) and divided these genes into three groups: genes involved in the regulation of GA biosynthesis, genes for GA signaling components, and genes that are downstream effectors. We shall meet some of these genes in Part B, where the patterning of specific cells and organs will be discussed. As the isolation of genes involved in GA modulation is at present an active field of investigation, it is to be expected that the list of genes in Table 1 will be extended in the future. It should be noted that the abovementioned list contains 12 genes that were revealed in Arabidopsis, where the genomic sequence of DNA is now known and freely available, and where forward and reverse genetic methods are well advanced.

An emerging issue in the impact of GA on morphogenesis is its controlled degradation by ubiquitination. In this process, the interaction with other plant hormones plays an important role. For example, it seems that auxin and ethylene can directly or indirectly affect this degradation of GA.

Table 1 Genes involved in GA Modulation of Morphogenesis

Gene	Species	Protein Function	Relation to GA Pathway	Mutant Phenotype(s)
GA biosynthesis regulators				
AGL15	*Arabidopsis*	MADS-box transcription factor	Induces GA2ox6	Reduced height, late flowering (OE)
DDF1	*Arabidopsis*	AP2-like transcription factor	Decreases GA levels	Reduced height, late flowering (OE)
FUS3	*Arabidopsis*	B3 transcription factor	Represses GA3ox	Cotyledons replaced by vegetative leaves (LOF); reduced height (OE)
KNAT1	*Arabidopsis*	KNOX	Represses GA20ox	Increased leaf lobes (OE); repressed by GA
POTH1	Potato	KNOX	Represses GA20ox	Reduced height, rounded leaves (OE)
RSG	Tobacco	bZip transcription factor	Represses KO	Reduced stem elongation (DN)
STM1	*Arabidopsis*	KNOX	Represses GA20ox	Defective or missing shoot apical meristem (LOF), enhanced by *spy*
GA signaling components				
DELLA	Multiple(b)	DELLA transcription regulator	Represses GA response	Reduced height, reduced fertility, delayed flowering (GOF); tall, slender phenotype in single-gene species (LOF)
PHOR1	Potato	Ubiquitin E3 ligase	Promotes GA response	Reduced height (AS)
PKL	*Arabidopsis*	CHD3 chromatin	Promotes GA response	Embryonic characteristics in root tissue (LOF)
SLY1/GID2	*Arabidopsis*/rice	F-box protein	Promotes degradation of GA repressors	Reduced germination and height, delayed flowering (LOF)

(*Continued*)

Table 1 (*Continued*)

Gene	Species	Protein Function	Relation to GA Pathway	Mutant Phenotype(s)
SNE	*Arabidopsis*	F-box protein	Promotes degradation of GA repressors	Rescues height of *sly1* (OE)
SPY	*Arabidopsis*, barley, petunia	*O*-GlcNac transferase	Represses GA response	Elongated stem, reduced fertility (LOF)
Downstream effectors				
GAMYB	Multiple	Transcription factor	Promotes GA-induced gene expression	Reduced α-amylase expression, reduced pollen development (rice LOF); male sterility (barley OE)
GL1	*Arabidopsis*	MYB transcription factor	Induced by GA	Lacks trichome development (LOF)
LUE1	*Arabidopsis*	Katanin p60-like	Promotes GA effects	Reduced stem elongation, delayed flowering (LOF)
miR159	Multiple	MicroRNA	Promotes degradation of GAMYB	Delayed flowering, reduced anther development (*Arabidopsis* OE)
SOC1	*Arabidopsis*	MADS-box transcription factor	Promotes GA effects	Early flowering, partially rescues flowering delay of *ga1-3* (OE)

(a) Mutant phenotype caused by: AS, anti-sense; DN, dominant negative; GOF, gain-of-function; LOF, loss-of-function; OE, over-expressors. (b) Includes *RGA*, *GAI*, *RGL1*, *RGL2*, *RGL3* of Arabidopsis, *SLN1* of barley, *SLR1* of rice, *d8* of maize, *RHT* of wheat, *Vvgai1* of grape, and *DWF2* of *Brassica*. *GL1*, *GLABROUS1*; *KNAT1*, *KNOTTED1*-like in *Arabidopsis thaliana*; *PHOR1*, *PHOTOPERIOD-RESPONSIVE1*. (From Fleet and Sun, 2005, where references are provided.)

Brassinosteroids

Brassinosteroids (BRs) were identified as plant hormones considerably later than other plant hormones. The history of the discovery of BRs was reviewed by Mandava (1988). Briefly, the first indications on the role of BRs in plant growth and differentiation were observed in the 1960s in Japan. Japanese investigators used the rice lamina inclination bioassay to evaluate levels of growth promoters in plant extracts. In this essay, leaf laminae are excised from etiolated rice seedlings and exposed to solutions containing growth-promoting compounds. The level of these compounds is then evaluated by the angle that is formed between the leaf blade and the sheet. This bioassay is very sensitive. Levels as low as 0.0001 mg/L of certain BRs can be detected by this bioassay. The Japanese investigators thus discovered a peculiar growth promoter in the leaves of the plant *Distylium racemosum*.

Later, investigators of the U.S. Department of Agriculture (USDA) initiated a project to look for growth promoters in plants and focused on pollen as a source of such substances. They opted to start with a vast amount of pollen. For that they recruited bees, which collected for them about 230 kg of rape (*Brassica napus*) pollen. Extractions, purification and crystallization yielded (in 1979) 12 mg of a crystallized steroid compound that was later coined *Brassinolide* (see Fig. 15). Subsequently, another plant steroid with plant hormone characteristics was isolated from insect galls of chestnut (*Castanea crenata*) and was termed castasterone. This was followed by the identification of additional BRs that differed in the substituents of their A and B rings and their side chains. The surge of BR research and the investigation of the impact of BRs on growth and differentiation in plants happened during the 1980s, although, for quite some time BRs did not catch much attention. Only in recent years has it become evident that BRs, and especially brassinolide, play a major role in plant growth and differentiation, and interact with other plant hormones in crucial regulations of plant growth and differentiation.

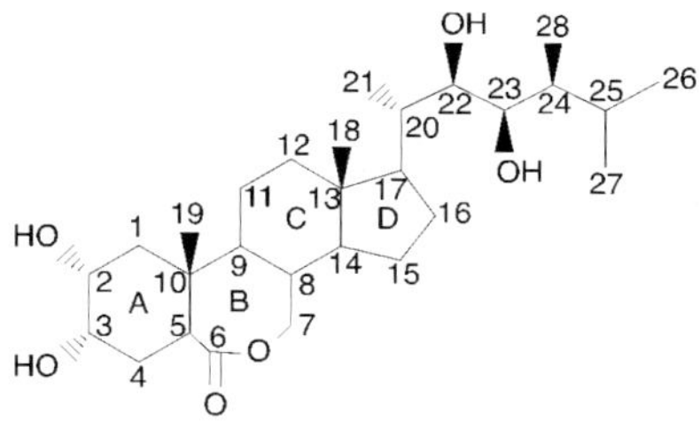

Brassinolide

Fig. 15. The structure of brassinolide, a commonly occurring Brassinosteroid (BR) with high biological activity, showing *numbered positions* mentioned in the text. In natural BRs, hydroxylation can occur in *ring A* at positions 3-, and/or 2-, and/or 1-; also found are epoxidation at 2,3-, or a 3-oxo-group. In *ring B*, alternatives are 6-oxo- and 6-deoxo- forms. In the side chain methyl-, ethyl-, methylene-, ethylidene-, or *nor-* alkyl groups can occur at 24-, and the 25-methyl-series is also represented. (From Clouse and Sasse, 1998.)

Effects of applied BR on growth, differentiation and on the metabolism of plants

With the availability of BRs, the application of these compounds to plants and plant organs became a common procedure, and many reports on the impact of these BRs were published. In brief, the range of effects of these applications was rather wide. It included elongation of shoots (especially in certain dwarf mutants), pollen tube growth, unrolling of leaves in Gramineae, reorientation of cellulose microfibrils, xylogenesis, and enhanced ethylene release. BRs were also found to cause proton pump activation.

In addition, less defined, but practically useful effects were reported (as reviewed by Mandava, 1988), such as the increase of yields in several vegetables, as well as in barley, wheat and potatoes. Even the yield of maize was claimed to be increased by spraying the ears and the silk with BRs, and some reports claimed that BRs increased the tolerance of crops to diseases.

Nevertheless, the existence of endogenous BRs in plants and the effects of BRs on plant growth and differentiation do not constitute proof that endogenous BRs play an essential role in the regulation of plant growth. Such proof had to wait for studies by Chory and associates (see review of Li and Chory, 1999), which started at the end of the 1980s, and shall be mentioned below.

Occurrence and biosynthesis of BRs

Once the rice lamina inclination bioassay and other bioassays, such as the bean second internode bioassay, for BRs became available, the search for endogenous BRs started to flourish. A favorite source of BRs was pollen grains, and Mandava (1988) listed 16 plant species in which pollen grains contain BRs. These included several plant families: dicots as well as monocots, and annuals as well as trees. It is possible that BRs exist in all pollen grains. It should be noted that bioassays suffer from possible effects of other endogeneous compounds that may promote BRs or suppress the BRs' effect in the bioassay. BRs were also found in other plant constituents, such as young leaves. The levels of BRs vary, but, in total, they are rather low (in the range of a few nanomolars). On the other hand, BRs were found virtually wherever they were looked for. The number of BRs found in plants is vast. At least 40 were identified, and these include, in addition to brassinolide and castasterone, steroids such as sitosterol, which is probably the most abundant sterol in many plant species, accounting in the latter for 50% to 80% of the total steroid content. However, sitosterol does not seem to be the normal precursor of physiologically active BRs in plants. This role is attributed to campesterol.

The biosynthetic pathway that leads to physiologically active BRs is still under investigation. It is also not yet clear whether the *in vivo* synthesis of BRs is taking place in all plant cells or there is commonly a long-distance (acropetal) transport. Clearly, in at least some cases, such as in pollen grains, developing seeds and cell cultures, there is a "local" biosynthesis. Symons and Reid (2004) found no evidence for long-range transport of BRs. The details of *in vivo* biosynthesis of BRs as brassinolide are quite complicated and not entirely clear yet, but some general characteristics of this biosynthesis did emerge by a combination of mutants that inhibit the pathways of synthesis, and by feeding experiments in which the blocks caused by these mutants were overcome.

The very early stages of the synthesis pathway of BRs are identical to those of other sesquiterpenes, meaning that geranyl diphosphate is converted to farnesyl diphosphate and, after several additional steps, squalene is produced. In further steps, campesterol is produced and the latter is considered as the source of plant BRs. From campesterol, the pathway may go either through an early C_6 oxidation pathway or a late C_6 pathway. In the former, the C_6 oxidation leads to castasterone and, from this steroid, through teasterone and typhasterol to castasterone. In the late oxidation of C_6, there are several conversions from campestanol through 6-deoxycastasterone, 6-deoxoteasterone, etc., until 6-deoxocastasterone is oxidized to castasterone. In both pathways, castasterone is converted to brassinolide. It seems that in some plants both the early and the late C_6 oxidations take place. The names and the structures of these steroids may well confuse readers, but to those who would like to go into more detail, the reviews by Clouse and Sasse (1998), Li and Chory (1999), and Crozier *et al.* (2000) are recommended. The article by Kim *et al.* (2005) provides a further update on BR biosynthesis. These reviews also indicate which mutants were found to interfere with steps in the pathways of brassinolide biosynthesis. Interestingly, an elaborate system involving many genes was found in Arabidopsis; this system establishes homeostasis of BR levels through feedback regulations (Tanaka *et al.*, 2005; He *et al.*, 2005).

Brassinosteroid perception and signaling

The identification of a receptor of BR that spans the plasma membrane and has an extracellular amino acid sequence that protrudes out of the cell membrane, to which BRs are bound, is a relatively recent finding. Much of our present knowledge about the perception and signaling of BRs in plants was gained by the studies of Joanne Chory and her associates at the Plant Biology Laboratory at the Salk Institute for Biological Studies in La Jolla, California. The proof for the binding of BRs to a specific receptor was reported relatively recently. However, the history that led to this proof is interesting, and shall be summarized here.

In her early studies (Chory *et al.*, 1989a, 1989b), Chory was interested in the photomorphogenesis of Arabidopsis seedlings. A mutation in the *DET1* gene (*de-etiolated1*, *det1*) caused the Arabidopsis seedlings, which were grown in the dark, to display many characteristics of light-grown wild-type plants. Studies of this mutant, as well as *det2*, *det3* and *det1-1* alleles of *det1* (Chory and Peto, 1990), led to the suggestion that *DET1* encodes a negatively acting regulatory molecule that is involved in light stimulus activities. Similarly, the *det2* mutant rendered dark-grown Arabidopsis seedlings to show many characteristics of light-grown plants. However, in this case the chloroplast development program was not initiated in the dark. The phenotype of the *det1* and *det2* mutants was found to be additive (Chory *et al.*, 1991), and *det2* also altered photoperiodic response in light-grown plants, indicating that the DET2 protein is also required during later stages of vegetative growth.

Until this stage, Chory did not handle BRs; but in a further study (Li *et al.*, 1996), it was revealed that *DET2* encodes a protein that shares sequences with the mammalian steroid 5α-reductase, and also that *det2* mutants can be ameliorated by the application of brassinolide. This indicated that BRs have an important role in light-regulated development in plants. Furthermore, the mammalian gene for the 5α-reductase could replace the DET2 of plants (Li *et al.*, 1997). While the *det2* mutant could be corrected by BR application, Chory and her associates isolated several dwarf/photomorphogenetic Arabidopsis mutants that could not be ameliorated by the application of BRs, meaning the defect is not in the biosynthesis of BRs.

After Clouse *et al.* (1996) isolated a BR-insensitive mutant, *bri*, Li and Chory (1997) revealed 18 mutated alleles of *BRI*. Cloning and sequencing this gene indicated that the gene encodes a protein with a putative leucine-rich repeat (LRR) receptor kinase. It was known that the extracellular LRR domains play an important role in receptors that are located at the cell surface. This finding led to a series of publications by Chory and associates on the perception and signal transduction of BR in plants (e.g. Friedrichsen *et al.*, 2000; Li *et al.*, 2001; Yin *et al.*, 2002; Nemhauser and Chory, 2004; Mora-Gracia *et al.*, 2004; Nemhauser *et al.*, 2004; Russinova *et al.*, 2004; Yin *et al.*, 2005; Kinoshita *et al.*, 2005). The investigation of the perception and signal transduction of BR plant hormones was also taken up by several other laboratories (see Tichtinsky *et al.*, 2003). It was found that the leucine-rich repeat (LRR) subfamily of receptors actually represent a large proportion of the plant receptor kinases (PRKs). These include CALAVATA1 (for stem cell maintenance); FLS2 (flagellin-sensing 2), which mediates response to bacterial pathogens; and the S-locus receptor kinase (SRK), the female determinant of self-incompatibility in Brassicaceae.

The BRI1 was found to interact with another LRR receptor-like kinase (RLK), called BRI1-ASSOCIATED RECEPTOR KINASE1 (BAK1), which is also essential for BR signal transduction. The association of BRI and BAK1 is enhanced in plants by BR, causing heterodimer formation between the two RLKs and phosphorylation of the numerous Thr residues in the extracellular binding domain. It should be noted that plants (e.g. Arabidopsis) have a vast number of (over 600) RLKs; hence, the latter probably control a wide range of physiological responses, only one of them being the response to BR.

The general structure of the BRI1 can be summarized as follows: BRI1 consists of an extracellular domain, a single-pass transmembrane segment and a cytoplasmic serine/threonine kinase domain. The extracellular domain contains a signal peptide, a leucine zipper motif, two cysteine pairs that flank 25 LRR motifs, and a characteristic 70-amino-acid "island" domain that is located between the 21st and 22nd LRR. The BRs are assumed to bind to this "island" (see Fig. 16A). On the basis of their own and others' results (Wang *et al.*, 2005a, 2005b), the Chory team suggested a model for BRI activation (see Fig. 16B) in which the binding of BR to the "island" of the BRI1 causes a transduction of the BR signal and leads to the

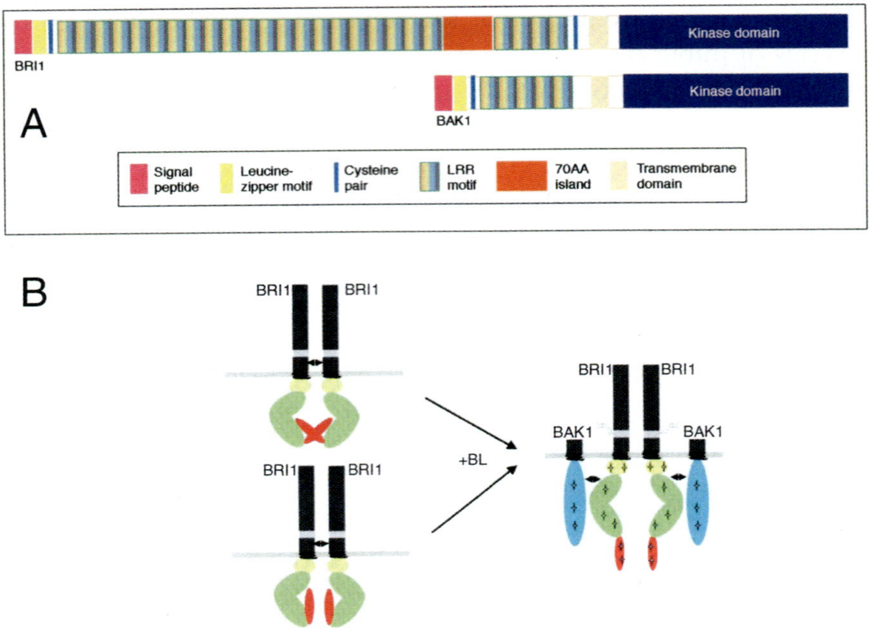

Fig. 16. **(A)** The structures of the two LRR-RLKs that are involved in BR signaling. BRI1 contains 25 LRR motifs and a characteristic 70AA island in its extracellular domain, whereas BAK1 contains only five LRRs and lacks the 70AA island and the second cysteine pair. **(B)** A model for BRI1 activation in the absence of BRs. BRI1 kinase is in a low basal activity state, in which BRI1 kinase is hypophosphorylated and kept in an autoinhibitory conformation by the CT domain, either in *trans* or *cis*. BL binding to the island domain of BRI1 induces a conformational change between the kinase domain through the JM domain, leading to phosphorylation (start) of residues in the CT domain. This phosphorylation could facilitate the release of its autoinhibition and allow further autophosphorylation, thus creating a fully activated receptor. Following the phosphorylation and activation of BRI1, BAK1 and BRI1 likely form a multimeric complex through their intracellular domains to transduce the BR signal. (From Wang *et al.*, 2005b.)

phosphorylation in the cytoplasmic domain. After this activation of BRI1, BAK1 and BRI1 are likely to interact and to form a multimeric complex through their intracellular domains. The cytoplasmic domain of BRI1 is then phosphorylated at multiple Thr and Ser residues. The BR signal is thus transduced to the cytoplasm and, from there, to the nucleus in which BR-reactive genes are consequently regulated.

The two publications of Wang *et al.* (2005a) and Wang *et al.* (2005b) were submitted by two different groups. Although the first author in both publications is X. Wang, the X stands for Xiaofeng and Xuelu, respectively. Confusing? It probably is confusing, since the Wang *et al.* (2005a) article is a report from the Steven Clouse team of North Carolina State University, while the Wang *et al.* (2005b) article is from the Salk Institute in California; moreover, the same Japanese investigator, Tadao Asami, participated in both teams.

Li (2005) updated the BR signaling from receptor kinases to transcription factors. He first stressed the difference between the signaling of animal steroid hormones that rely mainly on nuclear receptors to convey their signals into the nucleus. BR of plants have a transmembrane receptor kinase, spanning the cell's membrane, that initiates the phosphorylation cascade to transduce the BR signal through the cytoplasm and then into the nucleus. While the BR binds directly to the "island" of 70 amino acids, C-terminal flanking LRR are also vital for this binding. In the presence of BR, there is a heterodimerization of BRI1 and BAK1, but the role of this dimer in BR signaling is not yet determined. The receptors are then internalized, probably by endocytosis.

Two substrates for the activated BRI1 were revealed: transthyretin-like (TTR) and TGFβ receptor-interacting protein1 (TRIP1). These were identified in Arabidopsis, but their exact role in the signaling of BR awaits clarification. An additional protein, BIN2, is negatively regulating the transduction of BR. BIN2 can phosphorylate two proteins, BES1 and BZR1, tagging them for degradation by ubiquitination. Reduction of the BIN2, which is apparently caused by BR inhibition, may thus maintain BES1 and BZR1. In the absence of BR, both BES1 and BZR1 are phosphorylated and targeted for proteasome-mediated degradation. BES1 and BZR1 move into the nucleus where they are capable of direct binding to DNA, causing the control of expression in BR-reactive genes.

While much of the present knowledge on BR perception and signaling was derived from studies with Arabidopsis, other plants were also used in such studies, such as rice (Hong *et al.*, 2005), pea (Schults *et al.*, 2001) and tomato (Koka *et al.*, 2000). Studies in other plants suggested that the perception and signaling of BRs proceeds in similar lines in all angiosperms.

Interactions between BRs and other plant hormones

Studies on interactions between BRs and other plant hormones are relatively novel, but several investigations already indicate that such interactions do exist.

Nemhauser *et al.* (2004), of Chory's laboratory, reported on an interaction between auxin and BR. They described a shared auxin/BR pathway that is required for seedling growth. The response from one pathway required the function of the other pathway. Increased auxin levels saturated the BR-stimulated growth response and greatly reduced the BR effect on gene expression. The integration of the two pathways (of auxin and BR signaling) was found to take place downstream from the AUX/IAA in the auxin pathway and downstream from the BES1; hence, rather close to the final regulation of the auxin- and BR-responsive genes.

A negative interaction between BR and gibberellin was reported by the team of John Mundy at the University of Copenhagen (Bouquin *et al.*, 2001). These investigators used a BRI1 mutant and found that it displayed altered expression levels of genes that are differentially regulated by GA. It was suggested that BR and GA antagonistically regulate the accumulation of mRNAs of the GA-responsive genes.

We should not be surprised if these investigations are the swallows that herald the spring, and that interactions between plant hormones that regulate the expression of hormone-responsive genes are actually common phenomena.

Jasmonic Acid, Salicylic Acid, Polyamines, and Abscisic Acid

These four types of compounds are widespread in angiosperms and may be included in the group of plant hormones, but their known involvement in plant patterning is meagre, either because of lack of knowledge or because they (or at least some of them) do not directly regulate the differentiation of plants.

The recognition that jasmonates (JAs) are essential hormones in plants that participate in the regulation of defense responses, as well as in the regulation of development, emerged from recent studies. Interactions between JAs and other plant hormones (e.g. ethylene auxin, abscisic acid) were reported in recent years. However, our knowledge about JAs is still meagre. Readers interested in an update on JAs will find the reviews of Turner *et al.* (2002) and Lorenzo and Solano (2005) very useful.

Salicylic acid containing plant material was derived frequently from willow trees (of the genus *Salix*, hence the name of this compound). It was used since about 2000 years ago as a pain reliever. Towards the end of the 19th century, it was chemically synthesized, and later its acetylated derivative (acetylsalycylic acid) was synthesized by the German company Bayer and termed "Aspirin". Since then it has been in medical use to relieve pain and other maladies. Most people (but not all) can take aspirin without having side-effects. However, a certain proportion of people suffer (even severely) from aspirin-caused allergy or other side-effects. In plants, salicylic acids may retard the senescence of petals (when cut flowers are put in a vase), probably by the reduction of the conversion of ACC ethylene. While not considered a flower-inducing compound, with the application of salicylic acid to the medium of cultured duck weeds (e.g. *Lemna gibba*, *Spirodela polyrrhiza*, *Wolffia microscopica*), these plants may flower in noninduced (short days) photoperiods. The most impressive phenomenon involved in endogeneous salicylic acids concerns the thermoregulation of the voodo lily. In the flower of this plant, the temperature of specific floral members can rise at specific times to 14°C above the regular temperature. This is caused by a shift in metabolism and is attributed to salicylic acid. Salicylic acid treatment can mimic this vast rise in temperature.

While polyamines are found in plants at much higher levels than regular plant hormones, they are not known to regulate the evolvement of patterns in plants. Polyamines participate in several physiological responses, such as cell division, pollen tube formation, root initiation, embryogenesis, flower development, and fruit ripening; but they do not appear to have a hormonal role. Rather, they take part in several key metabolic pathways that are essential for efficient functioning at the cellular level.

For many years, abscisic acid (ABA) was considered a plant hormone that is involved in stress conditions, seed dormancy and stomata closure. However, in recent years, insights into the signal transductions of ABA revealed a dramatic change in the view of plant biologists on this plant hormone. More than 1000 genes were found that are differentially regulated by ABA (see Himmelbach *et al.*, 2003 for review and literature). ABA signaling is activated in the early stages of plant ontogenesis, as well as in processes that were recognized many years ago (e.g. salt and drought stress responses and closure of stomata). ABA is clearly involved in cell growth and cell division. ABA also interacts with JA in the control of cell cycle progression (e.g. Swiatek *et al.*, 2002). On the other hand, many details on the perception and signal transduction of ABA are still enigmatic. We shall "meet" ABA again in Part B, where the patterning of specific plant cells and organs shall be discussed.

Peptide Hormones

Until recent years, there was no awareness that short peptides, which are prevalent among animal hormones, also play important roles in plants. Hence, in the review of plant hormones (Crozier *et al.*, 2000) in the book *Biochemistry and Molecular Biology of Plants* (Buchanan *et al.*, 2000), peptides were not yet mentioned as plant hormones. It is now clear that indeed such peptides have vital roles in plants. The subject of peptide hormones is presently emerging, and considerable advances in understanding the roles of peptide signaling in plants are expected in the coming years. Matsubayashi and Sakagami (2006) reviewed the peptide hormones of plants. They furnished updated information on several kinds of these peptides. Some of these, such as *systemines*, *phytosulfokines* and *ENDOD40*, have no direct roles in plant patterning; while other peptides, such as those encoded by

CLV3, have roles in patterning. The first three shall therefore be mentioned only briefly, while more information shall be furnished on *CLV3*-derived peptides.

Systemines

These peptides are involved in defense mechanisms in plants and were studied mostly in plants belonging to the *Solanaceae* (e.g. tomato and tobacco). The tomato systemin (TomSys) is an 18-amino-acid peptide that is dervied from a 200-amino-acid precursor. While TomSys itself can move in the plant, it seems that the wound-induced release of TomSys activates jasmonic acid biosynthesis, and the latter acts as the main long-distance signal that induces the systemic wound response. The TomSys peptides can bind to leucine-rich repeat receptor-like kinases (LRR-RLK).

Phytosulfokine (PSK)

The PSKs are short peptides (e.g. YIYTQ), in which the tyrosine (Y) are posttranslationally sulfated. They are derived from precursors of about 80 amino acids. They were detected as the factors of "conditioning medium" required for the division of diluted cell cultures. Even nanomolar concentrations of PSK can induce cell division in nondividing sparsely plated cells of several plants (e.g. asparagus, rice, maize, *Zinnia*, carrot, and Arabidopsis). The induction of cell division in sparsely plated cells is related to my own investigations of over 30 years ago, when it was observed that diluted protoplasts, derived from leaf mesophyll or callus, will not divide unless cultured over a layer of (γ-irradiated) nursing cells as "feeder layer" (Raveh *et al.*, 1973). This procedure enabled obtaining calli from individual protoplasts that could then differentiate into mature and functional plants (e.g. tobacco, *Citrus* species, potato). The "feeder layer" technique was very useful in the derivation of plants from single protoplasts, or from protoplast cybrids, in my laboratory (e.g. Galun, 1981; Galun *et al.*, 1982; Vardi *et al.*, 1982; Aviv *et al.*, 1984a,b; Galun and Aviv, 1986; Perl *et al.*, 1990a, 1990b), as well as in other laboratories. However, the component in the conditioning medium, or the feeder layer, that was responsible for the cell division capability was unknown until 1996, when the PSKs were revealed (Matsubayashi and Sakagmi, 1996).

CLAVATA (CLV)

CLAVATA and its acronym *CLV* stand for genes that encode proteins involved in signaling in plant cells, such as genes involved in the regulation of shoot apical meristems (SAMs) that will be detailed in Chapter 8. The name of these genes was derived from the Latin word for "club-shaped" because floral meristems of *CLV* mutants often form numerous extra club-like carpels. Although the genes share the same acronym, different *CLV* genes code for very different proteins. Hence, *CLV1* encodes a receptor-like serine/threonine kinase with an extracellular domain composed of leucine-rich repeats, a transmembrane domain and a cytoplasmic kinase domain. *CLV2* encodes a receptor-like protein that is similar to *CLV1*, except that *CLV2* codes for a protein that has only a small intracellular domain and lacks the kinase. *CLV1* and *CLV2* probably form a heterodimer that serves as a receptor for a peptide encoded by *CLV3*. The following amino acid sequence, derived from the *CLV3* transcript (written by the single-letter designation) is considered to be the precursor for the peptide hormone that interacts with *CLV1/CLV2*.

MDSKSFVLLLLLFCFLFLHDASDLTQAHAHVQGLSNRKMMMMKM
ESEWVG.ANGEAEKAKT.KGLGLHEELRTVPSGDPLHHHVNPPRQP
RNNFQLP

The underlined sequence of amino acids is considered to be the CLE motif conserved in several peptides that were recognized as peptide hormones. Prolines of the CLE motif may be modified to hydroxyprolines after translation, resulting in the final ligand that interacts with the receptor-like kinases. In Arabidopsis, the *CLV3* peptide that serves as ligand contains 12 amino acids of the CLE motif, and two of the three prolines in this motif are hydroxylated (Kondo *et al.*, 2006; Ito *et al.*, 2006; Simon and Stahl, 2006, for overview).

It is clear that the product of *CLV3* can move through several layers of cells: in the SAM it moves from the (upper) L1 layer to the L3 layers. Hence, the designation of this peptide as a peptide hormone is fully justified.

This mentioned *CLV3*-derived protein is only one out of 31 other proteins which were already detected in Arabidopsis, and which contain the CLE motif near their carboxyl terminus. Hence, additional peptide hormones of a similar kind may be revealed in plants. Moreover, plants have a plethora of genes that encode receptor-like kinases. These are candidates for interactions with peptide hormones. It is expected that the studies aimed to reveal peptide genes involved in plant cell-to-cell communications that started recently will soon yield novel knowledge on peptide hormones in plants.

Chapter 3

THE CYTOSKELETON OF PLANT CELLS

The term *cytoskeleton* encompasses several entities that comprise a network of filamentous proteins and accessory proteins. It contains mainly three polymers: *intermediate filaments*, *microfilaments* and *microtubules*. Cytoskeleton is a convenient term because by using it we do not have to list all its various components. But, there is a warning: the term may be misleading, because there is no analogy between the "skeleton" of vertebrates and the "skeleton" in the cytoskeleton of plant cells. The two "skeletons" vary vastly. The skeleton of vertebrates serves to provide rigidity and stability to the animal, and is an anchorage to soft tissues such as muscles. The fibrous polymers of the plant cytoskeleton are far from stable and rigid. They are very dynamic and quickly polymerize and depolymerize. Each of the sections of these polymers has a very short half-life that can be as brief as a few minutes. Interestingly, while Ca^{2+} contributes to the strength and maintenance of the vertebrate skeleton, Ca^{2+} can cause depolymerization of the cytoskeleton's filamentous polymers.

In this chapter, the intermediate filaments shall be mentioned very briefly. The microfilaments, the microtubules and their respective accessory proteins will be handled in greater detail because they play important roles in shaping the plant cell and in cell division; consequently, they are vital components in plant patterning. The involvement of these filamentous polymers in cell division, in growth and in shaping the plant cell will also be discussed in Part B of this book, where the patterning of specific cells and organs shall be detailed. The review of Baskin (2000) is recommended as a general background for the plant cytoskeleton. It is amply illustrated and clearly summarizes this subject, although ample information has been added since this review was written.

Intermediate Filaments

The information on intermediate filaments (IFs) comes from animal cells. In these cells, the IFs serve as "bones" of the animal skeletons. The classes of IF are defined by the monomers (the subunits) of the respective polymers. These include lamin, keratin, vimentin and neural filament protein. In each case, there is a central "rod"-forming domain that is flanked by globular domains. The monomers dimerize, then tetramers are formed; the tetramers are attached "head to tail" to form the protofilaments and, finally, the latter are associated laterally to form the bundle of the final IF. The evidence that plant cells contain IF in order to add mechanical strength and maintain resilience is not conclusive. Possibly, the tasks of strength and resilience in plant cells were taken over by the cell wall. No genomic sequences that encode IF-like proteins were revealed in Arabidopsis.

Actin and Microfilaments

Characteristics of microfilaments

The microfilaments (MFs) are polymers composed of actin monomers. These monomers are globular proteins that are also coined "G-actin". In plants (and in other multicellular eukaryotes), there are families of actins encoded by respective gene families. The number of the family members varies. For example, about 10 gene families of actin were found in Arabidopsis, while there are more than 100 actin genes in Petunia. The reason for so many actin genes is still enigmatic. The actin subunits are polar, and when polymerization takes place these subunits are joined in a head-to-tail manner, giving the polymer itself a polarized direction. The single-actin molecule contains 375 amino acids and attains a U-shape, in which the upper side is the minus end and the bottom side is the plus end. ATP is inserted into the open cleft, and three such G-actin subunits complex to F-actin trimers in which the clefts are closed and the ATP changed to ADP.

In the following step, the trimers polymerize or are added to existing polymers. From *in vitro* experiments, it was deduced that the formin homolog domain 2 (FH2) has a decisive role in the seeding (polymerization) of MF. Dimers of FH2 stabilize actin dimers and then accept additional actin subunits at the plus end. For its function, additional proteins are bound to FH2 (Zigmond, 2004). An *in vivo* role in MF nucleation was also reported by Ingouff *et al.* (2005). Because the subunits have plus and minus ends, the (helical) polymer as a whole also has a plus ("barbed") end and a minus ("pointed") end.

The diameter of the microfilaments is about 10 nm. The addition of actin to existing filaments is mainly to the "barbed" plus end. The actin filaments are dynamic: actin is added (mostly) at the plus end, while actin subunits are depolymerized from the other (minus or pointed) end. Hence, even when the total length of the polymer is stable, the individual actin subunits move constantly from one end (mostly the plus) to the other end. This dynamic behavior of the microfibril is called "treadmilling". This is a treadmill without a horse (or a donkey or a man) to power it. Adding subunits is probably regulated by the concentrations of the incoming (ATP-containing) and outgoing (ADP-containing) subunits in the cytosolic surrounding of the filament. Due to the treadmilling dynamics, we may envision the subunits in the microfilament as a wagon of a train that moves along the filament (e.g. from the plus end to the minus end). Any "cargo" attached to a specific subunit will thus also move along the microfilament. Special proteins can mediate the attachment of cargo to the subunits of the filament. This provides the capability of microfibrils to transport material to different locations in the cell.

The critical concentration of actin required for assembly into the microfibrils is 0.2 μM, while the concentration of actin in the cytosol is much higher (0.1 to 1 mM). There is thus a tendency to polymerize. The cell has means to sequester the level of the hydrolized actin (that comes out of the minus end), and by that to reduce its concentration. The ADP of nonsequestered actin subunits is then converted to ATP. The dynamic equilibrium of the microfilament length is thus achieved by the addition and removal of the actin subunits from the respective ends of the microfibril. Otherwise, the microfibril may either extend its length or shorten it.

For the *de novo* assembly of microfilaments, a lag phase is required, during which the subunits assemble in trimers and associate correctly to form a template for further assembly. The initial assembly is termed "nucleation". In *in vitro* systems, the nucleation can be facilitated by the addition of "seeds" that are short sections of a performed polymer.

Accessory proteins of microfilaments

While actin and its microfilaments are highly conserved among all organisms in which they exist, this is not true for the accessory proteins (i.e. proteins that are copurified with the microfilaments). Among these actin-binding proteins (ABP), there is the family of myosins. The myosin is the force-producing component of animal muscles. The plant myosins VII and XI are endemic in plants. Their molecular masses are from 90 to 230 kDa and they contain a motor domain, a calmodulin-binding motif and a coil–coil region (see Reddy, 2001, for details). There are many classes of myosin in animals, but only very few in plants.

There are two additional accessory proteins on actin microfilaments: *fimbrin* and *α-actinin*. The fimbrin links actin filaments that have the same polarity into bundles. The α-actinin links chains of actin filaments that have an opposite polarity. When the link is to parallel filaments, the linker moves to one direction. This happens during cytoplasmic streaming. Another accessory protein, *filamin*, cross-links actin filaments at an angle to form a gel-like network. A typical plant accessory protein is the pollen allergen, *profilin*. The profilin binds to soluble (monomeric) actin, and by that lowers its level, reducing the rate of incoming subunits to the polymers (see Staiger *et al.*, 1997). Additional actin-binding proteins (ABPs) cross-link the cytoskeleton to other cell components, such as membranes, enzymes or components of signal transductions. Several animal ABPs are conspicuous in their absence from plant cells. Among these are the ABPs talin, vitronectin and vinulin (see review of Wasteneys and Galway, 2003, for a discussion on additional actin accessory proteins).

Some actins and ABPs of plants could be tissue-specific so that certain ABPs could have a cargo specificity. The bean (*Phaseolus vulgaris*) genome encodes only two profilin genes, of which one is only expressed in root nodules; but when the proteins were analyzed, several isoforms of this ABP were revealed that resulted from phosphorylations of tyrosine residues. On the other hand, Swoboda *et al.* (2001) detected the same profilin in root hairs and pollen tubes of tobacco.

Involvement of actin microfilaments in cell function and patterning

Before dealing with the involvement of MF in cell function and patterning, it should be noted that recently a role was suggested for actin in the very basic cellular activity: the transcription of genes. This subject was briefly reviewed by Visa (2005). It appears that in metazoa, and probably in all eukaryotes, there is a nuclear actin; this actin interacts with nuclear myosin (NM1) and also with other proteins to regulate the transcription by the three polymerases PolI, PolII, and PolIII.

In the following, we shall mention the cell activities and growth patterns in which MF are involved. We shall skip the details because many cases, such as those concerned with patterning of cells, will be discussed at some length in Part B. Also, when dealing with microtubules (MTs), in the section below, we shall return to MF because MT and MF interact in the execution of several cell activities. Such an interaction is exemplified during nuclear division and cytokinesis.

Actin is involved in tip and diffused growth of plant cells. The tip growth occurs in pollen tubes as well as in root hair cells. The extension of pollen tubes requires the movement of vesicles to the tip, where these vesicles reach the tip's plasma membrane. This vesicle movement is mediated by actin (MF). The cascade of molecular signals that activates the dynamics of the MF is beginning to be resolved.

The cascade includes the activation of formin (which belongs to a group of poly-L-proline-containing, PLP, proteins). The formin interacts with the protein, *Profilin*. The latter, when bound to G-actin subunits, will, in the presence of active formin, add these subunits to the barbed end of the dynamic MF (which may shed ADP-containing subunits from its pointed end). The formins of Arabidopsis contain domains that indicate that they are signal transducing proteins with an N-terminal region that perceives the signal. While there are similarities between animal, yeast and plant formins, the details in the respective domains of this protein differ considerably in plant formins (see Deeks *et al.*, 2002).

An important role in the capability of MF to shape plant cells was assigned to the Arp2/3 complexes. These proteins are assumed to be important for nucleating the actin monomers (G-actins) into the MF (F-actin). In mutants of Arp2/3, the fine filaments of the F-actin in the subcellular cortical regions are changed into denser active bundles. In epidermal leaf cells, this is correlated with the lack of cell lobes. In these mutants, the differentiation of leaf cell trichomes and the elongation of root hairs are also inhibited. Also, for yet unknown reasons, functional Arp2/3 seems to be important for the integrity of the vacuole membrane and, consequently, in maintaining the turgor pressure in plant cells (see Deeks and Hussey, 2003).

The metaphor of describing the MF (and the MT) as tracks or rails along which "cargo" is transported to the various cellular locations was detailed in the review of Vale (2003). In this review, the term "tool box" motors was coined. There are ample tool boxes in the cells, and it appears that the whole transportation system evolved early in the eukaryotic evolution — possibly it was initiated even earlier. It also seems that different tool boxes are utilized to move the different types of cargo. The cargo can be small vesicles, but also large subcellular organelles such as Golgi, chloroplasts and nuclei. The motor tool boxes used in plants differ from those in animals. For example, kinesins are amply used in animal cells, while plants use several myosins for the transportation of cargo and the transportation is mainly along MF "rails". The specificity of motor tool boxes in plants, for specific cargos, is still being clarified. Evidently, the "rails" (e.g. MF) can also move each other by sliding in opposite directions, as two adjacent microfilaments.

During mitosis and meiosis, both MF and MT play major roles. These roles will be noted below.

Tubulins and Microtubules

Characteristics of microtubules

The microtubules (MT) are built of α- and β-tubulin dimers. GTP-containing dimers are added to the helices of the tubulin polymer. Thirteen such polymers are wrapped together, so that in a cross-section a ring of 13 protofilaments can be seen. During the addition of dimers to the growing microtubule, the GTPs of the α-tubulin subunits are hydrolized at some distance behind the growing point. The tip of the plus end itself is thus rich in GTP (of the α-tubulin) and is regarded as GTP-capped, which stabilizes the tip. The diameter of the microtubule is 25 nm. When the supply of the tubulin dimers is low or when the hydrolysis of GTP reaches the tip, the protofilaments may peel apart and a disassembly of the microtubule will start. As in MF and MT, there is "treadmilling", and a dynamic process can maintain the length of a MT, shrink it or elongate it. MT are very dynamic polymers, elongating and shrinking and having a short half-life (usually 10 minutes, albeit in some cases as short as 1 minute). Two drugs are commonly used to study the role of MT. One is colchicine, which prevents MT polymerization; while the other is taxol, which binds to MT and stabilizes them. The dynamics of MT are schematically shown in Fig. 17.

The accessory proteins of microtubules

The mechanochemical enzymes that convert chemical energy to work are termed "motor proteins". Two such motor proteins bind MT: *dynein* and *kinesin*. These comprise large families of proteins. They use ATP as the energy source. They function as dimers and have a similar general structure. Their heavy chain contains a globular component (the motor head) that is attached to the microtubule,

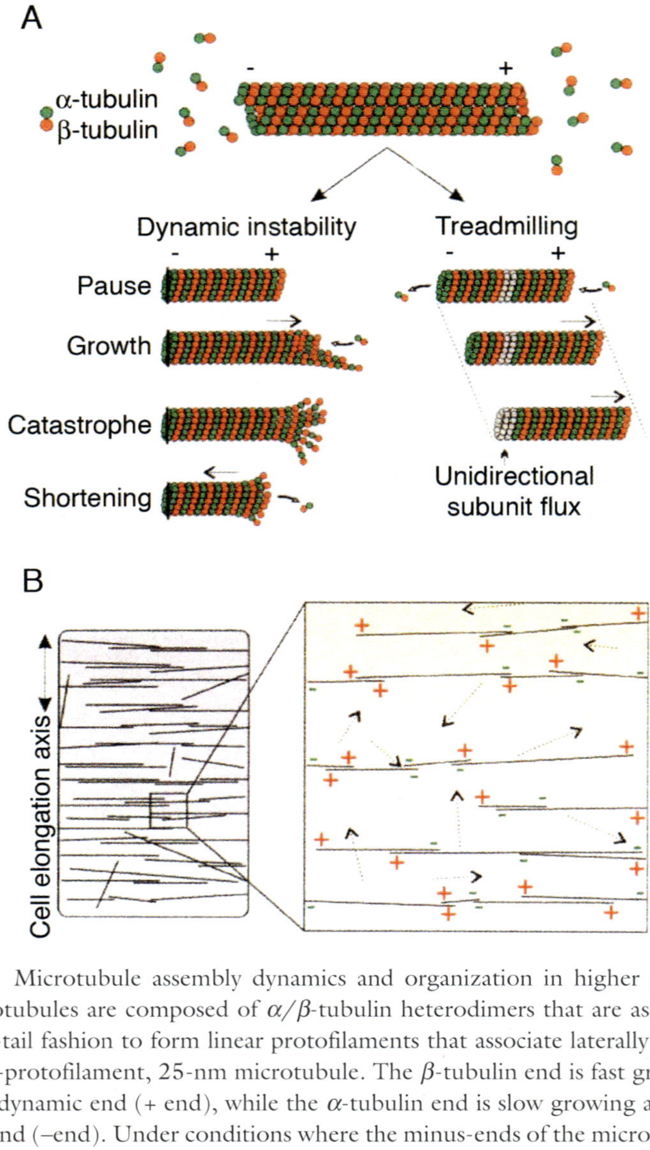

Fig. 17. Microtubule assembly dynamics and organization in higher plant cells. **(A)** Microtubules are composed of α/β-tubulin heterodimers that are assembled in a head-to-tail fashion to form linear protofilaments that associate laterally within the typical 13-protofilament, 25-nm microtubule. The β-tubulin end is fast growing and the more dynamic end (+ end), while the α-tubulin end is slow growing and the less dynamic end (–end). Under conditions where the minus-ends of the microtubules are anchored, only the freely exposed plus-ends are dynamic and exhibit periods of growth and shortening with stochastic transitions between these two phases (dynamic instability). If both ends of the microtubules are freely accessible to the soluble tubulin subunits, then the assembly dynamics may be marked by net polymerization at the

and an elaborated "tail" that can hold the cargo at its distal end. Dyneins and kinesins operate as dimers. Thus, the dimers use the two "heads" (that can also be regarded as "legs") for a firm attachment to the MT, and the combination of the two ends of the tails are used for holding the cargo. There are also light chains that support the binding of the cargo to the motor protein. Alternate binding of the two heads of kinesins to the MT can cause the "walking" of these motor proteins along the MT. Most of the plant kinesins walk towards the plus end of the MT. All the dynesins walk in the opposite direction. The analogy to biped walking is far-fetched — it is by strides of a two-headed entity, and the two heads are not mirror images (as legs) so that with each "step" each head rotates 180° in order to return to the right orientation for the binding to the MT.

A detailed description of the structure and function of plant kinesins and dyneins was provided in the review of Reddy (2001), and they were also illustrated in Baskin (2000).

The main functions of kinesins and dyneins

Microtubules and their motors, kinesins and dyneins, have important roles in cell division (nuclear division and cytokinesis) and several other cell activities. Their roles in cell division will be handled in a subsequent section. Here is a short description of their roles in the other activities. The role of MTs in the tips of pollen tubes and root

---

plus-ends and net depolymerization at the minus-ends, thereby leading to a directional subunit flux (shown by the gray subunits) and an apparent repositioning of the polymer over time (treadmilling). **(B)** Plant CMTs form arrays in specific patterns. In a rapidly elongating cell, the CMTs are predominantly oriented transverse to the elongation axis. The individual CMTs are relatively short, and they bundle into a higher order structure that spans the circumference of the cell. CMTs in an elongating cell arising in random orientations throughout the cells' cortex (as depicted by the dotted arrows), but discordant microtubule orientations do not persist for long. Furthermore, the parallel alignment of CMTs is not based on a unidirectional orientation of their plus-ends. Therefore, CMT nucleation and organization are not tightly coupled in plant cells. (From Dixit and Cyr, 2004.)

hairs was mentioned above; but in regular cells, the MTs are located just inside the plasma membrane in the cortical cytosol, at a distance of about 200 nm from the plasma membrane and in some places contacting this membrane. However, with the elongation of the cells, as happens in the roots, there is a reorganization of most of the MTs, and they are arranged perpendicularly to the direction of cell growth. This direction is parallel to the formation of cellulose fibrils. The MT thus probably have a role in producing these fibrils. When the growth (cell elongation) stops, this "transverse" MT changes once more to return to the normal orientation. Also, under experimental conditions, when protoplasts are isolated and cultured *in vitro*, the newly formed cellulose microfibrils parallel underlying MT. Such an association between cellulose microfibrils and MT was observed also in other cases of cellulose deposition, such as in the differentiation of tracheids.

It is assumed that the motor proteins associated with the MT are transporting the material required for the cellulose deposition. Another possible role of MT is in the transfer of extracellular signals. There are several indications that in plants, as was established in animal cells, the sensing of external stimulations, such as gravity, salt and pathogens, special proteins (integrins in animals) transfer the relevant information from the matrix in the cell wall through the cellular membrane to the MT. The plasma membrane keeps points of contact with the cell wall, even under plasmolysis, when the plasma membrane shrinks. At these contact points, analogs of integrins operate to transfer signals from the microfibrils in the wall to the cytoskeleton.

As for cytoplasmic streaming, there is evidence that it is operated by MF/myosins, but there could be exceptions in which MTs are active in cytoplasmic streaming.

The Cytoskeleton in Cell Division

There are two types of cell division in eukaryotic cells. One of these is the regular division that takes place in all vegetative tissues. This is termed *Mitosis*. The other division is confined to the gametes in the reproductive organs. In plants, the female gametes (macrospores) are produced in the megaspore mother cell during the process of megasporogenesis, which involves meiosis and results in the haploid megaspores. The male gametes (microspores) are produced by meiosis in the pollen mother cell. Ample information now exists on mitosis and meiosis of plants, including information on structures and molecular aspects of these processes. A background on these processes is required in order to understand the role of the cytoskeleton in mitosis and meiosis of plants. Such a background is beyond the scope of this book. The readers are therefore referred to two chapters in Buchanan *et al.* (2000), namely Chapter 11 on Cell Division and Regulation (Doerner, 2000) and Chapter 19 on Reproductive Development (Bewley *et al.*, 2000).

Here we shall be dealing mostly with mitosis in plants, because there is more knowledge on the involvement of the cytoskeleton in mitosis than in meiosis. The most prominent difference in cell division between plants and animals becomes apparent during the metaphase stage. In animal metaphase, the two poles to where the spindle MT and kinetochore MT are directed is tightly focused, while this spindle pole body (SPB) is rather diffuse in plant cells. The other difference is during the telophase stage. During this stage, a phragmoplast is formed between the two, recently divided, nuclei. Thereafter, the cortical MT reappear and the phragmoplast develops as a curtain between the two newly formed cells. This curtain is then the basis for the new cell wall between the daughter cells. In animal telophase, a contractile ring is formed between the two newly formed nuclei. This ring narrows until the two daughter cells are separated (cytokinesis).

The mitotic spindles

The mitotic spindle of plant cells becomes apparent during the prophase stage. An early indication for the entry of the cell into this stage is the disappearance of cortical MT and the formation of a circular *preprophase band* of MT that encircles the "equator" of the future spindle. In this band, MF are running parallel to the MT. The nuclear envelope disintegrates at prometaphase. When the spindle MTs are formed, their minus end reaches the poles while the plus end is towards the middle of the spindle. Some of these spindle MTs run almost from pole to pole, while others reach only the equator. Of the latter, some MTs form bundles that will later bind to the kinetochores of chromosomes. These are the kinetochore MTs.

The spindle MTs start to be formed before the nuclear envelope breaks down (this envelope breaks down in angiosperms and in animals, but is retained during mitosis of many fungi). They are very dynamic and their half-life lasts only about 1 minute. The half-life of kinetochore MTs is longer, but they also do not last during the whole mitosis. When MT are renewed, they grow from the poles towards the equator (extending at the plus ends). As for kinetochore MTs, they undergo active treadmilling and subunits are added (mainly) at the kinetochore side. This brings up an old joke, in which two children who are watching a parade ask from where soldiers are growing. One claims that soldiers grow from their heads because all stand on the same level, but some heads are higher than others. The other child disagrees. The pants of some soldiers reach their shoes, while in other soldiers one can see the legs between the shoes and the pants. Hence, soldiers grow from their legs.

At the end of prophase, after nuclear envelope breakdown, the spindle MTs initiate their interaction with the chromosome mainly by attachment of the kinetochore MT to the kinetochore. The protein/DNA composition of plant kinetochores is not yet fully clarified. The kinetochore splits into two sister kinetochores; each of these sisters is linked to MT that leads to opposite poles. The plus end of the MT is adjacent to the kinetochore, while the minus end points to the poles. The chromosome (actually the kinetochores of the chromosomes) then all move either away from the pole or towards the pole until they are all situated in one plate between the two poles. This is the metaphase plate of the chromosomes. Removal or addition of tubulin subunits, mainly at the attachment to the kinetochore, can move the kinetochore (and with it the chromosome) towards or away from the pole. This kinetochore dynamic will result in chromosome movements.

While situated at the metaphase plate, the chromosomes of dividing animal cells may oscillate, but still roughly maintain the plate location. No such oscillation is observed in dividing plant cells. The change in location takes place during the anaphase stage. Then the centromeres split and the two sister kinetochores, each with its sister chromatid (derived from chromosome replication) is pulled by the MT towards the poles. It is not fully clear how this movement is operated. Several nonexclusive ways were suggested. Minus-end-directed motor proteins may exert the required pulling force. The kinetochore MTs may depolymerize at either the kinetochore end or near the poles, shortening these MTs. The kinetochore MTs could be pulled towards the poles by motor proteins that are bound to other microtubules, or even to the nonmicrotubular matrix. The poles themselves may move apart from each other.

In animal cells, the concept of "polar wind" was suggested, in which MTs that are not bound to the kinetochore are also active in moving the chromosomes during mitosis. But, such a "polar wind" does not seem to apply to the plant chromosome movement. Evidence for that comes from the behavior of acentric chromosome fragments in plants. These acentric fragments do not move actively to the poles, but rather move in various directions during the stages of mitosis (i.e. equatorially in anaphase but polewards in metaphase and telophase).

There is evidence for the depolymerization of the kinetochore MTs to be a main force in chromosome movement during anaphase. Such evidence came from treating cells that are in mitosis with drugs (e.g. low levels of colchicine), thermal treatments and exposure to various levels of calcium (high calcium levels are correlated with depolymerization of MT). The evidence from calcium injection into stamen hairs indicated that very high calcium levels quickly depolymerized the kinetochore MT, stopping anaphase movement of chromosomes. However, lower doses of calcium accelerated this movement and probably mimicked the regulated depolymerization at the ends of the kinetochrome MT.

The cytoskeleton during cell division

The establishment of two separate cells that follow nuclear division is termed *cytokinesis.* In plants, cytokinesis involves the formation of cell walls between the two daughter cells. Notably, in plants (as in animals), nuclear division is not necessarily followed by cytokinesis. When a syncytium is formed, nuclei divide, but no cell walls are formed. This is typically the situation in the endosperm. The plane and the location of the new cell wall are of major significance in plant patterning. These planes and locations are marked at a rather early stage by the establishment of the *phragmosome.* This is a thin layer of cytoplasm that may be visible rather early, before the nuclear division (mitosis) is apparent. It can be detected already in the S-phase (DNA replication) in cells that will go into mitosis. The phragmosome contains both actin MF and MT, and both of these fibers also participate in moving the nucleus, which is about to undergo mitosis, into the appropriate location in the cell where its division will take place. Such phragmosomes were revealed in vacuolated plant cells, such as cambial cells or mature cells that were stimulated to divide. It is not clear whether or not phragmosomes also form in meristematic cells that lack large vacuoles.

The phragmosome is a temporary structure and its exact role in cytokinesis is not clear. Another structure is clearly relevant in the future formation of the cell wall between the daughter cells. This is the *preprophase band.* The term is misleading because it is usually formed during prophase. This band starts to be formed as a ring of actin MF and MT that encircles the prophase cells, inside of the plasma membrane, and marks the location where the future cell wall will be initiated. At a certain stage, during the disintegration of the nuclear envelope, there is a depolymerization of the preprophase band. The cortical MTs disappear, but the actin MFs are retained at the location of this band.

The actual partition between the two daughter cells during plant cytokinesis is initiated by the phragmoplasts. This cytoplasmic curtain is built of MF, to which myosin motors are bound, as well as MT. Existing spindle fibers as well as newly polymerized MF and MT join in the formation of the phragmoplast, and are arranged perpendicularly to the plane of cell division. The fibers have a polarity; their minus ends are towards the poles. The phragmoplast is built during an interaction between MF and MT at the equatorial plane. While the phragmoplast is initiated at the center of this plane, the formation continues centrifugally to meet the existing cell wall. The center is later cleared and a ring band is formed near the existing cell wall. The MTs involved in the construction of the phragmoplast and its changes are highly dynamic, and the half-life of each MT was estimated to be about 1 minute; but the process of phragmoplast plate and ring formation may take several hours.

How the phragmoplast contributes to the future cell wall that is being produced between the two new daughter cells is still being investigated. The phragmoplast harbors large quantities of vesicles that may contain the building stones for the new plasma membrane and cell wall, which are formed centripetally.

The roles of the cytoskeleton in plant cell morphogenesis and other cell activities

The roles of MT and MF in shaping plant cells were studied by several means, such as the use of drugs that specifically antagonize the formation of either of these polymers. For example, the changes that leaf epidermal cells undergo during the formation of trichomes (that include a stalk and three or four branches) was studied by disrupting actin polymers (MF) or MT by specific drugs. The results indicated that both MF and MT are required for the morphogenesis of trichomes. Genetic studies were also utilized to identify the role of cytoskeleton polymers in trichome morphogenesis. The general strategy was to assemble many mutants (e.g. in Arabidopsis) that show abnormal trichome phenotypes, and subsequently focusing on the genes involved in these phenotypes, to provide information on genes that are essential for the formation of the respective cytoskeleton polymers. In such studies, not only genes that concern MF and MT were revealed, but also genes that encode motor proteins such as kinesin-like proteins. The morphogenesis of trichomes shall be handled in Chapter 11 of this book.

The cytoskeleton seems to be also essential for cell-to-cell communication in plants. Although plant cells are surrounded by a cell wall, there are specific connections between the cytoplasms of adjacent cells. These are the *plasmodesmata*. Through these plasmodesmata, a symplastic network that connects between plant cells, is established. Proteins (up to a certain molecular mass) and RNAs can move through these plasmodesmata. The transport of proteins and RNAs via plasmodesmata involves molecular motors of the cytoskeleton. For example, actin polymer depolymerization drugs enlarge the plasmodesmata and consequently facililtate trafficking of macromolecules.

An interesting suggestion regarding cell-to-cell communication was made by Baluska *et al.* (2005a). These investigators noted that plants synthesize neuronal-like molecules that may fulfill some "intelligent behavior", such as the discrimination in root cells between self and non-self. This requires neuronal synapse-mediated exchange of information. These authors also suggested that synapses are involved in auxin transport, and that in this synaptic transport, actin and myosin are involved. While there were several experimental results that supported this suggestion, further evidence is required to substantiate this.

An active role of myosin and kinesin in the transport of macromolecules was indicated by certain experiments. Also, calcium channels are probably affected by cytoskeleton components. In this connection, it should be noted that calcium itself probably affects actin-based protein motors; free Ca^{2+} above 10 μM will cause fragmentation of MF and inhibit the motor function.

A detailed update on the involvement of cytoskeleton components in plant cell activities was provided by Reddy (2001). See also references to later studies on this topic in Hepler *et al.* (2001), Mathur and Hulskamp (2002), Wasterneys and Galway (2003), Smith (2003), Brembu *et al.* (2004), Lee and Liu (2004), Mathur (2004), and Dogterom *et al.* (2005).

While ample information on the cytoskeleton and its role in cellular activities in plants has been provided in recent years, and the knowledge of the involvement of MF, MT and their motor protein, increased considerably, we are still ignorant with respect to major issues that concern the patterning of plant tissues in which the cytoskeleton is involved. These issues will be handled in Part B of this book, but here is an example of a major unsolved issue.

Let us look at the differentiation of a trichome: first, there is the decision of a leaf epidermis cell that it shall differentiate into a trichome. How does such a determination take place? We then go further to the cell divisions and cell elongations that render the cell, which was destined to become a trichome, to actually become differentiated into such a structure. One requirement is that the cell divisions should be at the correct planes. How are the planes of cell divisions (the locations of the preprophase bands, the axes of the spindles and, consequently, the planes and locations of the phragmoplasts) determined? This latter question is still without an answer, although it is one of the fundamental questions on the control of patterning, as indicated in the Introduction (Chapter 1): how is a three-dimensional cellular structure derived from the linear information that is stored in the genome?

Chapter 4

PHYLOGENY, TAXONOMY AND GENE DUPLICATION

The inclusion of a chapter on phylogeny and gene duplication in a book on the patterning of angiosperms may not appear to be an obvious choice at first sight. The justification for doing so will become clear during deliberations in this and subsequent chapters, but already here, this justification will be handled briefly. Many coding sequences that served during the early evolved organisms for the syntheses of certain proteins were modified during evolution and presently serve for the syntheses of modified proteins. The latter are now used in novel tasks. I referred to this process of using "old" building stones, after due modification, for the construction of "new" buildings in the Introduction, where this metaphor was termed as "The Pillars of the Mosque of Acre". This phenomenon is prevalent in proteins that are structural components and enzymes as well as in transcription factors. All these proteins may have important roles in patterning. Being aware of the phylogenetic origin of these coding sequences and the taxonomic relationships of the plant species in which these sequences (in the original and the modified forms) occurred is helpful for understanding the control mechanisms of patterning.

We shall also see below that gene duplications, which are caused by any of several processes (e.g. polyploidization, chromosome arm duplications, unequal crossing over, inverse repeats of DNA sequences) are a common cause for the ability of organisms to synthesize novel proteins, increase versatility and alter cellular and/or organ differentiation. While this is true for all eukaryotic organisms, examples for this process will be provided mainly from angiosperms.

Phylogeny and Taxonomy are Relevant to Plant Patterning

In a rather educative review, Meyerowitz (2002) noted the relevance of phylogeny for understanding pattern formation. In this review, the comparison was mainly made between the two kingdoms of plants and animals. Several features of this review were handled in the Introduction and shall therefore only be summarized here. There is good evidence for the existence, about 1.6 billion years ago, of unicellular eukaryotes, which were the ancestors of both plants and animals. After the split between the two kingdoms, there was a long period during which both kingdoms retained their unicellular state. When plants and animals became multicellular, they had behind them hundreds of million years of separation. Hence, in the period during which multicellular patterning evolved, each of the two kingdoms developed its own mechanisms of patterning control. However, the "logic" of developmental control in the two kingdoms was similar.

Meyerowitz provided examples, such as the HOX homeobox genes of animals, that have a decisive role in segmentation (e.g. in the early embryo of flies). HOX homeobox genes have no role in the patterning of plants, and the two HOX genes that were revealed in plants serve in very different tasks. On the other hand, MADS box proteins are master regulators in plant patterning (e.g. floral organ identities) and no homologies for these proteins were found in animals. It appears that the two nonhomologous proteins had their origins in the common unicellular ancestor of plants and animals; but then, each kingdom recruited one type or another type of these proteins for their respective patterning.

A similar situation probably existed with respect to symmetry specification. In animals, the dorsal–ventral specification is controlled by very different proteins than the control of *abaxial–adaxial* symmetry of leaves (the abaxial side of leaves is the side that is away from the shoot, the "lower" side; while the adaxial side is the one that faces the shoot, the "upper" side). Again, the common logic of patterning is that there should be a protein that specifies symmetry. However, the specific proteins that fulfill this requirement are very different in the two kingdoms. The abovementioned review noted another "twist" with respect to symmetry control. In the dorsal–ventral symmetry of *Drosophila*, a TolI protein is involved. TolI proteins are also involved in the innate immunity of flies against pathogens. There does not seem to be a cognate of TolI that is involved in symmetry control of plants, hence TolI-like proteins probably have no role in plant patterning. However, plants do have certain proteins that share domains of TolI (leucine-rich repeats). These plant proteins have a role in the defense response against plant pathogens. It could be that TolI proteins originally evolved as means of cell-to-cell recognition and some TolI-like proteins fulfill this task in plants. Again, the "Pillars of the Mosque of Acre" metaphor comes to mind.

These examples from Meyerowitz (2002) are from two kingdoms that were branched off from their common ancestor very early in the eukaryotic evolution. But, there are also examples from taxa that are much more closely related. Evidence for the claim that specific changes in regulatory proteins are taking place in closely related plant species that lead to different patterning comes from the study of Barrier *et al.* (2001). These investigators studied genera of Asteraceae that are endemic in specific Hawaiian islands. Actually, some of these are unique to one of the islands.

The proteins of the floral regulatory genes (homologs of the *APETALA1* and *APETALA3* of Arabidopsis) were investigated in the geographically isolated (but taxonomically closely related) species of Asteraceae. The changes in floral patterns were correlated with the unique changes in the respective regulatory genes. It appeared that each geographically isolated species adopted its specific modified regulatory gene. Apparently, the rate of diversification in regulatory genes (e.g. transcription factors) is greater than the diversification of other genes. Was this the result of selective pressure in the specific environments in which the isolated species evolved?

Clearly, some basic mechanisms that involve chromatin remodeling have very ancient roots in eukaryotic evolution. Hence, the Polycomb group of proteins involved in epigenetics is similar in plants and animals. On the other hand, there are many genes that do not have their origin in the common eukaryotic ancestor of plants and animals, but entered into plants by horizontal transfer. Such genes came with the cyanobacteria-like cells that entered the ancestors of unicellular plants, causing the establishment of the chloroplasts. Some of these genes, which came with the cyanobacteria-like cell, "jumped" from the chloroplast genome and were integrated into the nuclear genome of plant cells. One may say that the latter genes used the chloroplast as a "Trojan horse" in order to establish themselves in the nuclear genome (among these are genes involved in the perception of red/far-red light, perception of ethylene signals, cytokinin perception, and perception of osmotic pressure). These genes, and their encoded proteins, have obviously no cognates in animals.

Phylogeny of Angiosperm — General Remarks

The exact time (in either years or geological period) in which angiosperms branched off from other plants (e.g. gymnosperms) is still a subject of dispute, but most probably it occurred in the late Jurassic period or the very early Cretaceous period. Most scholars in this field agree that there was a blooming diversification of angiosperm clades during the early Cretaceous. It is also agreed that a great deal of this early angiosperm diversity became extinct. In spite of this massive extinction, it was estimated (Heywood, 1993) that there are presently 260 000 angiosperm species.

The timing of splits can be evaluated (or rather roughly estimated) by several approaches. One approach is paleobotany, by which remnants (these can be pollen grains) found in well-dated geological layers are used to estimate the age of these remnants. A very different approach is by sequencing DNA regions, such as the space between the 18S and the 25S rRNA genes. Different approaches can lead to very different estimates. For example, the earliest angiosperm pollen grains were found in a layer of the very early Cretaceous, estimated to be 144 million years old (Friis *et al.*, 2005), while an estimate based on DNA sequences of the spaces between the rRNA genes is that already 170 to 235 million years ago there was the split *within* the angiosperms between monocots and dicots. When the age of this split was estimated on the basis of chloroplast DNA data (Wolfe *et al.*, 1989), the estimate of the monocot–dicot divergence led to a split date of 200 million years ago, and the branching that led to Brassiales was estimated to have happened 90 million years ago.

When mitochondrial DNAs are available for sequencing, these can also serve in phylogenetic estimates, but ironically estimates based on the sequencing of nuclear DNA, mitochondrial DNA and chloroplast DNA may yield rather different results. An example of such discrepancy was reported by Yang *et al.* (1999b) with respect to genera of the family Brassicaceae (to which the model genus Arabidopsis belongs). This brings up the story of "Mother Eve". In 1987, Wilson and associates (Cann *et al.*, 1987) studied the variability in human mitochondrial DNA and derived an estimate: that the original "mother" of all present humans lived about 150 000 to 200 000 years ago in Africa. The descendants of this "Mother Eve" spread first to Asia and then to all over the world.

Weinscoat (1987) wrote in the same issue of *Nature* (in News and Views) that the human phylogeny can also be based on the DNA variability of the Y chromosome to estimate the location and the time of "Father Adam". If this will be done, could it be that Mother Eve and Father Adam were spatially and temporally separated? My humoristic suggestion was that the daughters of Mother Eve mounted wild horses (*Equus przewalskii*) and reached the steppes of Mongolia. This suggestion was based on the claim of George and Ryder (1986) that *E. przewalskii* split from the rest of the *Equus* genus in Africa about 150 000 years ago and reached Mongolia. This wild horse is the progenitor of the riding horse. Luckily, the editors of the *Proceedings of the 6th International Citrus Congress*, in which I wrote on the taxonomy of Citrus cultivars (Galun, 1988) did not eliminate my humor.

While different approaches to constructing phylogenetic trees may result in different trees, in most cases, the various approaches complement each other. An example for the latter case is from my own laboratory (Perl-Treves and Galun, 1985; Perl-Treves *et al.*, 1985), in which the phylogenetic relationships among species of the genus *Cucumis* were studied. Estimates based on chloroplast diversity gave the same results as those based on nuclear diversity. Both approaches indicated that all the African species are closely related, but the cucumber (*C. sativus*) is an "outsider" and native of India. The split was estimated to have happened about 90 million years ago, at about the same time that India was separated from East Africa and moved towards the Asian continent. If we accept that the Indian subcontinent is moving at a velocity of 5 cm per year (or 50 km per million years), the distance between East Africa and the present India can be covered in 90 million years.

Now back to the phylogeny of angiosperms. In spite of the reservations indicated above, there is a fair agreement on the basic branchings of the phylogenetic tree of angiosperms. Figure 18 presents an example of such a tree. This tree was constructed by a team in the Missouri Botanical Garden (see Stevence, 2005; a constant update of the phylogeny of angiosperms is provided by the website http://www.mobot.org/MOBOT/research/APweb/). This team furnished a convenient source for the phylogeny of orders of seed plants as well as the relationships between families of these orders. Table 2 provides connections between angiosperm orders and taxa that serve as model plants or are cultivated genera. Similar phylogenetic trees of angiosperms were presented by numerous other authors (Soltis *et al.*, 1999; Magallon and Sanderson, 2001). Phylogenetic trees of specific families, such as Brassicaceae, are also available (Yang *et al.*, 1999a).

Amazingly, the overall taxonomic relationships among angiosperms were correctly estimated many years ago. This brings up another memory. In 1974, while I was collecting literature for my first book (Frankel and Galun, 1977), I was a guest of the late Professor J. Heslop-Harrison, who was then the Director of the Royal Botanic Gardens in Kew (London) that included the herbarium and a large Botanical Library. I was given a long desk in the library near a window that overlooked the tides of the river Thames. On my two sides were shelves loaded with books of and about the work of Carolos Linnaeus (1707–1778). I was already then amazed by the accuracy of the taxonomy of Linnaeus.

By merely considering morphological features of plants, he categorized plant species, genera and families rather correctly. The differences between the taxonomy of Linnaeus and the presently suggested taxonomies, which are based on molecular biology methodologies or on paleobotany, are rather small. This means that the ramifications of the

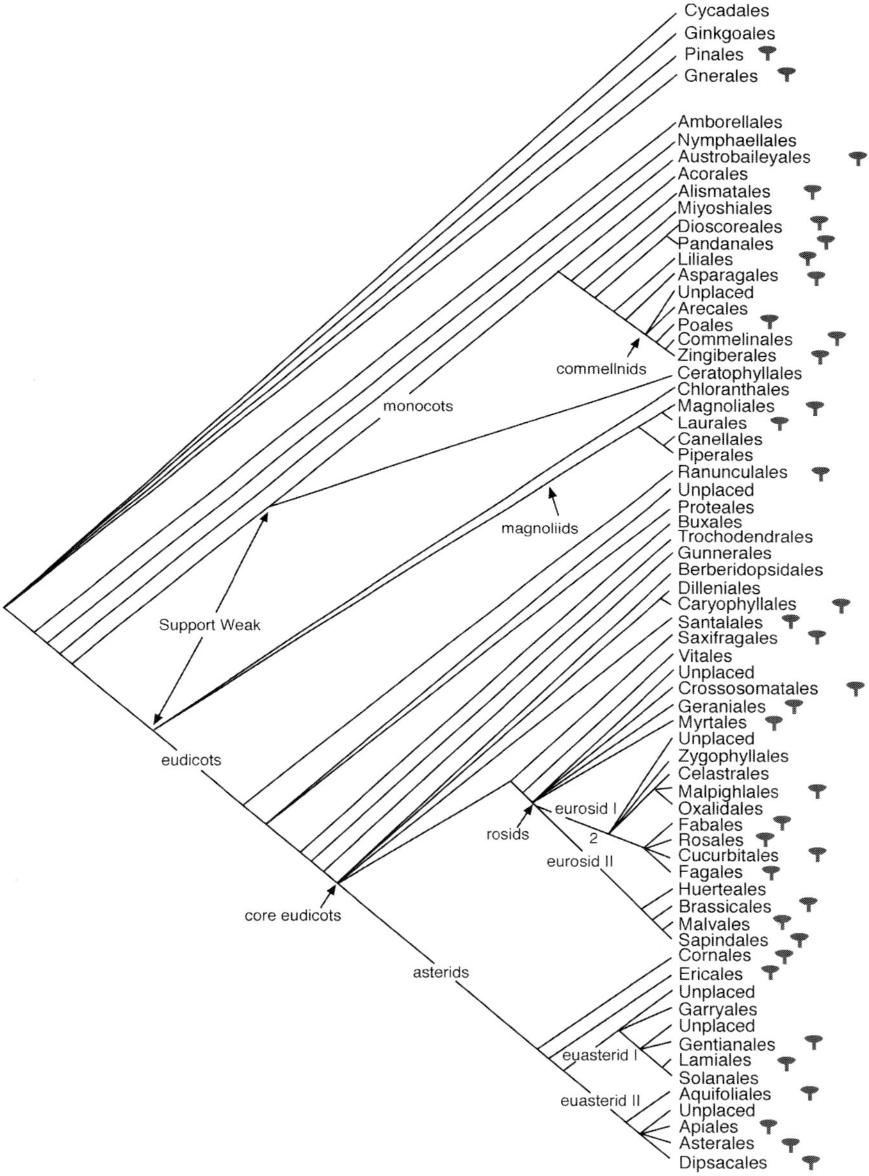

Fig. 18. Phylogenetic tree of seed plants. (From Stevens, 2005. Angiosperm Phylogeny website, version 6, May 2005, http://www.bobot.org/MOBOT/research/APweb/).

Table 2 Familiar Taxa and the Clades in which They are Placed in the Phylogenetic Tree shown in Fig. 18

Familiar Taxa/Model Organisms	Clade (Family or Order)	Larger Group
Nymphaea (water lilies)	Nymphaeaceae	
Magnolia	Magnoliales	Magnoliids
Orchidaceae (orchids)	Asparagales	Monocots
Poaceae (grasses): *Triticum*, *Oryza*, *Zea*	Poales	Monocots
Fabaceae (legumes)	Fabales	Eurosid I
Rosaceae (rose family): *Rosa*, *Malus*	Rosales	Eurosid I
Arabidopsis, *Brassica*	Brassicales	Eurosid II
Dianthus (carnation), *Spinacia*	Caryophylales	Eudicots
Cornus (dogwoods)	Cornales	Asterids
Hydrangea	Cornales	Asterids
Lamiaceae (mints)	Lamiales	Euasterid I
Antirrhinum	Lamiales	Euasterid I
Solanum (potato, tomato)	Solanales	Euasterid I
Nicotiana (tobacco)	Solanales	Euasterid I
Petunia	Solanales	Euasterid I
Apium (celery), *Daucus* (carrot)	Apiales	Euasterid II
Helianthus (sunflower)	Asterales	Euasterid II

(From Soltis *et al.*, 1999).

phylogenetic tree of plants, as seen by Linnaeus about 250 years ago, is still valid today. Notably, the work of Linnaeus predated the work of Charles Darwin (1809–1882) by many years, but the grandfather of Charles, Erasmus Darwin (1731–1802), was a contemporary of Linnaeus. It should be noted that Erasmus Darwin already expressed ideas on the evolution of organisms, hence he preceded his famous grandson in these ideas. “Old masters”? Indeed! All the three scholars mentioned above lived 72 ± 3 years. Not being disturbed by handling complicated instruments and elaborate experimental procedures, these “old masters” had ample time to exercise their mental facilities.

Gene and Genome Duplication

The relevance of genome and gene duplication in the evolution of organisms was already emphasized by Ohno (1970). While genome duplication (polyploidization) in plants was recognized as a very common phenomenon about 100 years ago, the awareness of the extent and frequency of gene duplication has increased substantially by the results of the sequencing (or even partial sequencing) of the genomes of angiosperm plants.

The synteny between the nuclear sequences of the fully sequenced Arabidopsis and sequenced regions in other species revealed an interesting "picture". Such comparisons were made between the Arabidopsis genome and the then available sequences from the soybean genome (Grant *et al.*, 2000), as well as between Arabidopsis and sequences from the tomato genome (Ku *et al.*, 2000). In both cases, it became evident that the two compared genomes had a common ancestor, but also that after the diversions between Brassicaceae and Leguminosae and between Brassicaceae and Solanaceae, the ancestors of Arabidopsis underwent genome duplications (polyploidizations) that were followed by massive losses of DNA sequences.

The phylogenetic trees (see below) indicate that the diversion between the tomato (Solanaceae) ancestor and the ancestor of Arabidopsis (Brassicaceae) occurred much earlier (over 100 million years ago) than the diversion between the ancestor of soybean (Leguminosae) and the ancestor of Arabidopsis. There are indications that one genome duplication occurred already in the common ancestor of all these mentioned taxa. The team of Steven Tanskley (Vision *et al.*, 2000) attempted to bring order into this subject, and to present a timetable for the large-scale duplication events that occurred in the ancestors of Arabidopsis.

While the validity of maps of duplicated blocks that were provided by Vision *et al.* (2000) is not questioned, the interpretations are open for discussion. The subject of nucleotide sequences in the duplicated pairs of sequences in Arabidopsis, was studied in more detail by Zhang *et al.* (2002). Further effort to estimate the timing of the whole genome duplication in Arabidopsis was made by Ermolaeva *et al.* (2003). These authors suggested that the possible date for such a duplication is about 38 million years ago (this means after the diversion of Leguminosae from Brassicaceae), but pointed also to the possibility that the duplication date was more recent.

A further study on the locations of duplicated genes in Arabidopsis was conducted by the team of Kennet Wolfe of the Trinity College in Dublin, Ireland (see: Woolfit and Wolfe, 2005). These authors used advanced approaches to reveal these duplications, which were based on amino-acid-derived sequences rather than on nucleotide sequences. The timing of segment duplications and their scale were evaluated in Arabidopsis. The ages of the duplications of blocks were estimated. The results indicated that there were two events of polyploidization: an old one, dated right after the monocot/dicot divergence; and a young polyploidization that happened before the Arabidopsis/*Brassica rapa* split, probably during the early emergence of the crucifer (Brassicaceae) family, 24 to 40 million years ago.

A wide-scale analysis of gene duplications was performed in 14 model plants, including major crop plants such as wheat, maize, cotton, tomato, potato and soybean. In all of the analyzed species, pairs of paralogous genes were detected, and the data led to a strong suspicion that events of gene duplications caused the formation of such pairs. The manner by which redundant genes were modified during evolution to better serve species functions was analyzed by Moore *et al.* (2004; see: Moore and Purugganan, 2005) in two paralogous gene pairs of Arabidopsis. These pairs control floral organ identity (SEP1 and SEP2) and seed shattering (SHP1 and SHP2), respectively. In Arabidopsis, tandem gene duplication indicated that polyploidization is not the only source of gene duplication in this model plant. Whether there was genome duplication (polyploidization) or duplication of genes, in the presently analyzed plants, up to 90% of the duplicated genes have returned to a (functional) single-copy state. The loss of DNA sequences and further rearrangements of the genomic DNA render the timing (the dates) of the various duplications a problematic task. The studies of Blanc and Wolfe (2004a, 2004b) revealed that the retention of duplicated genes in Arabidopsis is not at random: genes involved in signal transduction and transcription have been preferentially retained, while those involved in DNA repair were preferentially lost.

In the Middle East, it is said that the only sure thing about dates is that they grow on palm trees (rather than being estimated on the basis of phylogenetic trees). With this remark, we shall visit the Middle Eastern genera *Triticum* and *Aegilops*, and handle studies conducted by the team of Moshe Feldman and Avraham Levy of our department at the Weizmann Institute of Science (Ozkan *et al.*, 2001; Shaked *et al.*, 2001; and Eckhardt, 2001 for review of these studies).

It is widely accepted that most (almost all) existing angiosperm species underwent, during their evolution, one or more steps of polyploidization. This polyploidization can result from either doubling the chromosomes in the same species (autopolyploidization) or from mating between two different species (allopolyploidization). The polyploidization could then lead to losses of genomic sequences. Losses of noncoding sequences should not affect the vitality of the newly formed polyploid (unless it includes sequences that are transcribed to essential microRNA). The loss of redundant genes should also not adversely affect selection during evolution.

These basic features of duplication and losses of genomic sequences are actually common not only to angiosperms, but were also documented in animals. The Feldman–Levy team based their study on previous findings on the elimination of DNA sequences in polyploid wheats (Feldman *et al.*, 1997) and then focused on *Aegilops* and *Triticum*, in which the investigators had ample experience and cytogenetic knowledge, and in which species of three levels of ploidity are available (see Fig. 19 for phylogenetic relations in wheat-tribe *Triticeae*). Allopolyploidization can be achieved in these genera and the authors obtained numerous synthetic intergeneric allopolyploids between *Ae. sharonensis* and *T. monococcum*, as well as between *T. turgidum* (durum wheat) and *Ae. speltoides*. The fate of specific genomic DNA sequences was followed, and it was found that after such allopolyploidizations there is a massive (and rather quick) elimination of DNA sequences that was found to be typical for the parental pair and apparently occurred mainly in noncoding regions.

The first elimination was found already in the F_1 generation, but proceeded to F_2 and F_3. This process of DNA sequence elimination can lead, in nature, to novel alloploidic species with homeochromosomes related to the ancestral chromosomes. An intrageneric cross between *Ae. sharonensis* and *Ae. umbellulata* showed that 14% of the

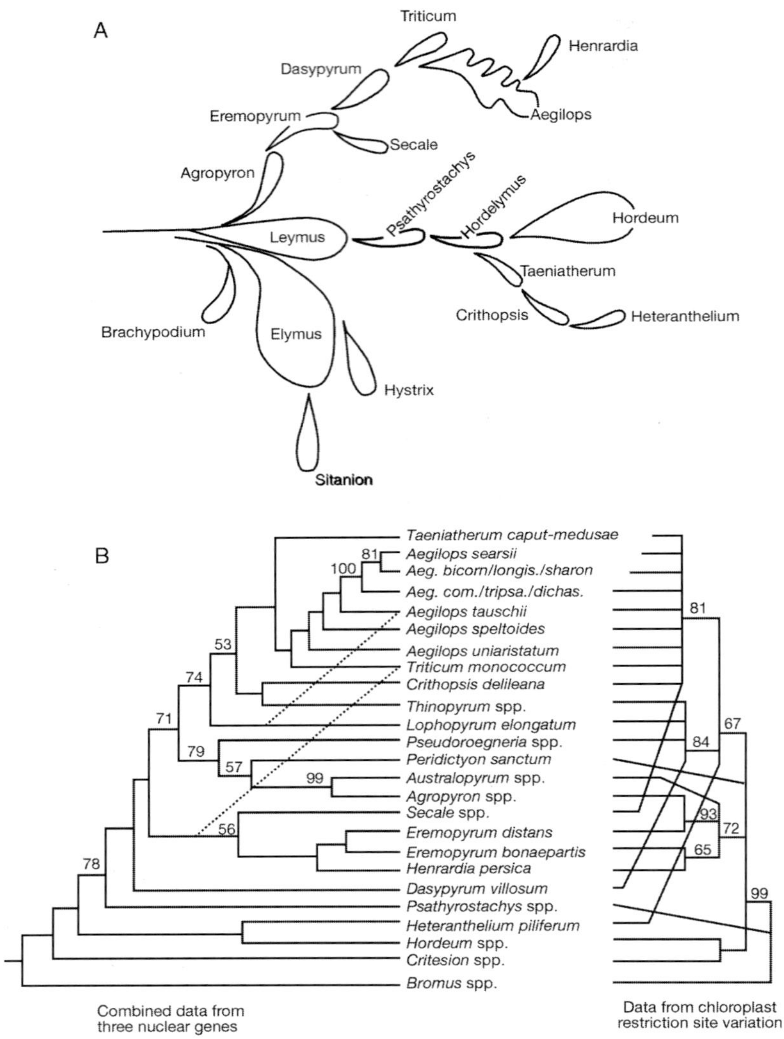

Fig. 19. **(A)** Diagram of relationships in Triticeae. (From Clayton and Renvoire, 1986.) **(B)** Evolutionary relationships in the grass tribe Triticeae. Relationships identified on the basis of three nuclear genes are shown on the left, and relationships identified on the basis of the chloroplast genome are on the right. Crossed lines indicate that there are many differences between the chloroplast and nuclear gene phylogenies. The differences among the various sources of phylogenetic data may indicate a history of introgression. Dotted lines indicate well-supported differences among the nuclear genes. Numbers above the branches indicate bootstrap support values. (Redrawn from Kellogg *et al.*, 1996).

analyzed loci from *Ae. sharonensis* were eliminated, compared to only 0.5% from *Ae. umbellulata*. Changes in cytosine methylation in the genomic DNA after synthetic alloploidization were also revealed in the progeny of the interspecific crosses, and concerned both codings and noncoding sequences. The authors suggested that these genomic duplications, which represent similar duplications in nature, render the resulting plants to a great plasticity and may lead to novel functions in the resulting plants. It is not yet known what mechanism causes this massive postpolyploidization elimination, but clearly polyploidization triggers an increase in the plasticity of plant genomes.

A later effort to unravel the evolution of angiosperm genomes by phylogenetic analysis of chromosome duplication events was made by Bowers *et al.* (2003). These authors compared Arabidopsis with other taxa, but rather than analyzing DNA sequences, they focused on the amino acid sequences encoded by the plants' genes. This study indicated that a genome-wide duplication postdates the divergence of Arabidopsis from most dicots, and the authors noted the propensity of angiosperms to undergo gene losses. However, what caused the later shrinking of the Arabidopsis genome, which is now believed to be derived from ancestors that underwent several rounds of chromosomal duplications?

Another example that indicates the relevance of gene duplication for the understanding of plant patterning comes from a study by the R. Carpenter and E. Coen team of the John Innes Centre of Norwich, UK. The study (Keck *et al.*, 2003) concerns the patterning of floral organs in Arabidopsis and *Antirrhinum* (which shall be detailed in the last chapter of this book). In the model of floral development coined as the "ABC model", there are gene classes for floral organ determination: A, B and C. With respect to B- and C-class genes, there is a clear conservation between Arabidopsis and *Antirrhinum*. However, previous studies failed to find in *Antirrhinum* a gene that is similar to the A-class gene, *AP2*, of Arabidopsis. A-class genes (e.g. *AP2* of Arabidopsis) are essential for the determination of the sepal whorl and, together with B-class genes, are essential for the determination of the petal whorl.

AP2 is also antagonistic to C-class genes (C-class genes are essential for carpel formation and, in combination with B-class genes, for stamen formation). While no gene similar to *AP2* could be found in *Antirrhinum*, *Antirrhinum* has two corresponding genes, *LIP1* and *LIP2*; these are similar genes, and when *both* are mutated, an A-class–deficient phenotype is obtained. But, these double mutants are not involved in the antagonism to C. Hence, *LIP* genes (unlike *AP2*) are not required for the repression of C.

The full story is rather elaborate and here I shall summarize the well-based conclusions. There is sufficient similarity between *AP2* of Arabidopsis and the *LIP* genes of *Antirrhinum* to indicate that these genes were derived from a common ancestor. In an ancestor of *Antirrhinum*, there was a gene duplication and the derived sequences underwent further modifications. The modified (*LIP*) genes retained their role as A-class genes, but both derived genes lost the antagonistic effect to C, and the latter role was assigned in *Antirrhinum* to different genes (*FIS*, *FLO*, *CHO* and *STY*). In Arabidopsis, the dual roles of an A-class gene as well as a repressor of C were retained in the same gene (*AP2*).

Another example does not come from angiosperms, nor does it involve patterning. It concerns the fermentation capability of yeast (*Saccharomyces cerevisiae*) in the production of alcoholic beverages. Aristotle would say that the purpose of yeast is to satisfy the pleasure of man. Yeast quickly converts the sugars of fresh fruits (and germinated cereal grains) to ethanol. In this process, sugar is converted to pyruvate. The pyruvate loses CO_2, producing acetaldehyde, and the latter is reduced by alcohol dehydrognease 1 (*Adh1*) to ethanol. The alcohol accumulates, but it is also toxic to microorganisms. Most microorganisms cannot grow at the level of alcohol accumulation performed by yeast. Hence, yeast has a competitive advantage in the consumption of fruits. However, yeast is also able to consume alcohol by exploiting the gene *Alh2*, which is a homolog of *Alh1*.

Thompson *et al.* (2005) studied these *Alh1* and *Alh2* genes and their common ancestor. They suggested that the ancestor, Alh_A coded for an enzyme that was not useful in the consumption of ethanol, but this enzyme was able to recycle NADH which is generated during the glycolytic pathway. An idea was forwarded that the ancestral Alh_A gene underwent a duplication, probably during the

Cretaceous period, and the duplicated genes evolved into *Adh1* and *Adh2*. The Cretaceous is the period when fleshy fruits arose on earth. This is a plausible and interesting idea, and the yeast/ethanol example is relevant to the question of what the roles of gene duplication are in acquiring new functions. However, I suggest that we do not take the question further (or backward) and rather pose the following question: what were yeasts and other fungal organisms eating before the emergence of plants?

Clearly, fungal organisms predated multicellular plants! Several guesses can be made. My favorite guess is that fungal organisms went into symbiosis with photosynthesizing unicellular organisms (e.g. cyanobacteria and/or unicellular green algae), establishing lichen-like associations.

A further study with yeast (*S. cerevisiae*) noted that gene duplication is the prime source of new genes (He and Zhang, 2005). These authors found that duplicated genes (that are retained in present-day yeast) code for longer proteins than the proteins encoded by nonduplicated genes. The former proteins have more functional domains and more *cis*-regulatory motifs than proteins encoded in single genes. Hence, in the yeast, gene duplication increases both the number of genes and gene complexity — two factors that affect the organism's complexity.

The consequences of gene and genome duplication in plants (actually angiosperms) were reviewed by two research teams (Moore and Purugganan, 2005; Adams and Wendel, 2005) in the same issue of *Current Opinion in Plant Biology* (see editorial by Wesseler and Carrington, 2005). Moore and Purugganan (2005) analyzed research results of investigations that were looking for the roles that gene and genome duplication play in plant evolution. Plants excel, among eukaryotes, in that the duplicated loci compose a large fraction of their genomes. This is probably a result of higher frequency of *retention* of the duplications in genomic segments, and because of polyploidizations. In spite of its tiny genome, about 90% of the Arabidopsis genome loci is inferred to be duplicated.

It is noteworthy that the rate of duplications of a given gene in Arabidopsis is estimated to be 2×10^{-3} in one million years, a rate that is similar to that estimated for *Drosophila*; but, the retention rate of these duplications seem to be much higher in Arabidopsis than in *Drosophila*. The reason for the much longer "half-life" of plant duplications than that of animals is not yet clear, but Moore and Purugganan (2005) described probable fates of the duplicate gene (see Fig. 20). An important aspect of these fates is that duplicated genes may acquire debilitating but complementary mutations that alter one or more subfunctions of the single-gene progenitor. This was described above for the *LIP* genes in *Antirrhinum* and the *Adh* genes in yeast, but there are numerous additional examples, such as duplication in the MADS box genes that control floral organ patterning in plants. We shall discuss the latter genes while dealing with floral patterning.

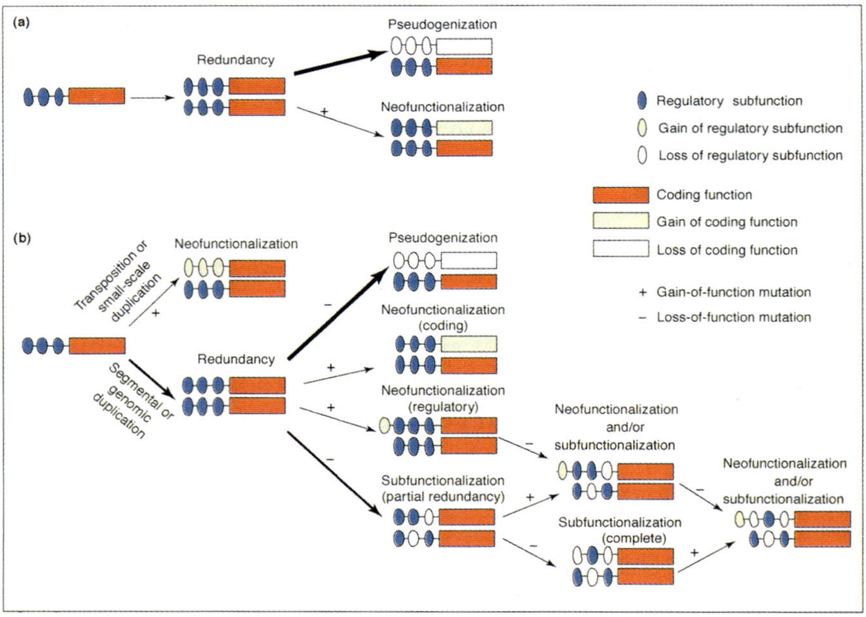

Fig. 20. **(a)** Classic Ohno (1970) model of duplication gene fates. Mechanisms of duplication and fates of genes are indicates. Thickness of arrows indicates relative frequency of possible fates. **(b)** Recent theoretical work supports a much more complex model for the fates of duplicate genes. (From Moore and Purugganan, 2005.)

The abovementioned authors (i.e. Moore and Purugganan) also concluded that inferring patterns of plant genome evolution requires a phylogenomic approach; genomic data on structural arrangements of extent genes must be compared to that from related taxa to determine the extent and timing of genome duplication events. Only when such comparative approaches are used will we be able to truly understand the contribution of single-gene and whole-genome duplications to the emergence and diversification of plant species.

The publication of Adams and Wendel (2005) that followed the review of Moore and Purugganan (2005), in the same issue of *Current Opinion in Plant Biology*, deals mainly with whole-genome duplication, i.e. the impact of polyploidization on genome alteration and the diversification of functions in plants that are derived from polyploidization. As it is assumed that most presently existing angiosperm species (and surely all cultivated plant species) underwent at least one round of polyploidization, one may deduce that the present diversity in plant species owes its existence to polyploidization.

We should also recall that several rounds of polyploidizations were inferred from sequence analyses. Recently, four (rather than two) rounds were assumed to have occurred in Arabidopsis. Each round leads to additional diversification of genes and may "give birth" to a new species. As indicated above, polyploidization can result from the duplication of one genome (autopolyploidization) or by interspecific crosses that result in allopolyploids. Both types of polyploids usually undergo a quick process of elimination of sequences (including elimination of genes) as well as diversification of one or both doubled sequences (that may result in new functions). What happens in reality depends on the selective advantage of the modified genes. In cultivated species, breeders have a considerable role in selecting plants with modified genes that have an agricultural advantage. Adams and Wendel presented a scheme of polyploidization in the ancestors of several cultivated taxa as well as in Arabidopsis (see Fig. 21). If we take the presently cultivated potato, it probably has a history of four rounds of polyploidization.

In addition to the regular genome duplication that results in autopolyploids, the allopolyploidization in plants is quite common.

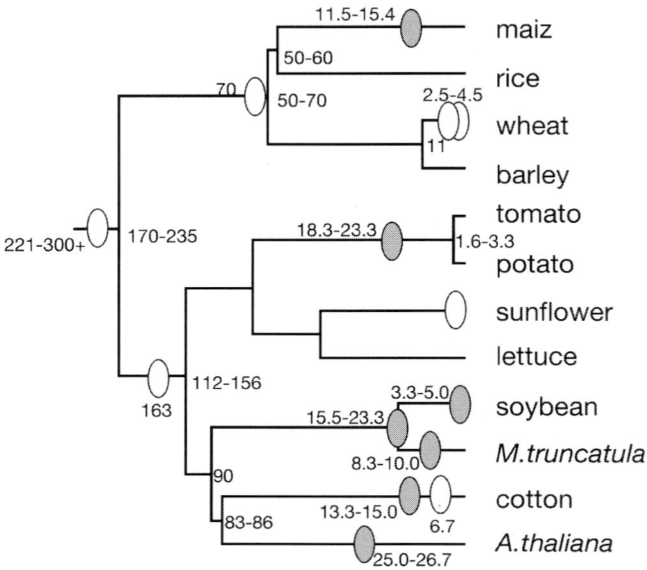

Fig. 21. Suspected large-scale duplication events presented in phylogenetic context. The phylogenetic tree represents the currently accepted phylogeny of the plant species. Branch lengths are not to scale. The shaded ovals represent suspected large-scale duplication events (polyploidy or aneuploidy) recovered in this study. White ovals correspond to suspected polyploidy or aneuploidy events inferred in previous publications but not evidenced in this analysis. Estimated dates of duplication and speciation events are given in million years when available. (From Blanc and Wolfe, 2004a.)

Many of the present species of crop plants were derived from allopolyploidization (e.g. tobacco, rapeseed, bread wheat, cotton). More than one round of allopolyploidization during the history of a present plant species is plausible. Such rounds may result in selective advantage, but they are causing confusion to taxonomists, who are eager to know the ancestors of the present cultivated species.

Further literature on polyploidity in angiosperms can be found in Blanc and Wolfe (2004a,b), Moore *et al.* (2005), and Adams and Wendel (2005). The advantages and disadvantages of polyploidity in plants were reviewed by Comai (2005).

The construction of phylogenetic trees, and especially the timing (providing dates) for the splits of branches in such trees, is based on reasonable assumptions, rather than on solid proof that can be validated by further experimental procedures. Clearly, evolution cannot be reconstructed. Hence, in a way, all these phylogenetic assumptions should be regarded as theories. This leads us to the work of Sir Karl Popper (1902–1994) on the logic of scientific discovery and, more specifically, to his lecture on "Science: Conjectures and Refutations" (published in London, 1963). Popper asked himself: when should a theory be ranked as scientific? (Socrates would have stopped Popper right here, asking: and what is a "theory" and what is "science"?)

Popper developed his theme by looking for a clear demarcation line between scientific theories and theories that are close to myths. For the former (scientific) theories, Popper has the example of Einstein's theory of gravitation. This theory could be tested by an experiment that would either refute the theory or validate it. For Popper, a genuine test of a scientific theory was a honest attempt to falsify it. This was actually done by the Eddington expedition (1919) which showed that, contrary to common intuition, assuming that light is not affected by heavy bodies, light was indeed attracted by the sun; hence, Einstein's theory was validated. By analyzing other theories, such as Marx's theory of history, Freud's psychoanalysis theory and Adler's individual psychology, Popper rationalized that these three theories are based on plausible arguments, and those who believe in them repeatedly found cases that were in agreement with these theories. But these theories could not be put to refutation tests, and therefore are separated from scientific theories by a clear line and categorized with other groups of theories (such as those of astrology) as pseudo-scientific and close to myths.

To Popper, induction, meaning inference based on many observations, is a myth. Only the falsity of the theory can be inferred from empirical evidence, and the latter inference is a purely deductive one. In short, according to Popper, the criterion for the scientific status of a theory is its falsifiability or refutability or testability. Phylogenetic tree theories will not pass Popper's criteria. On the other hand, one may ask if such trees, as well as other assumptions that do not pass Popper's criteria for scientific theories, do contribute to progress in understanding patterning. The probable answer is in the affirmative.

Processes Leading to Gene and Genome Duplication

The various means by which whole genomes or components of genomes (e.g. chromosomes, chromosomal arms, genes, coding sequences of genes) are duplicated constitute a vast subject that is beyond the scope of this book. I shall therefore merely list such means, and even this list will not be comprehensive.

<u>Genome duplication</u>: Genome duplication is the process that causes polyploidity. In plants, the common result of this duplication is the conversion of a diploid species into a tetraploid one. This process is apparently more common in plants than in animals. We do not know what mechanism is causing this type of genome duplication in nature. We do know that it is experimentally possible (at least in many plant species) to stop cytokinesis without stopping chromosome replication. This can be done by certain treatments (e.g. by colchicine). Clearly, genome replications without cytokinesis happen regularly in plants (e.g. in the endosperm and in the tapetum); however, these tissues do not contribute to the germ line and thus do not result in polyploidity of the progeny.

<u>Allopolyploidity</u>: This can result from interspecific crosses, so that the first progeny of such a cross contains two genomes. Subsequent generations may either maintain the two parental genomes or lose chromosomes (or fractions of chromosomes) from either or both parental genomes.

<u>Duplication of chromosomal arms</u>: This can happen due to chromosomal arm translocation under natural or experimental conditions (such as mutagenesis by X-ray irradiation). Plants with such duplicated arms may lose, simultaneously, a given chromosomal arm. Such cytological phenomena are more prevalent in polyploid plants (such as bread wheat), where there is a redundancy of homeologous chromosomes, than in diploid plants.

<u>Duplication of chromosomal regions</u>: Under this heading, there are several types of duplications, such as duplication of extensive

sections of chromosomes, duplication of genes, duplications of DNA sequences that are not genes (sequences that encode microRNAs) and duplication of parts of genes (e.g. of the coding sequence of a gene).

While we do not have full knowledge on how these types of duplications are controlled naturally, we are familiar with some mechanisms that can lead to these duplications. For example, dysfunction of the telomers can cause aberrations during mitosis and meiosis; in certain cases, continuous cycles of "breakage-fusion-bridge" (BFB, the phenomenon that was discovered by Barbara McClintock; see chapter 6 in Galun, 2003). Other possible causes of gene duplication are non-homologous crossing-over events during meiosis and other cytological irregularities during meiosis. Clearly, plants and other organisms have in each of their cells an elaborate means for the duplication of specific genes: genes encoding ribosomal RNAs. However, these duplications are not carried onto the gametes and thus are not involved in the types of gene duplications that we handle here. Incidentally, it should be noted that ciliate protozoa have a regular means to multiply certain genes in their macronuclei, but again, the macronucleus does not serve in formation of the next generation. The latter two examples indicate that organisms developed, very early on in their evolution, efficient means to replicate specific genes. The roles of transposable elements in gene duplication/modification were reviewed by Bennetzen (2005).

Chapter 5

REGULATION OF GENE EXPRESSION

The correct spatial and temporal regulation of gene expression has a prime role in plant patterning. Hence, the reader should be aware of the myriad means and mechanisms by which gene expression can be regulated. Obviously, the full range of gene regulation mechanisms at the levels of chromosome, chromatin, DNA, RNA and protein is far beyond the scope of this book. Therefore, in this chapter the various regulatory mechanisms will be noted, but not detailed. Where required, the reader is referred to the pertinent literature. For a general background, Chapters 6, 7 and 9 in the book *Biochemistry and Molecular Biology of Plants* (Buchanan, Gruissen and Jones, 2000), written by Sugiura and Takede (2000), Ferl and Paul (2000), and Spremulli (2000), respectively, are recommended.

Some processes that have an impact on the levels of certain kinds of gene expression are not directly involved in plant patterning. An example of these processes is the duplication of ribosomal genes. The nuclear organizer region in specific chromosomes contains many copies of the ribosomal RNA genes. Clearly, this abundance of genes is a means to increase the capability to construct ribosomes, but it does not regulate patterning specifically. Such kinds of regulation of gene expression shall not be handled in this chapter.

However, I shall list the regulations of gene expression at the following sequence:

Chromatin
↓
DNA
↓
Pre-messenger RNA
↓
Messenger RNA
↓
Protein

Chromatin

As in other eukaryotes, the nuclear DNA in plants is organized in chromatin (there is no chromatin in the chloroplasts and mitochondria of plants). For an overview and detailed literature on plant chromatin, readers can refer to Ferl and Paul (2001) and to the Appendix on Chromatin Remodeling in Galun (2003). The packaging of genetic information (DNA) in chromatin takes place in eukaryotes, while no such packaging exists in prokaryotes. This is a primary difference in the logic of the regulation of gene expression between the latter organisms and eukaryotes (see Struhl, 1999). The basic structure of the chromatin is a string of double-helix DNA that is "decorated" with nucleosomes. The nucleosomes are structures that are composed of a hectamer of histone proteins (two of each: H2A, H2B, H3 and H4), that is wrapped twice by 146 base pairs (bp) of DNA. Another histone, H1, resides outside of the nucleosome core. This basic structure was nicknamed "beads on a string". Between the "beads" is the double-stranded DNA ("string"). The space between the beads is variable (e.g. 20 to 200 bp). This nucleosome array is actually a fiber of ca. 10 nm in diameter. The described array may condense and form a *solenoid* of 30 nm in diameter, forming a compact structure in which the DNA is not accessible to enzymes involved in replication and transcription to RNA. It also furnishes protection against mutagenic effects.

Already in the late 1950s and 1960s, James Bonner regarded plant chromatin as a possible regulatory agent in gene expression (e.g. Huang *et al.*, 1960; Bonner *et al.*, 1961). It was postulated that acetylation/methylation, phosphorylation and possibly other changes in the amino acids of histones would "open" the chromatin and thus enable transcription of genes. After a lot of sacks of pea seeds were germinated by the Bonner team, the shoots were separated by a Maytag washing machine and huge quantities of chromatin were extracted and analyzed from chromatin modifications. This subject went into hibernation for many years, and only in recent years have the roles of chromatin remodeling in gene expression been clarified (see Appendix in Galun, 2003, for an overview of chromatin remodeling). Special attention was paid to the methylation (for silencing) and demethylation of lysine-9, as well as acetylation (for activation) and deacetylation of lysine-9 and lysine-14 of histone H3 (see Grewal and Elgin, 2002; Grewal and Rice, 2004; and Peters and Schubeler, 2005, for a general background). Verbsky and Richards (2001) reviewed the chromatin remodeling in plants and, in a more recent review, Guymarch *et al.* (2005) surveyed the involvement of chromatin remodeling in regulating the expression of genes involved in plant patterning, such as *SHOOT MERISTEMLESS* (*STM*) and *KNOTTED1-LIKE FROM ARABIDOPSIS THALIANA1* (*KNAT1*). It was recently found that methylation and demethylation of histone are rather dynamic processes, providing another means for the regulation of transcription (Peters and Schubeler, 2005).

M.A. and A.Y.M. Matzke were pioneers in analyzing the impact of RNAi on transcription. In a detailed review on the RNAi-mediated pathways in the nucleus in animals and plants, Matzke and Birchler (2005) noted that diced dsRNA can lead to heterochromatization (and consequently, silencing transcription). Also, specific loci in the euchromatic regions of the plant chromatin can be affected. They suggested a scheme in which specific siRNAs guide histone methyltransferases to a specific site on the chromatin. The methylated H3 is then bound to an additional protein(s), such as HP1, to maintain the silencing of this site by the inhibition of transcription of the DNA in this site. If such a process takes place during defined stages of the differentiation of the plant, the RNAi-mediated silencing of chromatin has a potential role in plant patterning. This scheme requires the "homing" of the specific RNAi to specific sites that are characterized by the DNA sequence inside the chromatin. It is not clear how the RNAi finds its specific target. While during most of the duration of the cell cycle the chromatin is packed in a compact manner, the chromatin exposes the DNA during the S-phase (i.e. during DNA replication); could it be that the RNAi utilizes this phase to find its target? If this is so, then the RNAi–chromatin remodeling effect on patterning should require at least one round of nuclear division.

In *Drosophila*, there are groups of genes that are termed Polycomb group (*Pc-G*) genes and trithorax genes. These two gene groups repress or derepress, respectively, transcription from certain regions of the genome. There are indications that the control of gene expression is mediated by changes in the chromatin in these regions. The effects of *Pc-G* and *BX-C* are epigenetic and are transmitted through cell division to the progeny cells. Goodrich *et al.* (1997) found that the *CLF* gene of Arabidopsis that is required for stable expression of a floral homeotic gene encodes a protein with an extensive sequence that is homologous with a *Drosophila* protein. The latter is encoded by the polycomb-like gene *enhancer of zetse*. In their review on the impact of polycomb group proteins in plants, Hsieh *et al.* (2003) discussed how homeotic gene expression that regulates spatial and temporal changes in patterning can be changed by the modification of histones (e.g. deacetylation and methylation). It should therefore not surprise us if further investigations will reveal patterning roles for polycomb-like and trithorax-like genes in plants. Clearly, a mechanism that will either silence or reactivate groups of genes, which are required only during specific stages of plant differentiation (e.g. only during embryogenesis or only during floral differentiation) but are shut off during all other stages, will be of benefit in the regulation of plant patterning. However, does such a mechanism that reversely changes euchromatic to heterochromatic chromatin really exist?

There is an emerging awareness on the interaction between the histones in the chromatin and the DNA that is complexed with it in the same chromatin regions. This interaction is apparent between methylation of the lysine 9 in the H3 and methylation of cytosine in the DNA. The latter methylation also causes silencing of genes. An interaction between the methylation of histone and DNA methylation was reported, for example, in the fungus *Neurospora crassa* (Tamaru and Selker, 2001), where the methylation of DNA was dependent on the methylation of H3 at lysine 9 in the chromatin. The interaction between histone methylation and DNA methylation was reported by Gendrel *et al.* (2002), who found that the same Arabidopsis gene, *DDM1*, which is required for DNA methylation, is also essential for the methylation of lysine 9 in H3. This brings us to the role of DNA in the control of gene expression.

DNA

Clearly, the nuclear DNA in eukaryotes is not merely a linear array of coding sequences. A very rough calculation shows this rather clearly. If we assume that 10^3 base pairs are required for encoding an average protein, then for 30 000 plant genes, the total coding sequences should be about 3×10^7 bp; while many plants' genomes contain over 10^9 bp. Also, by more accurate calculations, coding sequences in the plant genome comprise only about 1%–2% of the total DNA sequence. This is not an exact number; in some plants the percent of coding sequences is much higher (Arabidopsis), while in other plants (e.g. lilium) it is lower.

What are then the noncoding regions? First, a gene is not only coding, it has several other components that, together, are commonly much more extensive in sequence length than the coding region (see Fig. 22). These include enhancer regions and a promoter region, as well as transcribed, but not translated, regions. Beyond the latter regions is the region that serves for the transcription of pre-mRNA. The latter RNA commonly has extensive sequences that are spliced and do not serve for the coding of proteins. Thereafter, are termination sequences. In addition, the genome is not composed only of genes, but quite to the contrary: genes usually constitute a minority of the whole genome's length.

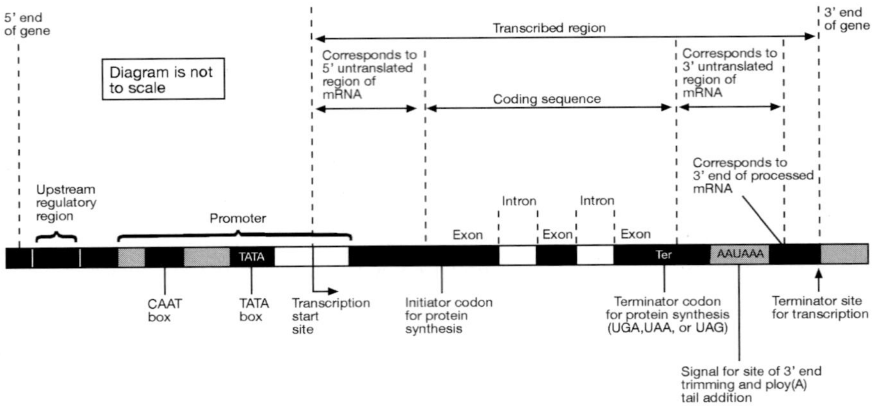

Fig. 22. Structure and organization of a eukaryotic gene. A gene is divided into several sections. The transcribed region functions as a template for the synthesis of RNA, which is then edited and translated into the protein product of the gene. The transcribed region is interspersed with noncoding sequences that partition the region into coding sections (exons) and noncoding sections (introns). The transcribed region is flanked on either side by noncoding sequences that play a role in the regulation of the gene. Most of the regulatory sequence elements are in the 5′-flanking region. The first 1000 bp or so of the 5′-flanking region is referred to as the gene promoter, as it contains sequence motifs important for the "promotion" of transcription. These sequence motifs are called "*cis*-acting elements." The most highly conserved *cis*-element is the TATA box, which is usually found within the first 50 bp of the transcription start site. The TATA box coordinates the recruitment of RNA polymerase to the gene. (From Ferl and Paul, 2000.)

What are all the other sequences in the plant genome? There is information on some of them, as various kinds of active and nonactive transposable elements (in maize, they occupy about half of the genome's length), but we still lack knowledge on all of these nongenic sequences that in the past were coined as "junk". We also have no good evidence for the reason that in some plants the genomes are huge; while in other plants, which probably contain a similar number of genes, the genome is by far smaller.

In recent years, it has become clear that sequences that encode microRNAs are rather abundant (see Galun, 2005a). However, how many such sequences exist in the plant genome? There is evidence for several hundred such sequences, but there could be even more. Only recently, Bentwich *et al.* (2005) found that in man, where 220 microRNAs were previously detected, there are good indications for the existence of about 800 microRNAs. The microRNAs of plants have an important role in regulating the levels of proteins, among them proteins that are involved in the control of patterning. Hence, part of what was dismissed in the past as junk DNA is presently considered an important component of the regulation of patterning. Examples for the roles of microRNAs in plant patterning will be provided in Part B of this book.

In summary, the role of DNA in gene expression is manifested in quite a lot of ways. These ways shall be abbreviated in the following:

- A gene may undergo duplication so that the transcription can proceed, in parallel, from two or more identical genes.
- The genes of eukaryotes (including plants) are organized as schematically shown in Fig. 22. Starting from the 5′ end, there is a region where *enhancers* can bind; downstream of this region is the *promoter* region that spans towards the 3′ up to, or a little beyond, the site where the transcription is started. Then there is the transcribed region. The transcribed region has a 5′ side that is not translated into proteins as well as a 3′ end part that is not translated, but contains the signal for a poly(A) tail.
- The enhancer and promoter regions are of prime importance for patterning because these regions have motifs that bind the RNA polymerase and a large group of other proteins that, in concert, regulate the level of transcription. The latter proteins are themselves synthesized in response to specific external and/or internal cues. By that, specific cells are activated spatially and temporally to express specific genes in these cells (see below).
- The *enhancer-binding* region can be located at quite a distance from the transcription start (up to several kbp upstream).
- The promoter is a rather complex region. It has motifs for binding many *cis*-acting proteins. The most crucial of these is the *TATA box*, which is essential for binding of the RNA polymerase (in genes encoding proteins, this is *PolII*).

- Another "box" that is crucial in many genes is the *CAAT box*, which is close to and upstream of the TATA box. The protein-encoding genes have at least two additional motifs for binding regulatory proteins.
- One of the additional boxes is the *G-box* (5′ CCACGTGG 3′), which is required for the recognition of environmental cues. Yet another box may be required in order to cause the activation of the respective genes. An example of such a situation is the binding of the abscisic acid-responsive element ABRE, which is required to confer ABA responsiveness.
- The binding of the PolII to the promoter requires interactions with a great number of DNA-binding proteins, and only the correctly organized complex of proteins will initiate the transcription.
- The latter proteins, termed transcription factors (TF), enter the complex in a rather orderly manner, briefly: the TFIID binds to the TATA box; TFIIA and TFIIB join the TFIID; thereafter come TFIIF and PolII; and, finally, TFIIE and TFIIH bind to the DNA and to the complex.
- While this may seem complicated, in reality it is much more complicated, because the TFIIs themselves have protein–protein binding motifs. For example, TFIID can bind additional regulatory proteins (as enhancers) and thus change configuration and activity. Also, some TFs may not bind the DNA directly, but rather act by binding to other proteins in the complex, and the latter may require binding with the former protein in order to bind to DNA.
- Now enter the *homeobox proteins* (e.g. MADS boxes of plants). The homeobox proteins of (probably) all multicellular eukaryotes contain a common DNA-binding box of 180 bp, while the regions outside of this box are variable and typical for each individual homeobox. The homeoboxes bind to the appropriate motifs on the DNA (by a helix-turn-helix binding) and, through their binding with transcription factors, the regulation of specific protein-encoding genes can take place.
- The coding region of the gene starts with the AUG codon (for methionine) and is terminated with one of three terminator codons (UGA, UAA or UAG). In this region, there are sequences

that will be translated (*exons*) and sequences that will be spliced (*introns*). There are signals in the DNA for the borders between exons and introns and between introns and the next exons.
- The function and role in the regulation of gene expression of the DNA region that is downstream of the translated region are less known than the functions and the roles of the upstream regions of the genes.

The rather elaborate system that was briefly presented above is especially important for the correct spatial and temporal expressions of genes that regulate patterning. The control of expression of genes that encode housekeeping enzymes requires relatively less elaborate mechanisms.

Pre-Messenger RNA

The primary transcript from nuclear genes that encodes proteins spans from several nucleotides upstream of the first codons in the translated region to several nucleotides downstream of the signal (commonly AAUAAA) for the addition of the poly(A) tail. As indicated above, the pre-mRNA undergoes splicing, during which introns are eliminated. Introns may already exist upstream of the translated region and the pre-mRNA contains signals (commonly of two nucleotides) on each side of the border between an intron and an exon, but the signals are possibly more elaborate. The splicing of plant mRNA is not yet fully clarified. The splicing is also an elaborate process; hence, mutations in components of the splicing complex and/or in the signals for splicing can affect the splicing. When this happens, a different protein will finally be translated. The poly(A) tail is essential for the export of the mRNA from the nucleus to the cytoplasms; it is also relevant for the rate of translation (probably by interaction with the upstream end of the mRNA and the translating complex), and it can regulate the longevity (half-life) of the respective mRNA. Consequently, the tail is relevant for the level of gene expression.

A recent study with mice indicated that the poly(A) tail of certain transcripts may, under given conditions, be eliminated; this happens while the pre-mRNA is still in the nucleus. Such a pre-mRNA will be retained in the nucleus, and no mature mRNA will be released to the cytoplasm. However, under different conditions (stress), a 3′ UTR polyinosine motif is cut-off; the poly(A) tail is thus formed and the transcript, in the form of mature mRNA, is released to the cytoplasm, where it is translated (Prasanth *et al.*, 2005, and preview by Bass *et al.*, 2005). This is schematically shown in Fig. 23. Whether or not similar phenomena also exist in plant pre-mRNAs is still an open question. However, the possibility that pre-mRNAs undergo modifications that affect the respective gene expression should not be dismissed.

The sequences of the introns are also not junk, at least not all of them, because some introns are coding for pre-microRNAs (see below) and some mature microRNAs have obvious roles in plant patterning (as detailed in Galun, 2005a and 2005b, and some more information on plant microRNAs is provided in the next section on Messenger RNA).

Messenger RNA

We shall focus our attention on nuclear-derived messenger RNAs (mRNAs). The genomes of mitochondria and chloroplasts (as well as plastides that lack chlorophyll) also have transcripts that are processed to mRNAs, but the proteins encoded in these mRNAs are commonly components of enzymes and are not known to be directly involved in formal plant patterning. These enzymes are composed of nucleus-encoded and organelle-encoded polypeptides. On the other hand, organelle-genome-derived mRNA cannot be dismissed from any role in patterning. In many plant species, there are phenotypes that are male sterile (MS). The phenotype of MSs range from defective pollen grains to complete lack of stamens. At least in some cases, as in cybrids between *Nicotiana* species, the MS phenotype is a result of incompatibility between proteins encoded by the nuclear genome of one species and proteins encoded by the mitochondrial genome of another species (see Breiman and Galun, 1990, for a detailed review on this subject). How exactly this incompatibility leads to the MS phenotype is not fully understood, but clearly the organelle-derived mRNA is involved in the process that causes MS.

The awareness of the role of plant microRNAs (plant-miRs) in patterning started only in 2002. Actually, the existence and activity of these

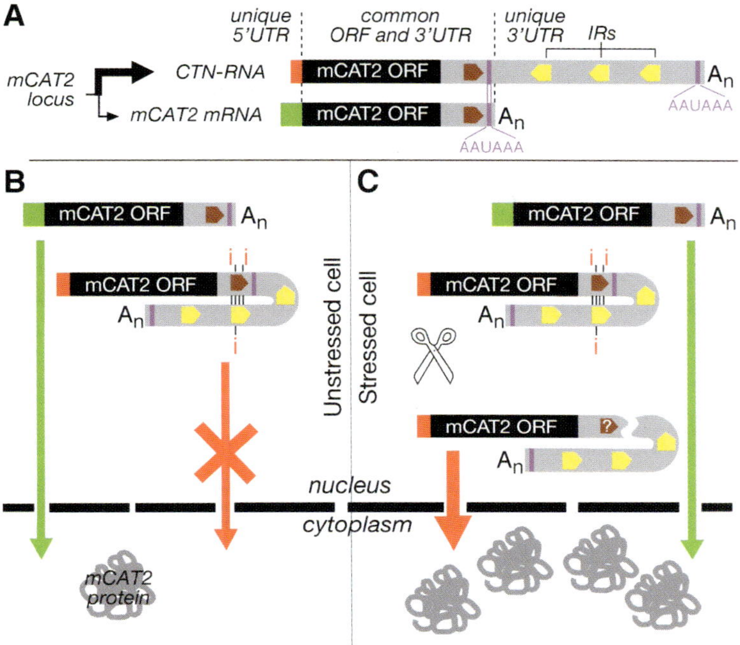

Fig. 23. The different fates of mCAT2 transcripts. **(A)** Two polyadenylated (A_n) transcripts of the *mCAT2* locus are shown. The more abundant CTN-RNA and the *mCAT2* mRNA differ only in their 5′-UTRs (orange and green, respectively) and an extended 3′-UTR unique to CTN-RNA. The proximal and distal polyadenylation sites are indicated (blue) as well as the forward (brown) and reverse inverted repeats (IR, yellow). **(B and C)** In both unstressed (B) and stressed (C) cells, the *mCAT2* mRNA is exported to the cytoplasm and translated to yield mCAT2 protein (gray). In contrast, CTN-RNA is retained in the nucleus in unstressed cells but cleaved in stressed cells to yield a transcript that is exported and translated. Inosine residues from RNA editing are indicated (orange "i"); the question mark signifies that the cleavage site has not been mapped, so it is possible that the forward repeat is in the other fragment. (From Bass *et al.*, 2005.)

small RNA sequences in eukaryotes is a recent discovery. However, once miRs were discovered, they attracted intensive investigation. The subject of RNA-imposed silencing of gene expression was detailed in my recent book (Galun, 2005a) and in a review (Galun, 2005b) that focused on RNA silencing in plants. The mature miRs are single-stranded RNAs of about 20–24 nucleotides. MiRs are derived from nuclear genome transcripts (pri-microRNAs) that fold into hairpin structures and are further processed (to pre-microRNAs) until the mature microRNAs are formed. These miRs enter a protein complex that is termed RNA-induced silencing complex (RISC), between PAZ and PIWI proteins, and these bind to compatible sequences of specific mRNAs. The binding may happen between the miR and the coding sequence of the mRNA (these are the common binding sites in plants). Consequently, the RISC will cut the mRNA at the coding region and destroy the mRNA. The binding can also occur at the 3′ untranslated region (UTR) of the mRNA and cause the inhibition of translation of the bound mRNA. The latter binding is probably more common in animals than in plants.

It is also possible that the RISC with the microRNA (between the PAZ and PIWI) will impose histone 3 methylation of a specific sequence in the chromatin. While the roles of plant miRs in affecting mRNAs and methylation (of chromatin and DNA) are still under active investigation (see examples in Bao *et al.*, 2004a; Ronemus and Martienssen, 2005; Eshed and Bowman, 2004; Kidner and Martienssen, 2005; Matzke and Birchler, 2005; Yu *et al.*, 2005; Sontheimer, 2005; Voinnet, 2005; Tomari and Zamore, 2005; Chan *et al.*, 2005), it is already very clear that the microRNA–mRNA interaction is an important means of regulating gene expression in plants, and this means of regulation should be taken into account while dealing with plant patterning.

Before being transferred from the nucleus to the cytoplasm, the 5′ end of the transcript is capped: in this process, the terminal phosphate is removed from the 5′ triphosphate and replaced by a 7-methyl guanosine and the 2′ hydroxyl groups on the first and second ribose are methylated. Hence, the mature mRNA emerge from the nucleus equipped with a cap and a poly(A) tail. It is then ready to serve, in the ribosome-mediated process, as a template for the translation of proteins. The quantity of the protein translated from a mature mRNA can now be affected mainly by two factors. One is the efficiency of the ribosome-mediated translation (e.g. distance between ribosomes on the mRNA, efficiency of the translation complex, and supply of required amino acids and tRNAs that bring the amino acid to the translation complex). The other factor is the longevity (half-life) of a given mRNA.

The mature mRNA may, right after its release from the nucleus, associate with free ribosomes in the cytoplasm, or it may associate with membrane-bound ribosomes. In the latter case, the nascent polypeptide will cross the membrane (e.g. into the lumen of the endoplasmic reticulum). The questions of whether and how patterning can be regulated by processes that operate between the stage of mature mRNA and the initiation of translation remain unanswered. However, for one important phenomenon that was mentioned above, short siRNAs (e.g. of some mature microRNAs) may bind to a complementary sequence in the untranslated 3′ end of the mRNA and cause inhibition of translation. This latter effect is indeed mRNA- and microRNA-specific and can affect patterning.

Protein

The initiation of translation from an mRNA template, the further translation and termination of translation, the processing of the translated polypeptides (e.g. correct folding and formation of disulfide bonds), as well as the controlled degradation (ubiquination) of proteins, are all rather elaborate mechanisms that shall not be detailed here. However, these mechanisms are relevant to patterning. Details on these mechanisms can be found in molecular biology texts, and these mechanisms in plants were described in the chapter by Spremulli (2000), in the book edited by Buchanan *et al.* (2000).

Synthesis of polypeptides

There are several processes during the translation of polypeptides from the respective mRNA templates that can be regulated. One of these is the initiation of translation. During the latter process, proteins termed *eukaryotic initiation factors* (*eIF*) play a major role. There are at least six main types of eIFs (1 to 6) and half of these have subtypes (e.g. eIF4 and eIF4A). For example, of eIFs' role, it is noted that the subtypes of eIF4 (eIF4E, eIF4G, eIF4B and then also eIF4A) bind to the 5′ cap region of the mRNA and closely downstream of the cap. When there is no correct binding (e.g. lack of complementation between an eIF and the RNA), the initiation will be inhibited. Also, there are eIFs that are essential for binding the small (40 S) ribosomal subunit to the mRNA and for the correct recognition of the first codon (AUG) of the coding region of the mRNA. Again, correct complementation between additional eIFs and the respective sequence in the mRNA are required to join the large (60 S) ribosomal subunit to the initiation site on the mRNA. Clearly, mismatches between certain eIFs and the mRNA, as well as mismatches between two different eIFs, can cause inhibition of polypeptide synthesis. Components that can operate in a spatial- and temporal-specific manner can thus regulate the level of synthesis of certain proteins and, consequently, the patterning can be affected.

Trimming and folding

In the early days of molecular biology, there was an assumption that the amino acid sequence of a protein would lead directly to the correct three-dimensional shape of this protein. Now, we know that an immature protein may pass through stages during its maturation. Most newly synthesized polypeptides must be cleaved (commonly the first amino acid, methionine, is removed from the nascent chain). After the polypeptide reaches its destination, it may undergo additional processings, such as being folded correctly in order to serve its purpose. True, the "address" to the subcellular location is "written" in the polypeptide (an analogy to an address of an e-mail), but then other cell components have to take part in the voyage (e.g. cytoskeleton polymers) and in assuring the correct trimming and folding. The trimming is accomplished by the appropriate proteolytic enzymes. In some cases, the vast majority of an immature polypeptide (e.g. prosystemin) is trimmed and only a short chain (e.g. the 18-amino-acid-long wound-response hormone systemin) is retained.

Correct folding, trimming and intramolecular interactions in the immature protein are probably mutually dependent processes that are not yet fully understood in many plant proteins. Clearly, correct folding is an essential mechanism. Hence, this folding deserves special attention.

A prime role in the correct folding is attributed to *molecular chaperones*. These are proteins that bind and stabilize the conformation of other proteins. Chaperones facilitate several processes, such as correct folding, oligomer assembly, subcellular localization and protein degradation. While not forcing specific foldings, chaperones prevent incorrect interactions within the polypeptide as well as other incorrect interactions, and by that increase the yield of correctly formed three-dimensional proteins that are functional. As detailed in Spremulli (2000), the invention of chaperones, as well as their subtypes, the chaperonins, is rather old.

Chaperones and chaperonins exist in prokaryotes (bacteria), meaning they originated several billion years ago. Plants have bacteria-like chaperones in their chloroplasts and mitochondria, as well as novel cytoplasmic chaperones and chaperonins. The chaperones can exercise their effect at several stages of the polypeptide foldings. For example, Hsp70 chaperones may bind (e.g. to hydrophobic patches) of the nascent polypeptides after about 100–200 amino acids have emerged from the ribosome (class I chaperones). The Hsp70-bound polypeptide may then either be transferred to different chaperones (i.e. Hsp60, class II chaperone) to control the final folding, or the Hsp70-bound and partially folded polypeptide is further folded without the aid of Hsp60 (in a process that requires ATP energy). It is assumed that the binding of the 100–200-amino-acid-long nascent polypeptide, to Hsp70, prevents undesired aggregation. It is also possible that first a Hsp40 chaperone binds to the nascent polypeptide and after its binding, Hsp70 joins. The Hsp40 will then be released. Thereafter, further chaperones and/or chaperonins will assure proper folding. There is a fair amount of knowledge of the chloroplast homologs of the bacterial chaperones GroEL and GroES. These are barrel-shaped protein complexes that prevent illegitimate folding of chloroplast proteins. Similar chaperones are most probably also active in plant mitochondria. There is no evidence yet for the direct involvement of chloroplast and/or mitochondrial proteins in the control of patterning. The well-known GroEL and GroES family will therefore not be handled in this book.

As for cytoplasmic folding of proteins, there are two major routes for nascent polypeptides that emerge from the ribosome. In cases where there is a signal recognition site on the polypeptide, it interacts with SRP; consequently, the respective ribosome is led to the outside of the endoplasmic reticulum (ER). The polypeptide will then cross the ER membrane and enter the lumen of the ER, where folding can take place. The insertion of polypeptides into membranes of the ER is of specific relevance to the plant hormone ethylene (see Chapter 2). The putative receptor of ethylene is located in the ER. Ethylene and its interactions with other plant hormones have roles in plant patterning. If no such signal exists on the released polypeptides, a nascent-chain-associated (NAC) particle will bind and the ribosome will not bind to the ER. The polypeptides released from such free ribosomes may then bind to chaperonins as the TRiC complex. The latter is a barrel-shaped complex that has similarity to GroEL/GroES, and also a remote similarity to the proteosome of ubiquination. The TRiC causes the correct folding of some proteins.

Several proteins, such as those that are secreted to the outside of the cell, contain cysteines in their amino acid chain. Disulfide bridges can be formed between these cysteines, causing primary folding. This folding may be changed further by bisulfide isomeration, forming the final three-dimensional protein. The enzyme causing this change in the S bridges between cysteines is termed *protein disulfide isomerase* (PDI). Such PDIs are located in plant chloroplasts and in the ER. The isomeration is a rather elaborate and multi-step process that ends with the most thermodynamically stable conformation. In addition to the disulfide bond isomerations, there can also be other changes in bonding between amino acids of the same polypeptide. There is, for example, the isomeration (*cis* ↔ *trans*) of bonds between proline and another amino acid ("X-Pro" bonds) by enzymes termed peptidyl prolyl isomerases. Such enzymes are instrumental in causing the required final conformation of some proteins.

In Part B of this book, it will become clear that many proteins that are involved in patterning interact with other proteins as well as with nonprotein polymers and plant hormones. These interactions require the correct three-dimensional conformation of the controlling polymers. Hence, the issues of trimming and folding deserve due attention.

Protein degradation

Just as the synthesis of proteins, especially of those proteins that have the ability to control cellular processes, is important for patterning, so is the degradation of proteins that fulfilled their task. The latter should be eliminated specifically. There are many means to degrade proteins. However, among those means that lead to very specific degradation of proteins at the required site and time, ubiquitination is the most important process of protein degradation. The process of ubiquitination was described in Chapter 2 (on plant hormones) and shall not be repeated here. However, it should be noted here that while the proteosome/ubiquitin way of protein degradation is reminiscent of the operation of a waste disposal utensil, only the former is environment-friendly: the breakdown material after ubiquitination can be swiftly reused for protein synthesis, and the ubiquitin itself is recycled for further ubiquitinations. As for proteosomes, recent work with yeast (*Saccharomyces cereviseae*) revealed another role for these complex particles: enhancement of the transcription of certain genes (Lee *et al.*, 2005). Whether or not plants also recruited proteosomes for additional tasks that are far removed form the waste disposal task is still an open question.

Chapter 6

SYSTEMS BIOLOGY

The May 5th (2005) issue of *Nature* opened with a question by the editor: "What is the difference between a live cat and a dead one? One scientific answer is 'systems biology'. A dead cat is a collection of its component parts. A live cat is the emergent behavior of the system incorporating those parts". In the case of the Schrödinger cat, we cannot know if the cat is alive or dead. Erwin Schrödinger (1887–1961) put his virtual cat in a box together with a bottle of poisonous gas, which would have killed the cat if a device connected to the bottle registered the radioactive decay of a radium atom. There is uncertainty as to whether or not this atom will decay during a given time frame and, consequently, there is uncertainty as to whether or not the cat is dead until an observer tries to determine in which state the cat is. Virtual cats serve not only to explain systems biology and quantum mechanics. There is the rather nasty maxim: "There are many ways to skin a cat." Another virtual cat served Albert Einstein in a public lecture, when he was asked how wireless telegraph works. His answer was that a cable telegraph is like a very long cat; you pinch its tail in San Francisco and it sounds "Miyau" in New York. The wireless telegraph is the same, but without the cat.

In the literature on systems biology, one can find the term *purpose* — the term used for assigning a purpose to the evolution of biological systems. I hold that purpose is a concept that is limited to humans; it does not constitute the driving force that created the universe, nor the creation of life on our planet. These creations resulted from the selection of successful interactions between components that did not lead to dead ends. Uncertainties and chances played a role in the retention of successful interactions. Once an interaction was retained, further interactions could take place. Fields of knowledge in areas such as genetics and biochemistry result from the direct acquisition of information about natural phenomena, while (as we shall see in this chapter) systems biology is clearly a human invention that may vastly improve our understanding of nature.

Although systems biology is an evolving approach, it is a rather complicated one. This chapter will merely provide a rough outline of it. The chapter will therefore be short. I shall provide the early history of systems biology, briefly relating how this approach evolved and providing a summary scheme. The main problems in establishing useful systems biology investigations will be presented. I shall also refer to a small number of examples for the execution of systems biology studies and end this chapter with the evolution of the usefulness of systems biology to plant biologists.

For readers who wish to obtain detailed information on systems biology, the following references are recommended. Green (1999) and Ideker *et al.* (2001) provided a background for this subject. The publications of Kitano (2002a–c, 2004) furnish an in-depth handling of conceptual, technical and mathematical aspects of systems biology. Ge *et al.* (2003) reviewed the development of strategies of systems biology and their relevance for the systems approach to modular biology. O'Malley and Dupre (2005) reviewed the fundamental issues in systems biology and advocate an active involvement of philosophers in this field of endeavor. The relevance of systems biology to plant biology was discussed by Belostotsky and Rose (2005). The subjects of "standards" for systems biology, with appropriate lists of acronyms, subjects and links as well as the development of components of data exchange standards were updated by Brazma *et al.* (2006).

The Evolution of Systems Biology

The application of mathematical tools in the study of pattern formation can be traced back many years. A frequently cited example is the work of the English mathematician (in a way the inventor of computers), Alan Turing, who developed the mathematics to study the flow of (a hypothetical) morphogen during the differentiation of Hydra (Turing, 1952). As for plant patterning, Green (1999;

a postmortem publication revised by Jacques Dumais) reviewed the combining of molecular and calculus-based biophysical paradigms in organ and cellular patterning of plants. It is rightfully claimed that progress in the acquisition of scientific knowledge is correlated with the development of tools for investigation.

With respect to systems biology, the tools were of very different kinds. On the one hand, the tools were from the realm of molecular biology; on the other hand, mathematical tools and computation techniques were being developed. The biological tools provided means of DNA sequencing that were rendered more efficient by mathematical and computation approaches. This has led to the sequences of hundreds of genomes of microorganisms, metazoa (including mammals such as man, mouse and rat) and plants (both dicots, such as Arabidopsis, and monocots, such as rice). Other tools enabled efficient analyses of specific transcripts (see Schmid *et al.*, 2005; Clarke and Zhu, 2006) and syntheses of proteins (see Phizicky *et al.*, 2003; Aebersold and Mann, 2003), as well as interactions between proteins (see Fields and Song, 1989; Li *et al.*, 2004). Moreover, high-throughput analytical procedures were developed that could handle thousands of gene expressions in a short time by small research teams. Also, the size of samples required for these analyses was drastically reduced until individual cell contents could be subject to analyses. Towards the end of the 20th century, the methodologies were ready for the next leap of investigations: to use mathematical and computer science approaches in order to get more biological sense out of the accumulating biological data, to make predictions and to perform further analyses with due perturbations to evaluate these predictions, etc.

Fashionable terms were used by those who considered themselves to be at the cutting edge of molecular biology. *Genomics* was used for the sequencing of genomes. *Transcriptome* became the term for all the transcripts (in an organism, an organ, or a cell, at a specific time), likewise, terms such as *proteome*, *interactome*, and *phenome* became useful and received the group name "omics".

In the following, a scheme for the evolution of systems biology is presented.

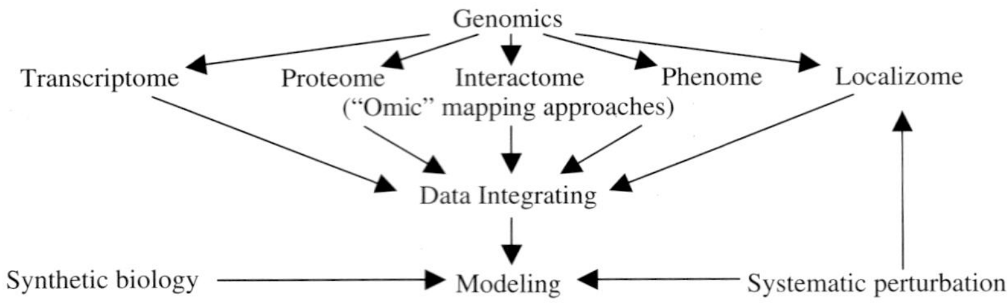

Omic scheme of systems biology.

Integrating "omic" information. With the availability of complete genomes and transcriptome sequences (genomics), functional genomic and proteomic ("omic") approaches are used to map the transcriptome (complete set of transcripts), proteome (complete set of proteins), interactome (complete set of interactions), phenome (complete set of phenotypes) and localizome (localization of all transcripts and proteins of a given organism). Integrating omic information should help to reduce the problems caused by false-positives and false-negatives obtained from single omic approaches, lead to better functional annotations for gene products and functional relationships between them, and allow the formulation of increasingly relevant biological hypotheses. Computational methods can then be used to model biological processes based on integrated data. The resulting models can be tested either by synthetic biology (*de novo* design and generation of biological modules based on suspected network properties) or by systematic perturbation, or both. Systems biology strategies can thus be viewed as a combination of omic approaches, data integration, modeling and synthetic biology (from Ge *et al.*, 2003).

Interestingly, while many reported investigations during the 20th century already used the systems biology approach (see below), the term "systems biology" appeared in titles of publications only in the early 21st century.

An Early Example of the Systems Biology Approach

Ideker *et al.* (2001) listed over 15 cases that were analyzed by systems biology approaches up to the year 2000. They concerned a wide range of organisms, from phages through bacteria, fungi, several metazoa and up to mammals; but did not include plants. Among these, the network of bacterial (*E. coli*) chemotaxis is frequently cited as a notable example. In this model system, the bacterium moves in a chemical gradient by means of a biased random walk, alternating swimming straight ("running") and random reorientation ("tumbling"). In order to run, the bacterium rotates its flagellar motor counterclockwise; a tumble occurs when the motor rotation is reversed. While moving, the bacterium senses gradients of chemical attractants or repellents as changes in concentration over time. Increasing the attractant or decreasing the repellent results in an increase in the durations of the runs in the appropriate direction. The system is schematically shown in Fig. 24. One interesting feature of this chemotaxis is

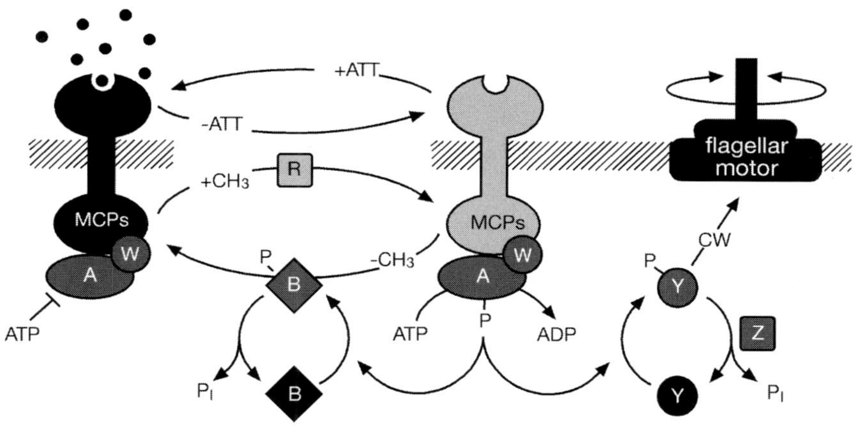

Fig. 24. The bacterial chemotaxis system. Transmembrane attractant receptors, known as methyl-accepting chemotaxis proteins (MCPs), form a signaling complex with CheW and a kinase, CheA, that autophosphorylates and transfers phosphates to a response regulator, CheY. Phospho-CheY interacts with flagellar motor proteins to induce clockwise rotation (tumbling). The CheZ protein promotes CheY dephosphorylation. CheA transfers also phosphates to the CheB MCP methylesterase. The activated phospho-CheB demethylates MCPs; this demethylation diminishes the kinase activity of the MCP-CheW-CheA complex as part of the adaptation mechanism. CheR is a constitutive MCP methyltransferase. In the presence of increasing attractant, CheA autophosphorylation is inhibited, counterclockwise flagellar rotation (running) is extended, and subsequent MCP methylation allows adaptation to the higher attractant concentrations. When moving through decreasing attractant, CheA autophosphorylation is stimulated. This, in turn, promotes phospho-CheY induced tumbling and then adaptation through demethylation of MCPs. (Figure from Spiro *et al.*, 1997; legend from Ideker *et al.*, 2001.)

the robustness of the system: the output of the system (e.g. the level of activity of the flagellar motor) is relatively insensitive to the input (e.g. attractant level); the same tumbling frequency will take place even in a wide range of attractant concentrations.

Barkai, Alon and associates analyzed the bacterial chemotaxis network by a systems biology approach (Barkai and Leibler, 1997; Alon *et al.*, 1999 and see also Alon, 2003). They first analyzed the sensitivity of the network to variations in biochemical parameters and found that the network is very robust in a wide range of biochemical parameters (such as the input of ligand/attractant levels). The simultaneous change of each rate constant (even severalfold) will cause only a single percent deviation from the perfect adaptation of the network. Their conclusion was based on thermodynamic and mathematical analyses, and a two-state model of chemotaxis was suggested.

This model could be mathematically reduced to a simple activity-dependent kinetic represented by respective equations. The robustness emerged as a crucial entity. The robust properties did not depend on the exact values of the network's parameters and should be relatively insensitive to the influence of a subsystem. Hence, it should be possible to explore some of the principles underlying cell function without knowledge of the molecular level. The initial approach of this team was further examined. They used experiments that focused on how response and adaptation to attractant signals vary with systematic changes in the intracellular concentrations of the components (e.g. use of bacterial strains having different levels of proteins such as CheB, CheY and CheZ; see Fig. 24) of the chemotaxis network. The investigators found that some properties, such as steady-state behavior and adaptation time, showed strong variation in response to varying protein concentrations. In contract, the precision of adaptation was robust and did not vary with protein concentration.

An Overview of the Systems Biology Approach

The definitions of *systems biology* by various scholars, while rather different, may lead the reader to an understanding of what is meant by this term. An interesting definiton was provided by Geoffrey North

(2005): "... Systems biology is a child of genomics. A characteristic feature is the 'top-down' approach — faced with large data sets generated by genomic analyses, systems biologists aim to gain insights by considering the 'system' as a whole, often represented in a rather abstract way, as a system of interacting components, shorn so far as possible from unattractive details". Kirschner (2005) of the Department of Systems Biology of the Harvard Medical School hesitates to provide a definition of systems biology. He claimed that one should await the words of the first great modern systems biologist. But still, he wrote: "Systems biology is the study of behavior of complex biological organizations and processes in terms of the molecular constituents . . . [It] attempts all of this through quantitative measurements, modeling reconstruction, and theory". Aderem (2005) of the Institute for Systems Biology in Seattle, WA, has another definition: "Systems biology is a comprehensive quantitative analysis of the manner in which all the components of a biological system interact functionally over time". Aderem emphasized that the analysis is executed by an interdisciplinary team that is also capable of developing required technologies and computational tools.

Several systems biology experts mentioned three basic concepts that are required for understanding complex biological systems: *emergence*, *robustness*, and *modularity*. By *emergence*, it is meant that there are emergent properties that are not demonstrated by the parts and cannot be predicted from the analyses of the parts lent to characterize the whole system. By *robustness*, it is meant that the system maintains relative stability in spite of (genetic and environmental) perturbations. The meaning of *modularity* for systems biology can be roughly descrcibed in a way that a module in a network is a set of nodes that has strong interactions and a common function.

As outlined in the scheme shown on page 148, a systems biology study starts with a global collection of data (the "omics"), such as DNA sequences, RNA and protein measurements, protein–protein and protein–DNA interactions, and biomodules, as well as signaling and gene regulatory networks. The data are also collected after genetic and environmental perturbations. The information can be obtained from specific cells, organs, organisms or even a population. The data are then processed by mathematical and computational procedures for intelligent interpretation and hypotheses are formulated. Means should then be found to present these interpretations to the scientific community and to enable further testing of the network by additional systems biology approaches.

Important Aspects of Systems Biology

In order to render the system-oriented approach effective, several aspects of systems biology have to be considered. Such considerations were reviewed thoroughly by Kitano (2002a–c). Several of these considerations will be mentioned in the following. First, the system-level understanding involves four *phases*: (1) system structure identification (i.e. to understand the structure of the system and collect all the required data); (2) system behavior analysis (the analysis of the system behavior and dynamics using computational methods); (3) system control (following the system behavior after exposure to genetic and environmental perturbations); and (4) system design (this is the final phase in which it is intended to comprehend the system).

The *measurement* issue should not be underestimated. Only the most exact and comprehensive data obtained from high-throughput analyses, which will be handed over to the computations, will yield meaningful results. An effective systems biology study requires a comprehensive *software platform* (e.g. storage of database, cell and tissue simulator, flux–balance analysis software, hypothesis generator and experiment planning software, and data visualization software).

The concept of robustness gained much attention by investigators that used the systems biology approach. Biological robustness was thoroughly reviewed by Kitano (2004), who pointed out that robustness is a property that allows a system to maintain its function despite external and internal perturbations. There are many examples of robust systems. In plants, the maintainance of the architecture at the shoot's apical meristem, as described in Chapter 8 below, is a robust system. One of the common means to maintain robustness is the existence of metabolic loops and feedback. However, other factors also contribute to it, such as buffering the perturbations and the establishment of a "bowtie" structure, where there are various inputs on the one hand and outputs on the other hand; these are connected in a "knot" of conserved core processes. While robustness is an advantageous feature of biological systems, it also has drawbacks: it complicates the system and may break down when unexpected perturbations are encountered.

Data Representation and Communication Standards

Because high-throughout technologies generate large amounts of data that have to be stored and communicated among systems biologists, it is imperative that the contact with these standards are widely known. Hence, Brazma *et al.* (2006) reviewed the topic of standards for systems biology. The following list of standards, with their respective acronyms and links, is an example of lists presented in this review.

List of Standards: Acronyms and their Definitions

Acronym	Full Name	URL
General		
XML	Extensible Markup Language	http://www.w3.org/XML
HTTP	Hypertext Transport Protocol	http://www.w3.org/Protocols/HTTP
UML	Unified Modeling Language	http://www.uml.org
RDF	Resource Description Framework	http://www.w3.org/RDF
OWL	Web Ontology Language	http://www.w3.org/TR/owl-features
SGML	Standardized Generalized Markup Language	http://www.w3.org/MarkUp/SGML
CORBA	Common Object Request Broker Architecture	http://www.omg.org/gettingstarted/corbafaq.htm
LSID	Life Sciences Identifier	http://lsid.sourceforge.net
OBO	Open Biological Ontology	http://obo.sourceforge.net
Genomics		
HUGO	Human Genome Organisation	http://www.hugo-international.org
GO	Gene Ontology	http://www.geneontology.org
MGED Society	Microarray Gene Expression Data Society	http://www.mged.org

(*Continued*)

(*Continued*)

Acronym	Full Name	URL
MIAME	Minimum Information About a Microarray Experiment	http://www.mged.org/Worksgroups/MIAM/miame.html
MAGE-ML	MicroArray Gene Expression Markup Language	http://www.mged.org/Workgroups/MAGE/mage-ml.html
MAGE-OM	MicroArray Gene Expression Object Model	http://www.mged.org/Workgroups/MAGE/mage-om.html
MO	MGED Ontology	http://mged.sourceforge.net/ontologies/index.php
MAGE-stk	MicroArray Gene Expression software toolkit	http://www.mged.org
MAGE	Combines MAGE-OM, MAGE-ML and MAGE-stk	http://www.mged.org/Workgroups/MAGE/mage.html
MISFISHIE	Minimum Information Specification for *in situ* Hybridization and Immuno-histochemistry Experiments	http://mged.sourceforge.net/misfishie
Functional genomics		
FuGE	Functional Genomics Experiment	http://fuge.sourceforge.net
FuGO	Functional Genomics Investigation Ontology	http://fugo.sourceforge.net
SysBio-OM	Systems Biology Object Model	http://cebs.niehs.nih.gov/content_sysbio_om.html
OME	Open Microscopy Environment	http://www.openmicroscopy.org
MIACA	Minimum Information About a Cellular Assay	http://miaca.sourceforge.net
MIARE	Minimum Information About an RNAi Experiment	http://www.rnaiglobal.org/gti.html
Pathways		
SBML	Systems Biology Markup Language	http://sbml.org/index-psp
CellML	Cell Markup Language	http://www.cellml.org
BioPAX	Biological Pathways Exchange	http://www.biopax.org

(*Continued*)

(*Continued*)

Acronym	Full Name	URL
MIRIAM	Minimum Information Requested in the Annotation of Biochemical Models	No link
Proteomics		
HUPO	Human Proteome Organisation	http://www.hupo.org
PSI	Proteomics Standards Initiative	http://psidev.sourceforge.net
PSI-MI	Proteomics Standards Initiative Molecular Interaction	http://psidev.sourceforge.net/mi/xml/doc/user
MIAPE	Minimum Information About a Proteomics Experiment	http://psidev.sourceforge.net
(MS, MSI)	(Mass Spectrometry, Mass Spectrometry Informatics)	
PEDRo	Proteomics Experiment Data Repository	No link
PEML	Proteomics Experiment Markup Language	No link
MCP	Molecular and Cellular Proteomics	http://www.mcponline.org
Metabolomics/metabonomics		
SMRS	Standard Metabolic Reporting Structure	http://www.smrsgroup.org
ArMET	Architecture for Metabolomics	http://www.armet.org
CCPN	Collaborative Computing Project for the NMR Community	http://www.ccpn.ac.uk
Standards organizations		
ANSI	American National Standards Institute	http://www.ansi.org
IEEE	Institute of Electrical and Electronics Engineers	http://www.ieee.org
IETF	Internet Engineering Task Force	http://www.ietf.org
OMG	Object Managament Group	http://www.omg.org
W3C	World Wide Web Consortium	http://www.w3.org

From Brazma *et al.* (2006).

It is already clear that systems biology caused an excitement among scientists and established a sink for considerable financial investment as well as an attractant for outstanding brains, even from outside the field of biological sciences. This should constitute a receipe for success. Traders commonly praise their merchandise — thus, Kitano (2002b) wrote about systems biology: "... its potential benefits are enormous in both scientific and practical terms". But Ronald Plasterk (Professor of Developmental Genetics at Utrecht University in the Netherlands) is less enthusiastic about systems biology, and in an interview (*Curr. Biol.* 15: R861, 2005) expressed "mixed feelings" and said, "All biology is systems biology. We do not need chemists and smart physicists to come and tell us that biological mechanisms can only be understood in light of the complete organisms". Future years will show whether the success of systems biology will be of benefit only to those engaged in this approach or the benefits will reach wider circles, such as plant biologists interested in plant patterning.

Part B

THE PATTERNING OF PLANT COMPONENTS

The sporophytic phase of the life cycle of plants starts with the zygote. The zygote is the result of mating between a gametophytic sperm nucleus and the nucleus of a gametophytic egg cell (in parallel, the triploid endosperm is initiated in the angiosperms). The genomes of the zygotes encompass the information that is required for the plant's patterning and activities (e.g. metabolic activities and responses to environmental cues). Only little is known about the roles of interactions between the nuclear genome and its products and between the organelles' genomes and their products in plant patterning. I shall therefore concentrate on the nuclear genome and its products.

In Part A (Chapter 5), I handled the various stages where the expression of genes can be regulated (chromatin, DNA, pre-mRNA, mRNA, protein conformation and protein degradation). For correct patterning, the regulation of gene expression should be specific with respect to a given cell; for that to happen, the cell should be able to perceive its location. Moreover, we should also consider the possible flow of hormones and polymers (as proteins and nucleic acids) from other cells to the cell where the proper regulation of gene expression should take place. Thus, the nuclear genome can be considered omnipotent with respect to its (about 30 000) genes; however, there are "checks and balances" that limit the realization to only a fraction of this potency in a given cell and during a given phase of development.

This brings us to a passage from the Hebrew Bible, Proverb 25:

תפוחי זהב במשכיות כסף

(in "free" translation: "Apples of Gold Encased in a Silver Network").

Maimonides (1135–1204), in his "Guide to the Perplexed", chose this metaphor to emphasize that the unenlightened look at a passage from the Bible superficially and see only the "silver", while the enlightened look deeper and reveal the "gold". I shall use the same metaphor for a very different purpose. I shall highlight a logic of patterning: there is a pleiotropic potential (the "gold"), but this gold potential is encased by an intricate "silver network" that permits the pattern formation in only one or another route. Here is an example that may be helpful to the reader. A linear sequence of amino acids may fold into the required protein configuration only after due interaction with an appropriate chaperone. Misfolded proteins may be degraded (e.g. by ubiquitination). Thus, the "silver network" has a prime role in affecting the final shape of the "apples of gold".

I have arranged the seven chapters of Part B on the basis of several considerations. The order is the following: Embryo → Shoot Apical Meristem → Root → Transition from Vegetative to Reproductive phase → Leaf → Flower. Clearly, the patterning of plants can be presented by a different order of chapters, but the present order was intended to render the book palatable to readers who are not specialists in this subject. To start handling angiosperm patterning with the newly established zygote means to postpone the detailed discussion on the development of the male and female gametophytes as well as the process of pollination and double-fertilization (see Raghavan, 2003). However, the main features of the embryo sac (the female gametophyte) and the pollen tube (the male gametophyte) shall be provided in Chapter 7. We shall note already here that the available structural studies indicate that, while the variability in patterns among the sporophytes of angiosperms is vast, there are only minor differences in the differentiation of the male- and female gametophytes among angiosperms. For example, in some plant families, there is only one sperm nucleus in the pollen grain and this nucleus divides once in the germinated pollen tube. In other families, this nucleus divides already before the maturation of the pollen grain. As for the arrangement of the cells within the embryo sac, there is also some variability among plant families.

Chapter 7

PATTERNING OF THE ANGIOSPERM EMBRYO

The Transition from the Gametophytic to the Sporophytic Phase

The transition from the gametophytic phase to the sporophytic phase in angiosperms is heralded by "fireworks". The pollen tube that is led towards the embryo sac by chemoattractants, which are released by the two synergid cells, approaches one of the two synergid cells (the degenerated one) after passing the filiform apparatus. The pollen tube then discharges its content into the synergid cell with a "big bang". This content includes the two sperm cells. In each sperm cell, there is a nucleus and a small amount of cytoplasm. One of the two sperm cells (probably the one that contributes its nucleus to the fusion with two nuclei of the central cell) may contain more organelles than the other sperm cell (the one that contributes its nucleus to the fusion with the nucleus of the egg cell). Almost instantaneously with the penetration of the tip of the pollen tube into the degenerated synergid cell, there is an "explosion" in the synergid cell and a breakdown of its content. Details and illustrations of the process of fertilization were provided by Bewley *et al.* (2000) and Lord and Russell (2002). However, the main features of the fusion process of one sperm nucleus with the egg cell nucleus, and the other sperm nucleus with the two central nuclei, were provided already 30 years ago by W.A. Jensen (see Fig. 3.15 in Frankel and Galun, 1977).

An experimental procedure to convert the male gametophyte directly into a haploid plant that is morphologically similar to the sporophyte emerged by chance. As a postdoctoral fellow at the Biology Division of the California Institute of Technology (Caltech) in

Pasadena (1960–1962), I cultured floral buds of cucumber and, under some conditions, the anthers of these buds could pass *in vitro* the meiosis phase (Galun *et al.*, 1962, 1963). Satish Maheshwari, who was also a postdoc in the same division, intended to carry this study further, and upon his return to the University of Delhi used anthers of *Datura* to obtain meiosis *in vitro*. He then observed that such cultured anthers can yield embryo-like structures that could develop into plants. Further analyses indicated that these were haploid plants that developed (without the zygotic stage) into functional haploid plants. The system was then developed by many investigators, and it became evident that microspore and/or anther cultures are proper means to obtain haploid plants in species of many plant families (especially in the Solanaceae). The popularity of this androgenesis resulted from the possible benefit of this procedure for genetics and breeding. Androgenesis of heterozygous plants will result in hemiploids, and the doubling of the chromosome in these hemiploids (which frequently happens without any treatment) will provide diploid homozygotes. In practice, this was possible, but it was rarely used in actual plant breeding. On the other hand, androgenesis provided a clear message: embryo differentiation is not restricted to the cells derived from the zygote.

Let us return to the natural fertilization and formation of the embryo. A detailed recording of the approach of the pollen tube to the embryo sac, and the double fertilization that will result in the diploid embryo and the (usually) triploid endosperm, was made possible by *in vitro* studies. In *Torenia fournieri* (Scrophulariaceae), the embryo sac protrudes from the microphyle, and thus is accessible to direct observations and cell ablation (by laser UV irradiation). The method of dealing with living embryo sacs and their fertilization was initiated by T. Higashiyama and collaborators (Higashiyama *et al.*, 1998, 2000, 2003), and then an *in vitro* procedure was developed.

The results of the experiments, in which these methods were applied, indicated that the migration of the pollen tube, starting from the germination of the pollen on the pistil's stigma until the approach of the pollen tube at the embryo sac, can be divided into several stages. The first stage is the movement of the pollen tube in the style until it reaches the ovule. This first stage is apparently not affected by the female gametophyte (the embryo sac), but further targeting of the pollen tube requires the existence of a wild-type female gametophyte. Both the passage through the funiculus and the entrance via the microphyle require guidance (e.g. funicular guidance and microphyle guidance). Studies with Arabidopsis mutants indicate that there may exist two diffusible chemoattractants (probably proteins) that are released from the embryo sac and are involved in these guidances. There also seems to be a mechanism (akin to animal egg cell fertilization) that assures that, after one pollen tube penetrates the embryo sac, the entrance of further pollen tubes is barred. The synergids are probably the source of the chemoattractants because the ablation of one synergid cell reduces this attraction of the pollen tube, and there is no attraction at all when both synergids are ablated.

The gene *MYB98* that encodes a transcription factor involved in pollen tube guidance is specifically expressed in the synergids (Kasahara *et al.*, 2005). However, what about embryo sacs that lack synergids (as in Plumbaginaceae)? Here, I would like to note that the cell composition of the embryo sac in Arabidopsis (i.e. three antipodal cells, a central cell, two synergids and one egg cell) is not identical in all angiosperms. For example, there are many antipodal cells in maize. In the text of P. Maheshwari (1950), 10 types of mature embryo sacs are listed. Only one of them, the *Polygonum* type, is represented in Arabidopsis. The procedure, which is based on the isolation of embryo sacs from *Torenia fournieri*, was used by Y. Fu and associates (e.g. Fu *et al.*, 2000, 2001; Fu and Yang, 2001) to further study the penetration of the pollen tube into the embryo sac. These studies focused on the changes in actin organization during fertilization. Also, the movement of the sperm nuclei themselves is apparently mediated by actin filaments and microtubules (MTs); but how these microfilaments and MTs are "ordered" to move the nuclei towards their target is still unclear. It was found that once the fusion of the egg cell nucleus and the sperm nucleus takes place, there is a significant change in the transcription of genes (Sprunck *et al.*, 2005).

The nuclei of the sperm cells and the vegetative nucleus of the pollen tube differ considerably. The difference is in the chromatin structure (e.g. in the composition of the histones and the density of the chromatin) and in their gene expressions. Are there also differences between the two sperm nuclei that either fuse with the egg cell nucleus or with the two nuclei of the central cell? Interestingly, the fusion between the nuclei of the two gametes (the sperm cell nucleus and the egg cell nucleus) will take place only when both nuclei are in phase G_2 of the cell division cycle. A phase difference between the two sperm cell nuclei could determine which of the two nuclei will fuse with the G_2 egg cell nucleus. An indication for a difference between the two sperm cells was found by Singh *et al.* (2002) in *Plumbago zeylanica*. In *Plumbago*, the two sperm cells differ. One contains more mitochondria, while the other contains more plastids (both cells are relatively poor in cytoplasm). The nucleus of the plastid-rich sperm cell of *Plumbago* fuses with the egg cell nucleus, while the nucleus of the mitochondria-rich sperm cell fuses with the central cell nuclei. The mitochondria-rich cells had a higher expression of ubiquitin than the plastid-rich sperm cells. It should be noted that *Plumbago* belong to the minority of genera in which the organelles are also paternally transmitted. Hence, such differences may not represent a general phenomenon in angiosperms.

In the male gametophytes of maize, the two sperm cells can be distinguished by B chromosomes. Quite frequently, the two B chromosomes are not divided equally in the two sperm cells. One sperm cell nucleus may receive two B chromosomes, while the nucleus of the other sperm cell will not contain any B chromosomes. Genetic studies indicated that the nucleus with two B chromosomes fuses with the nucleus of the egg cell. It thus seems that the targeting system is able to sense (at least in some cases) a difference between the two sperm cells.

From the Zygote to the Mature Embryo

A short structural description of the differentiation from the formation of the zygote (by the fusion of the haploid sperm nucleus with the haploid egg cell nucleus) up to the mature embryo, is as follows:

- The zygote divides (asymmetrically) into an apical cell and a basal cell;

- The apical cell divides three times in three planes to form an eight-celled embryo proper. At the same time, the basal cell, which is longer than the apical cell, divides transversely to yield the hypophysis adjacent to the embryo and several suspensor cells away from the embryo;
- The initial embryo proper (preglobular stage) consists of a small number of cells surrounded by the protoderm and is ball-shaped; it is connected by the hypophysis to the suspensor, and the latter connects the embryo to the ovule;
- Further cell division in the embryo proper leads to a central ground meristem surrounded by a procambium cell layer, which leads to the establishment of the globular embryo;
- After additional cell divisions, the embryo reaches the heart-shaped stage; further growth of the heart-shaped embryo leads to the torpedo stage; subsequent cell divisions and differentiation result in the mature embryo.

In recent years, much of the study on embryo differentiation has been performed with Arabidopsis. As this embryogenesis is very similar in all the dicots of angiosperms, the sequence of events that was described above provides a fair background to embryogenesis in all dicots. Much less knowledge exists on the embryogenesis of monocots.

Laux *et al.* (2004) of the University of Freiburg, Germany, reviewed the genetic regulation of embryogenic pattern formation in angiosperm. Their detailed review is based primarily on Arabidopsis embryos, and their illustrations are provided in Fig. 25. The genetic regulation starts at the earliest stage of embryogenesis. Already, the egg cell and the zygote express mRNAs of the WUSCHEL HOMEOBOX genes that encode the transcription factors WOX2 and WOX8 (Haecker *et al.*, 2004). Of these, WOX2 is specific for the early apical and WOX8 for the early basal embryo development. Figure 25B indicates which genes are expressed during embryo differentiation, starting from the 16-cell proper embryo stage until the late heart stage, as well as the spatial locations of these expressions. We should take into account that the various transcription factors are capable of intercellular communication, and in some cases, one factor may suppress the other

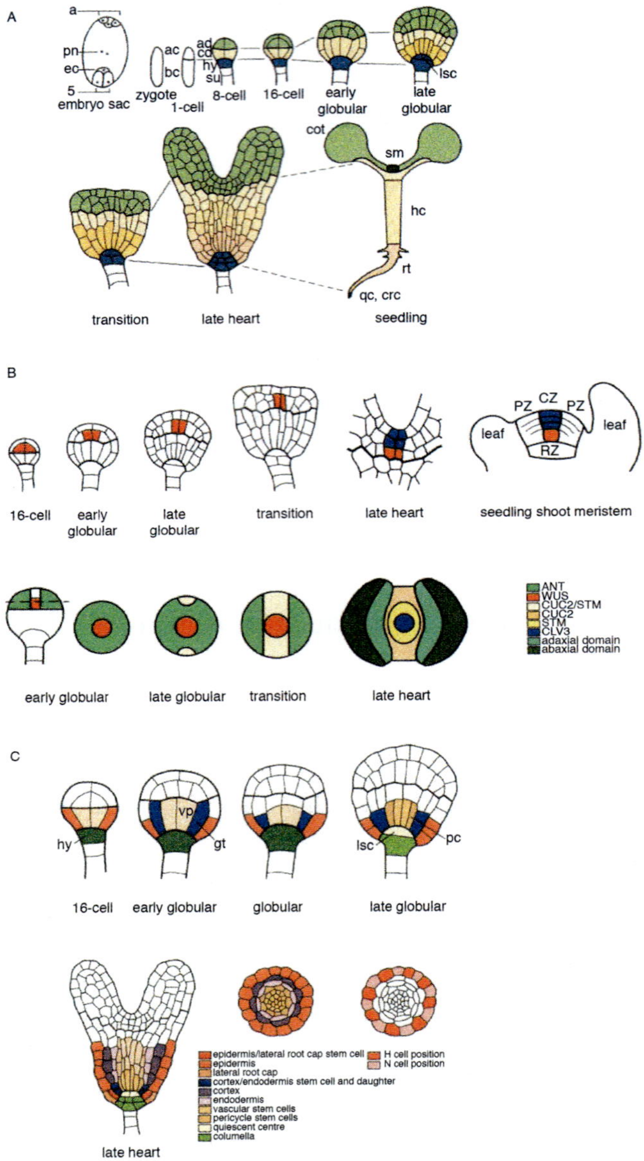

Fig. 25. **(A)** Apical-basal Arabidopsis embryo development. Schemes of longitudinal median sections. The upper and lower thick lines represent clonal boundaries between the descendants of the apical and basal daughter cells of the zygote and between the apical and central embryo domains, respectively. a, antipodes; ac, apical

factor in another group of cells. As was found in animal patterning (e.g. *Drosophila* embryos), the boundaries of transcriptional domains are progressively refined into more detailed region-specific expression patterns, in which intercellular interactions play an important role.

The perception of a cell's position has a major role from the first division of the zygote. As the egg cell is positioned in a specific location in the embryo sac, this division is always in the same plane (see Fig. 25A). Once this axis is set, all further planes of cell divisions in the embryo proper and in the basal cell follow suit.

In Chapter 5 above, several means of regulation of gene expression were listed. One of these that came to light relatively recently is gene silencing by microRNA (see Galun, 2005a, 2005b for details). From several studies, such as the mutation in the *DICER-LIKE1* (*DCL-1*) gene, which is required for microRNA-induced gene silencing, it became evident that microRNA-derived gene silencing is crucial for normal embryogenesis.

Returning to the first asymmetric division of the zygote, we already are faced with a plant hormone (auxin) that is involved in

---

daughter cell; ad, apical embryo domain; bc, basal daughter cell; cd, central embryo domain; cot, cotyledons; crc, central root cap; ec, egg cell; hc, hypocotyl; hy, hypophysis, lsc, lens-shaped cell; pn, polar nuclei; qc, quiescent center, rt, root; s, synergids; sm, shoot meristem; su, suspensor. (From Laux *et al.*, 2004.) **(B)** Development of the apical embryo domain. The top row shows schemes of longitudinal median sections. The upper and lower thick lines represent clonal boundaries between the descendants of the apical and basal daughter cells of the zygote and between the apical and central embryo domains, respectively. The bottom row shows cross-sections of the same stages as indicated by the dashed line at left. CZ, central zone; PZ, peripheral zone; RZ, rib zone. The expression domains of early genes in the apical region are shown in color as indicated. (From Laux *et al.*, 2004.) **(C)** Development of the radial pattern. The top row and the illustration at bottom left show schemes of longitudinal sections; the other illustrations in the bottom row show schemes of cross-sections through a root. The upper and lower thick lines represent clonal boundaries between the descendants of the apical and basal daughter cells of the zygote and between the apical and central embryo domains, respectively. Cell types are shown in color as indicated. Vascular and pericycle cells are shown in lighter colors than stem cells. gt, ground tissue; hy, hypophysis; lsc, lens-shaped cell; pc, pericycle; vp, vascular primordium. (From Laux *et al.*, 2004.)

the patterning of this early stage of embryogenesis. Genes of the *PINFORMED* (*PIN*) family, such as *PIN7*, are required for auxin efflux from cells. When the latter gene is mutated (*pin7*), the auxin efflux is inhibited and the divisions in the apical cell are inhibited. Another efflux-affecting gene, *PIN1*, determines the intracellular efflux of auxin and may affect the shape of the embryo cells. The wild-type of *PIN7* and *WOX2* proteins are required for normal divisions in the apical cell. It appears that the formation of the early-stage embryo causes a suppression of embryo formation from the suspensor cells. If the embryo is ablated, secondary embryos can form from the suspensor.

A comprehensive update on the roles of auxin in plant embryogenesis was provided by Jenik and Barton (2005). There is evidence indicating that auxin can affect the apical–basal axis patterning. Localized surges in auxin may cause specific ubiquitination of transcription repressors and thus control embryo patterning. Several gene products are expressed very early in the embryo's differentiation, but their manifestation, as evident by structural results, appears at a later phase. This is probably true for the *gurke* (*gk*) mutant. In the presence of this mutation, the differentiation of the cotyledons and the shoot apex between them will not happen.

Another noteworthy experimental result emphasizes the "memory" of patterning. A temperature-sensitive mutation in Arabidopsis can cause a temporary inhibition of embryonal primary roots. After shifting back to permissive temperature, the suppression that is caused by this mutation is relieved and a root differentiates from the same site. However, the differentiation of this novel root is apparently similar to the differentiation of postembryonic roots. This is reminiscent of the "memory" in *Fucus* embryos to form apical hairs and rhizoids at the proper locations after they are released from osmotic inhibition, as mentioned above (Galun and Torrey, 1969; Torrey and Galun, 1970).

Between the transition and the heart stages of the embryo, an important feedback loop is established that, in principle, is carried on to the shoot apex throughout the vegetative growth of the dicot shoot. This loop will be elaborated on in the next chapter; here, it will be summarized. A *central zone* (*CZ*) is formed in the cleft of the heart-stage embryo. This zone contains cells that will divide slowly, and it harbors the stem cell niche with the stem cells in the outermost of three cell layers. This is the *organizing center* (*OC*). The product of the *WUS* gene in the OC maintains the stem cells in an undifferentiated (dividing) state. The stem cells express the CLAVATA3 (CLV3) protein, which is the probable ligand of the CLV1 receptor kinase signaling pathway. CLV1 limits the size of the OC by *restricting WUS* expression. This loop adjusts the size of the stem cell pool in the developing embryo and subsequently in the shoot apex. The CZ is surrounded by cells that have a potential to divide rapidly, forming lateral outgrowths. However, this outgrowth is suppressed by the product of another gene: S*HOOTMERISTEMLESS* (*STM*). Only when the *STM* product is reduced will the outgrowth take place. While the structural manifestation of the different cell types is observed only at the late heart stage, the expression of the controlling genes is already initiated in an earlier stage of embryo differentiation. Mutation in another gene, *gurke* (*gk*), will prevent normal shoot meristem and cotyledon initiation, as mentioned above. The expression of WUS starts early: in the four subepidermal apical cells of the 16-cell embryo. Upon further divisions of these cells, the *WUS* product is distributed in a programmed manner in the embryo. For the correct spatial expression of *WUS* in the heart-shaped embryo, the CLV3 protein is required.

In the apical domain of the early embryo, the *CUP-SHAPED COTYLEDON* (*CUC*) genes (*CUC1*, *CUC2*, *CUC3*) assure that the outgrowth is limited to the embryonal cotyledons. The *CUC* genes may thus control bilateral symmetry.

Here, I would like to draw the reader's attention to names of genes that are assigned by investigators, especially those dealing with Arabidopsis genetics and molecular biology. The names of these genes do not necessarily describe a specific phenotype that results from the mutated gene. The names of the genes can be rather long, written in capital letters and in italics, commonly have abbreviations (commonly of two, three or four letters). Hence, gene names can confuse the novice, but are efficiently serving the experts in the field.

Now, back to the *CUC* genes. *CUC1* and *CUC2* are not equally expressed in the embryo; however, the expression is restricted to certain locations (as seen in the lower side of Fig. 25B). Moreover, for the spatial expression of *CUC1* and *CUC2*, the expression of two additional genes, *MONOPTERUS* (*MP*) and *PIN1*, is required, indicating that auxin is involved in this spatial regulation. Furthermore, *STM* promotes *CUC1* activity together with *PIN1* and is required for correct spatial expression of *CUC2*. Another gene, *AINTEGUMENTA* (*ANT*), is expressed in a ring around the circumference of the apical domain of the globular embryo (see Fig. 25B) and then is restricted to the incipient cotyledonary primordia, helping the outgrowth of embryonal cotyledons.

Once the cotyledonary primordia of the embryo are established, their future asymmetry with respect to differences in the adaxial and abaxial sides evolves. This asymmetric patterning seems to affect the shoot meristem. The gain-of-function *PHABULOSA* (*PHB*) gene is active in the promotion of the adaxial cell fate. When mutated, this gene can increase the size of the embryonic shoot meristem. Also, the *REVOLUTA* (*REV*) gene is probably involved in the shaping of the shoot meristem. In addition to *PHB* and *REV*, also another gene *ZWILLE* (*ZLL*) is involved in the patterning of the embryonic shoot apex and cotyledonary outgrowth. Ectopic expression of *ZLL* at the abaxial side of cotyledon primordia will transform these primordia into shoot axes! Normally, the pattern of *ZLL* expression in the developing embryo undergoes programmed spatial changes and *ZLL* is an essential component of gene regulation by RNA silencing. Thus, programmed and spatial silencing of regulatory genes may play a central role in embryonic patterning.

The Ups and Downs of the Differentiating Embryo

We shall return to the early stages of embryonal differentiation (see Fig. 25A) and look at the 8-cell and the 16-cell stages. The cells of these embryos have either of two destinations. Cells that are away (distal) from the suspensor will become the bases of the future cotyledons, while cells that are adjacent (proximal) to the suspensor will form the hypophysis and the embryonal root. The root cells include those proximal cells that constitute the root apical meristem. There are indications that steroids are involved in the differentiation of cells in the center of the embryo, since a gene involved with steroid metabolism, FACKEL (FK), is expressed in these cells. However, whether or not the plant hormones, brassinosteroids, have a decisive role in the elongation of central cells towards the differentiation of vascular tissue is not yet clear.

What happens in the basal (root side) of the early embryo is both complicated and yet to be fully understood. However, some structural features and gene activities shall be described here. Let us start from the hypophysis at the early globular stage (see Fig. 25C). At this stage, the hypophysis undergoes an asymmetric cell division, forming a small lens-shaped cell and a larger more basal cell. The lens-shaped cell will form quiescent-center (QC) cells (that are not dividing or have a very low cell division activity). The lower daughter cell of the hypophysis will give rise to the columella stem cells. This asymmetric division of the hypophysis is marked by the expression of two genes: *SCARECROW* (*SCR*) and *WOX5*. These expressions are then maintained only in the lens-shaped cell; this asymmetric expression may have a role in establishing the difference between QC and stem cells.

The early root differentiation is affected by two genes that are involved in auxin response: *MP*, which encodes a putative transcription factor that is related to auxin response factors, and *BODENLOS* (*BDL*). There is a probable interaction between *MP* and *BDL*. The *MP* gene is kept inactive by binding to *BDL*. *BDL* can be degraded by IAA signaling. When this happens, *MP* will be activated and affect its target genes. Defects in these two genes (i.e. *mp* and *bdl*) cause phenotypes already in a very early stage, after the first asymmetric division of the zygote. In the *mp* and *bdl* mutants, the normal octant stage is not achieved. Mutation in *WOX9* causes a slightly later abnormality: the hypophysis fails to yield the lens-shaped cell, inhibiting root formation. Mutants of other genes involved in auxin response, such as *auxin resistance 6* (*axr6*), also cause root differentiation defects. The *AXR6* encodes a subunit of a ubiquitin protein ligase that is a component of auxin-mediated degradation of specific proteins. Clearly, correct ubiquitination is essential for normal embryogenesis; when a subunit protein (*RPN1*) of the 26S proteosom is defective, embryo differentiation is impaired (Brukhin *et al.*, 2005).

Radial Differentiation in the Embryo

The start of the radial differentiation of the embryo is already visible during the transition from the 8-cell to the 16-cell stage. The cells in the eight-cell embryo undergo tangential cell division, yielding the outer *protoderm* that envelopes the inner cells (see Figs. 25A, 25B and 25C). Further divisions then establish a layer of cells, which in the lower side of the embryo (that leads to the development of the hypocotyl and the root), can be recognized (from the outside inwards) as *epidermal initiating cells*, *cortex*, *endodermis* and the central vascular initials. The vascular primordial cells will then differentiate into central vasculature and surrounding *pericycle* cells. Further divisions yield epidermis and the lateral root cap.

Which genes are involved in these radial patternings? The answer is that this is not known yet. However, there are clues. One suggestion was that the first signal for the tangential cell division originates already in the cell wall of the zygote. In support of this possibility is the expression of two Arabidopsis genes: *ARABIDOPSIS THALIANA MERISTEM LAYER1* (*ATM1*) and *PROTODERMAL FACTOR2* (*PDF2*), which are expressed very early in the development of the embryo in all the cells that possess an outer cell wall that is contiguous with the cell wall of the zygote. At the eight-cell stage, when the tangential cell division takes place, further expression of these genes is retained only in the protoderm. Thereafter, expression is limited to the epidermis. Interestingly, when both *ATM1* and *PDF2* are mutated, the embryo will have an aberrant surface, and the seeds germinating from such double mutants will lack an epidermis, even in the leaves. On the other hand, if only one of these two genes is mutated, no aberrant phenotype was observed. It should be noted that *ATM1* and *PDF2* encode homeodomain transcription factors that contain the five-amino-acid START domain, which is implicated in lipid/sterol binding.

Now we should be ready for additional genes that control the evolution of the vascular cells (xylem and phloem), the pericycle, the endodermis and the cortex. There is the *SHORT ROOT* (*SHR*) gene, which encodes a putative transcription factor and is expressed in the vascular cylinder and the pericycle, and by that regulates differentiation in the surrounding tissue. This is caused by activating the expression from another gene, *SCR*, which encodes a related transcription factor. These factors can move short distances. Hence, *scr* and *shr* mutants have only a single cell layer instead of the normal two layers: endodermis and cortex. Then there is another gene, *FASS* (*FS*). In the mutant of this gene (*fs*), there is an increased number of radial cell layers. The mutant *fs* is able to rescue the *scr* and *shr* mutants, re-establishing the normal number of cell layers. On the other hand, in the presence of *shr*, there is only the cortex but no endodermis, and *fs* does not reestablish the endodermis in *shr* plants. There are, therefore, indications that the *SHR* product promotes asymmetric cell division (via the *SCR* product) and takes part in specifying endodermal identity. Johnson *et al.* (2005) revealed a gene (*AtDEK1*) that is required for embryonal epidermis formation. Loss of function of the product of *AtDEK1* was correlated with loss of epidermis.

After considering the elaborate regulatory mechanisms and the interactions between genes that pattern cell differentiations such as those mentioned above, as well as considering other mechanisms that are narrated below in Part B of this book, the reader may be amazed. How can it be that the modest information stored in the tiny Arabidopsis genome is capable of executing the elaborate patterning process? The secret is probably that the genome not only codes for about 30 000 products, but also includes information of "where" and "when" to synthesize these products. An old anecdote comes to mind. A driver is crossing the Death Valley desert when suddenly the engine of his car stops operating, and the driver cannot repair it. He is desperate, but a man passes by and suggests to call a blacksmith who lives five miles away. The driver brings the blacksmith, who is carrying only a heavy hammer. The blacksmith hits the engine with his hammer, and the engine springs back to life. When asked about payment, the blacksmith asks for $101.00. "Why this sum?" asked the driver. The answer of the blacksmith was that $1.00 is for the hit and $100.00 is for the knowledge of *where* to hit.

Asymmetric Cell Division and Ramification of Developmental Fates

We noted above that asymmetric cell divisions take place in several stages in embryo development. Already the zygote undergoes asymmetric division, and such divisions also occur in later stages such as in the hypophysis. The great difference in cell size and shape among the two daughter cells provides a clear signal for the different future fates of these cells. However, such differences in shape and/or size do not condition the ramification of fate in two adjacent cells. Clearly, two cells of similar size and shape that are descendants of the same mother cell have respectively different neighbors, and can thus be differentially exposed to morphogenes or to other cell–cell communications with neighboring cells. The latter may determine the fate of one daughter cell in a different manner than the other daughter cell.

There are several types of candidates that may serve as agents of cell–cell communication, and plant plasmodesmata are the common portals for this communication. The cell-to-cell route via plasmodesmata was found to be involved in embryo development (Kim and Zambryski, 2005). Proteins (up to a given size), RNAs and plant hormones can thus enter a cell and exercise their patterning control. As for angiosperm embryo cells, knowledge accumulates on the signaling by auxin. Already at the daughter cells of the zygote, the apical daughter cells seem to have a higher auxin level than the basal cell. The transport of auxin from the basal to the apical cell is presumably mediated by *PIN7* (see Chapter 2 above for auxin metabolism and flow). At a later stage in the globular embryo, auxin appears to flow in a reverse direction: from the apical side, basipetally. This flow is probably mediated by the products of three *PIN* genes: *PIN1*, *PIN4* and *PIN7*. When lateral organs (i.e. cotyledons) are formed, the auxin seems to flow towards the tips of these organs. Jenik and Barton (2005) reviewed the subject of the role of auxin in embryo morphogenesis. The transport of auxin during several stages of embryogenesis is schematically presented in Figs. 26A and 26B, from this review. While auxin levels can be evaluated in tissues, there are as yet no reliable procedures to determine auxin levels in individual cells, let alone intracellular distribution of auxin. Therefore, I used the words: "appears to". It is expected that additional plant hormones also participate in the patterning of the developing embryo.

V. Sandaresan and associates (Pagnussat *et al.*, 2005), a team consisting of 10 investigators from UC Davis (CA), the Republic of Korea, and the Institute of Molecular Cell Biology in Singapore, conducted a large-scale search in Arabidopsis for mutants affecting the female gametophyte. They employed Ds transposon insertional mutagenesis and obtained 24 000 insertion lines. These were searched for mutated phenotypes, defective in the female gametophyte, that appeared during embryo sac development and had defects in early-embryo differentiation that followed fertilization with wild-type pollen. Nearly half of the identified mutants were those that were defective after fertilization. This indicated that genes expressed in the female gametophyte play a role in the early development of the embryo, although their embryo sac had an apparently normal phenotype.

Although efforts to identify genes that are required for the normal differentiation of the female gametophyte were already conducted several years earlier (Feldmann *et al.*, 1997), the study of Pagnussat *et al.* (2005) was the first large-scale effort in this field. The T-DNA used by Pagnussat *et al.* (2005) included a gene for kanamycin resistance (*kanR*). Selfing of plants derived from insertion should thus segregate (*kanR:kanS*) at a 3:1 ratio, unless part of the progeny will not produce seeds. Actually, the investigations frequently obtained a different ratio (e.g. 1.5:1).

Out of the candidates for mutations in the female gametophyte and embryo differentiation, 130 were further characterized (see Table 3). Among these were mutants with defects in the embryo sac; mutants that had apparently normal embryo sacs but did not respond to pollination, termed *UNE* (unfertilized embryo mutants); mutants that lacked pollen tube guidance, termed *MEE* (maternal effect embryo sac mutants); and those mutants in which embryo development was arrested (*EDA*). Interestingly, a considerable fraction of the female gametophyte

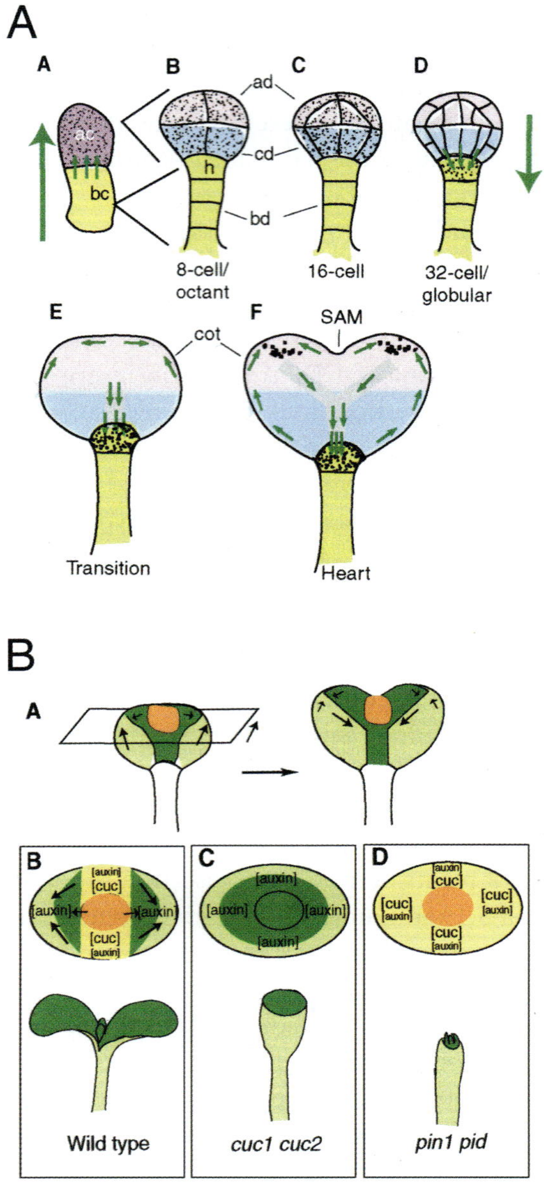

Fig. 26. **(A)** Auxin transport relative to early events in Arabidopsis embryo patterning. **(A)** An early Arabidopsis embryo, consisting of an apical cell (ac) and a basal cell (bc). Arrows indicate the direction of auxin transport; stippling indicates regions

mutants were also defective in the male gametophyte (pollen and pollen tube). This may indicate that some genes are essential for the normal development and function of both gametophytes. Several of the characterized mutations appeared to cause defects in known genes, such as *MEA*, *FIS2*, and *FIE* (causing male and female imprinting; see Guitton *et al.*, 2004 and Luo *et al.*, 2000, for details). Further studies with specific mutations that were characterized phenotypically by Pagnussat *et al.* (2005) will probably serve in future studies aimed to characterize

---

with high auxin levels. **(B)** Eight-cell/octant-stage embryo. (Cell numbers used to stage embryos reflect the number of cells in the apical cell lineage). The apical domain and central domain both derive from the apical cell and each consists of four cells. The basal domain derives from the basal cell. **(C)** A 16-cell stage, early globular embryo. **(D)** In a 32-cell stage globular embryo, auxin transport has shifted direction, and auxin now accumulates in the hypophyseal lineage. The hypophyseal lineage is derived from the hypophysis (h) — the suspensor cell closest to the embryo proper. This lineage gives rise to a portion of the root meristem, specifically the quiescent center and the central columella with associated stem cells. **(E)** A transition stage (transitioning between globular and heart stage) embryo. Auxin transport in the apical domain is directed towards the center of the cotyledon primordia (cot). **(F)** An early heart-stage embryo, showing the emergence of cotyledons and a cleft where the shoot apical meristem (SAM) will form. Gray indicates regions of vascular development. (From Jenick and Barton, 2005.) **(B)** Partitioning of the apical domain into shoot apical meristem and cotyledons. **(A)** Transition-stage and early heart-stage Arabidopsis embryos showing the direction of auxin transport (*arrows*) and the partitioning of the embryo into cotyledonary and meristem domains. **(B)** Cross sections (as shown in A) through the apical domain of a wild-type embryo, showing the region that will develop into the shoot apical meristem, intercotyledonary zones, adaxial domain (top) of the cotyledon, and abaxial domain (bottom) of the cotyledon. The meristem and intercotyledonary zones have low auxin levels and high CUC levels, whereas the opposite holds for the cotyledon primordia. **(C)** Cross-section of the apical domain from a cup-shaped *cuc1 cuc2* cotyledon mutant. No separation is made between the cotyledons and no shoot apical meristem is made in the central region of the embryo. **(D)** Cross-section of the apical domain from a *pin pid* embryo. These embryos lack the PIN1 auxin transporter. They also lack the PID kinase that appears to be responsible for positioning PIN1 on the apical side of the cell. Lack of high auxin levels in the cotyledon primordia in these embryos would allow CUC accumulation throughout the apical domain, thus preventing cotyledon outgrowth. (From Jenick and Barton, 2005.)

Table 3 Summary of Phenotypes Observed in Insertion Mutant Lines with Transmission Deficiency through the Female Gametes

Phenotype Observed	Number of Lines
Embryo sac with one functional megaspore	1
Embryo sac arrested at two-nuclear stage	5
Embryo sac arrested at four-nuclear stage	3
Embryo sac with varying defects in nuclei number and position	7
Unfused polar nuclei	18
Mutants showing embryo sacs arrested at varying stages of development	8
Defects in fertilization	18
Arrested at one-cell zygotic stage	
Endosperm development abnormal or arrested	56
Endosperm development normal	6
Embryo development arrested	8
Total	130

(From Pagnussat *et al.*, 2005).

the mutations at the molecular level by utilizing additional approaches, such as reverse genetics. It is thus expected that the knowledge on embryo patterning will be increased in the future.

Remarks on Phenomena that are Related to Embryogenesis

As an epilog to the chapter that deals with structural and molecular genetic aspects of embryo patterning, I shall have remarks on several phenomena that are related to embryogenesis. One remark is that in a book that deals with a scientific subject, there should be "drawers", although in nature the life cycle of plants is semicontinuous. This means that there are no clear distinctions between the differentiation before the embryo and the differentiation of the embryo itself. True, the fusion of the nuclei of the egg cell and of the sperm cell is a "starting point" manifested by the zygote, but this is not necessarily so: under some conditions, an embryo can differentiate without going through the zygote stage.

At the other end, it is common that after the mature embryo is established, there is a resting period. The embryo will develop further only after the seed (or the caryopsis) is moved to environmental conditions that will induce germination. However, even for this transition, there are exceptions. One such exception was recorded by me (but never published). While being active in breeding melons, melon growers in the Jordan Valley complained that in a year with an exceptionally warm spring (the temperature in the Jordan Valley can reach 55°C in some years), none of the melon fruits could be marketed; the seeds in all of them germinated inside the fruits. To trace this phenomenon, I grew melon and cucumber plants in acclimatized greenhouses, but put heating devices around individual fruits at several stages of fruit maturation. I found that indeed, when fruits were heated to 42°C during 2 weeks before maturation, all the seeds (in melons and cucumbers) did not attain any resting period, but rather went on and germinated within the fruit.

Double fertilization and the differentiation of the endosperm

The double fertilization in angiosperms was noted in previous sections of this chapter. Briefly, while the nucleus of one of the sperm cells of the male gametophyte fuses with the nucleus of the egg cell to result in the zygote, the nucleus of another sperm cell fuses with the two nuclei of the central cell of the embryo sac to initiate the endosperm. Ueli Grossniklaus and associates studied the effects of products of genes (e.g. *MEA*, *FIE* and *FIS* as noted above; also see Grossniklaus *et al.*, 2005 for review) on maternal imprinting and the development of the endosperm. Among these genes, *MEA* (*MEDEA*) and *FIE* encode polycomb proteins and the *FIS* genes are only expressed from the maternal allele. It was proposed that the proteins encoded by *MEA*, *FIE* and *FIS2* form a complex in which *FIS2* is bound to the DNA, and that the complex recruits a deacetylation complex which causes specific remodeling of the chromatin at the site of the protein-complexed DNA. The specific impact of the *FIE* on the development of the endosperm in Arabidopsis was studied by N. Ohad and associates (Mosquna *et al.*, 2004). They

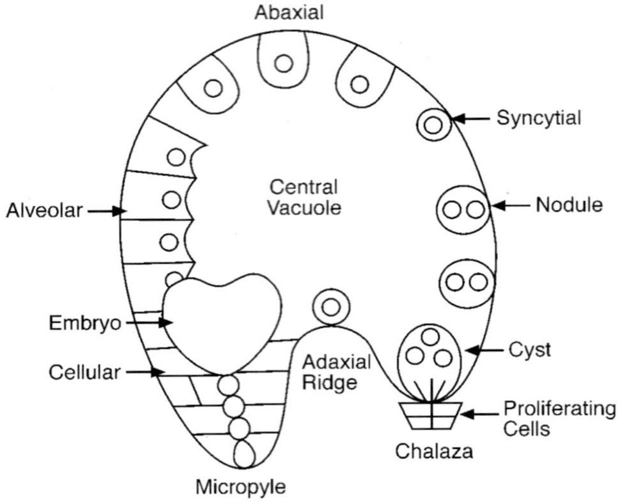

Fig. 27. Diagrammatic summary of endosperm development. In ovules containing heart embryos, all the stages of endosperm development can be seen, with the most advanced in the micropylar chamber. The first periclinal divisions are next to the embryo and followed by alveoli and syncytial endosperm in the central chamber. In the chalazal chamber are the endospermic nodules and a coenocytic cyst against the chalazal proliferating cells. (From Brown *et al.*, 1999.)

found that when the FIE protein was eliminated, the central cell of the embryo sac starts to divide and forms an endosperm without the fusion with the nucleus of a sperm cell. The retinoblastoma protein (rRp) that has a role in the cell division cycle is involved in the FIE-regulated endosperm development, but the details await further studies.

The details of endosperm differentiation shall not be provided here, and they seem to vary in different plant families. Only an example is provided in Fig. 27 from Brown *et al.* (1999). The authors are also a good general source for information on endosperms in angiosperms. We should note that in some plants such as legumes, the endosperm serves as a nutritional source for the developing embryo, while the nutrients for the germinating seedling comes from the cotyledons. In cereals, the caryopsis is mostly occupied by the endosperm. Hence, endosperms are not only the nutrient source for the germination seedling, they are also the major source of human nutrition (e.g. grains of wheat, rice and maize). The seeds of some plants, such as castor oil seeds, contain an endosperm and expanded (but thin) cotyledons.

Xenia and metaxenia

The terms xenia and metaxenia are used for different phenomena by different authors. The phenomena related to these terms are rather widespread and known since antiquity to horticulturists: the source of the pollen has a decisive effect on the characteristics of the fruit. While the parental imprinting that affects the endosperm is imposed by the maternal parent, metaxenia is imposed by the paternal parent. The phenomena have practical implementations. The fruits' (dates) quality of the dioecious palm tree (*Phoenix dactylifera*) depends strongly on the pollinator. Hence, good pollinator palms have to be selected. In several pepper cultivars, the pollination with hot pepper cultivars can render regular sweet peppers to very hot ones. The fruit character of vine cacti, such as *Selenicereus* spp., is strongly influenced by the species that supplied the pollen. This self-incompatible genus requires hand pollination in order to yield good crops; however, due to xenia, the choice of the proper pollination is important in the cultivation of these vine cacti. While the phenomena of xenia and metaxenia are well known, the genetics and molecular mechanisms that cause these phenomena are not yet fully understood.

Chapter 8

THE SHOOT APICAL MERISTEM

Background

An early scientific description of the shoot apical meristem (SAM) of angiosperms was provided by the German scholar Caspar Friedrich Wolff in his PhD thesis (submitted to the University of Halle, 1759). It is referred to in the Preface of this book, where I used an excerpt from his thesis as a motto. After analyzing the shoots of plants, Wolff came to the conclusion that all the above-ground components of plants are shoots and shoot derivatives. The shoot-derived leaf initials can either further develop to mature leaves or undergo metamorphosis and form flowers. This basic concept was also accepted a few years later (probably independently) by Wolfgang von Goethe. Hence, since about 250 years ago, it became evident that the "fountain" from which all above-ground plant components are derived is the shoot apical meristem (SAM). Scholars were obviously fascinated by several facts. First, in spite of a great similarity between the embryos of diverse plants (especially within all dicots and within all monocots), the SAMs derived from these embryos are the sources for very different mature plants. Furthermore, the SAM itself maintains the same basic structure from germination throughout the growth of the indeterminate vegetative shoot. In some tree species, this can go on for hundreds or (rarely) thousands of years.

We shall see below that the investigation of the SAM and the understanding of the controlling elements that cause the projections of plant components from the SAM were advanced with the advancement of the "tool kit" that became available. As in other themes of biological investigations, naked eye observations were first supplemented by microscopical analyses using histological methods. Then genetic approaches, such as the evaluation of phenotypes induced by mutations, were applied. Further investigations took advantage of molecular genetics. Genetic transformation was recruited to evaluate the impact of defined genes by using either or both the promoters and the coding sequences of genes, and the genes could be defined by nucleotide sequences and the encoded products (e.g. proteins, microRNAs). In examples that are described in this chapter, we will see that additional tools were employed, such as the ability to produce transgenic plants with elaborate combinations of hybrid transgenes and use of confocal microscopy to observe *in vivo* cell differentiation, as well as the use of mathematic algorithms in the evaluation of patterning and gene interactions. Thus, the sophisticated tool kit rendered the SAM a subject of advanced investigations. In spite of the enormous advantages furnished by the availability of tools that promoted the understanding of the patterning of SAM, several major questions still do not have answers. In many cases, it is not clear how a cell perceives its location relative to other cells. Also, while it is now possible to evaluate the levels of specific proteins and specific mRNAs in individual cells, such quantitative information on certain major plant hormones (e.g. auxin) is still lacking.

In one of my previous books (*Transposable Elements*, Galun, 2003), I already noted that scholars engaged in molecular biology have a flair for mythology and fairy tales. Hence, transposable elements were given names such as *Poseidon*, *Penelope*, *Neptune*, *Ulysses* and *Sleeping Beauty*. Investigators of plant patterning did not lag behind and gave mutations names such as *Medea* and *Dornröschen* (the latter is the Sleeping Beauty from a fairy tale of the Grimm brothers, coined by Kirsch *et al.*, 2003).

This chapter starts with a structural description of SAM, and handles morphological and histological events that take place at the SAM during the growth of the shoot and its lateral outgrowths. A special section of this chapter is devoted to the CLV-WUS loop, which controls the maintenance of a small group of stem cells at the summit of the SAM. Our understanding of this loop has advanced vastly in recent years due to clever use of the available tool kits. The chapter also notes the known involvement of plant hormones in processes that occur at the SAM, and the impact of the regulation of protein production from specific genes by microRNAs is reported. Recent studies point to chromatin remodeling as an additional factor in SAM patterning, and these studies will be briefly reviewed. Lateral organs (leaves) are initiated from the SAM in typical phyllotaxis orders. Some understanding of the regulation of these orders is emerging and will also be noted in this chapter. Several subjects that are related to SAM patterning shall be handled in subsequent chapters. Among these are the asymmetry of developing leaves, the transition from the vegetative to the reproductive SAM and the patterning of floral organs.

The SAM was investigated in only a handful of the several hundred thousand angiosperm species. Among the dicots, the investigators focused mainly on Arabidopsis, tobacco, petunia, tomato and snapdragon; while maize served as the main model plant for monocots. In recent years, a very high proportion of the investigations used Arabidopsis as the model plant for studies on the SAM. The initial choice of Arabidopsis is rather obvious, but then there was the "avalanche effect": the more molecular biology information was added to Arabidopsis, the more favorable this plant became for subsequent investigations.

In several reviews of SAM (e.g. Sussex, 1989), it is claimed that there is a fundamental difference between the differentiation of animals (e.g. flies, birds, mammals) and the differentiation of angiosperms. In the former, the basic body features are already established in the mature embryo; while in the latter, there is very little similarity between the mature embryo and the mature plant. It should be noted that at least part of this difference is based on semantics. We call the young sporophyte in the seed "embryo"; the same name as the animal embryo, and then we point out that in spite of the same name, they are rather different entities. As for the mammal embryo that is claimed to be equipped with the basic body features of the adults, one should recall that after birth important differentiation processes are going on, such as the additional nerve connections in the brain and secondary sexual changes in the female mammals. As for the difference in patterning strategy between animals and plants, we should not be surprised: the evolution of these two kingdoms was already separated at the unicellular stage, and each kingdom evolved the multicellular differentiation in an independent manner.

Early Studies of the SAM

As noted above, the growth of the angiosperm shoot was already analyzed about 250 years ago. Some progress in the understanding of projection of lateral organs and stem tissue from the SAM was made during the 18th and the 19th centuries; however, a significant advance in understanding the structure and differentiation of the SAM took place from the first half of the 20th century. Investigations that caused these advances used the emerging methodologies, such as mutations, induced polyploidization and refined histological procedures. The history up to the 1950s is summarized in books and reviews (e.g. Snow and Snow, 1931; Gifford, 1954; Wardlaw, 1957; Steeves and Sussex, 1989; Sussex, 1989; Evans and Barton, 1997; Kerstetter and Hake, 1997). Thus, readers interested in this history will find the information in these books and reviews.

Some of the findings during this early period should be noted. Satina and collaborators already used polyploid cells as markers of cell layers in the apex in the 1990s. They produced chimeras (in *Datura*), in which a cell in one of the three cell layers in the SAM was polyploid. They then followed the spread of polyploid cells and found that there are three separate layers that commonly do not mix: L_1 (the upper epidermis layer); beneath this a layer of L_2 cells; and a deeper population of cells, termed L_3 (see Fig. 28). Also, albino cells could serve in chimeras in a similar manner to trace dynamics of cell division in the cell layers. It was then found that these layers are characteristic to most (all) dicots and at least some monocots. It was further found that in some cases a cell from one layer joined another layer, but then the former cell acquired the characteristics of the layer to which the cell joined. Such a change in position can happen when, for example, a cell in the L_1 layer divides parallel to the surface (rather than the regular perpendicular division). When this happens, the lower daughter cell joins the L_2 layer. It was deduced that very few (probably three) cells at the summit of the apex (of Arabidopsis) are the primary source

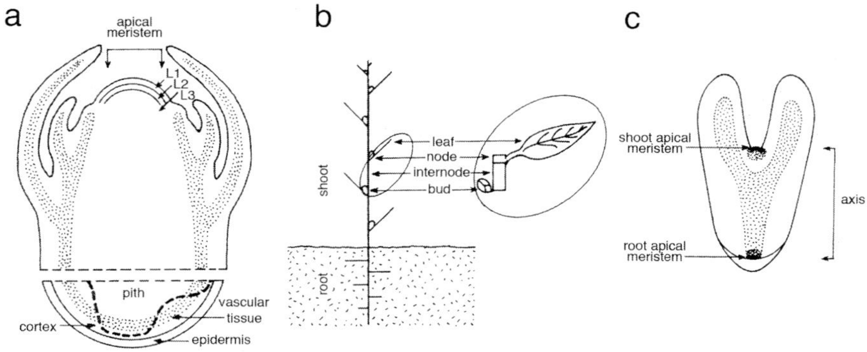

Fig. 28. Organization of the shoot meristem. **(a)** Longitudinal section of the shoot tip of a dicotyledonous angiosperm (dicot). The apical meristem is surrounded by developing leaf primordia. (Below) Partial transverse section of a stem showing the pattern of tissues that differentiate from the apical meristem. The heavy dashed line indicates the variable boundary between L_2 derivatives outside the line and L_3 derivatives inside the line. **(b)** Diagrammatic representation of a flowering plant showing shoot and root organization. At the right, a unit phytomer of the shoot is shown. **(c)** Embryo of a dicot showing location of the shoot apical meristem and the root apical meristem at opposite poles of the axis. (From Sussex, 1989.)

of all the cells of a layer, and that by the anticlinal cell division, the cells are "pushed" sideways and downwards to form the outer layer of cells in the growing shoot. Furthermore, it became evident that the fate of cells is determined by their *position* rather than by a fixed genetic program in the cells' genome.

Studies during the second half of the 20th century added another tool to the investigation of the SAM: microsurgical procedures (see Sussex, 1989, for details and references). In several cases, the SAMs that underwent such procedures that did not destroy all the meristem cells at the SAM, when followed by *in vitro* culture, resulted in the reconstitution of the SAM, and the SAM returned to its normal shape. Such results further supported the conclusion that the fate of SAM cells is determined by position rather than by strict epigenetic determination.

To what extent is the SAM autonomous with respect to its growth and differentiation? No unequivocal answer could be furnished by early studies. However, at least for a certain period, *in vitro* — cultured SAMs could develop further on a medium supplemented with auxin. We note below that auxin is clearly not the only plant hormone that has a role in cellular processes in the SAM. Cytokinin (CK), gibberellins (GA) and brassinosteroids (BR) are also involved. We shall also see that these plant hormones do not have to travel a long way: auxin and cytokinin are synthesized in the SAM, and GA is synthesized in the incipient leaf outgrowths.

The mature embryo of dicots differs considerably from the mature embryo of monocots (e.g. maize as reported already by Randolph, 1936). However, with the initiation of germination, the SAM of maize becomes rather similar to the SAM of dicots, although the initial SAM of the former plants is situated laterally to the embryo (opposite of the scutellum). The maize SAM is then formed between the coleoptile and the first leaf.

The diameter of most angiosperm SAMs is in the range of 100 to 200 μm, but can be much smaller (50 µm) or much larger (900 μm). The dome of the SAM may be rather flat, as in sunflowers, or finger-shaped, as in maize. Commonly, the SAM has the shape of a half-ball (e.g. Arabidopsis). Some investigators divided the SAM into the outer tunica (e.g. L_1 and L_2 in Arabidopsis) and corpus (L_3 in Arabidopsis).

Most investigations of the SAM were performed with Arabidopsis. Therefore, the SAM of this model plant is reviewed here, but, whenever needed, the SAMs of other plants shall be noted. Thus, here is the first note. The mature maize embryo already contains five leaf initials, while only the initials of two leaves (cotyledons) are contained in the mature embryo of Arabidopsis. Two pairs of opposite leaves are added during the initial growth of the shoot of the Arabidopsis seedling. Thereafter, additional leaf initials are formed at a less-than-180° (commonly 137°) angle from the previous leaf initial. These later leaves are thus situated in a spiral *phyllotaxis.* A later section shall deal with the control of phyllotaxis. During the early studies on the SAM, it could not be determined which signals determine the location of a new leaf outgrowth, but clearly those cells that are involved in the outgrowth of a new leaf should be able to perceive their location on the SAM.

One of the advantages of Arabidopsis over other model plants is the ease with which mutations can be produced and maintained (as homozygous, when viable, or as heterozygots). There are several means to induce such mutations and, as the sexual progenies can be obtained within a few weeks, a plethora of such Arabidopsis mutants were obtained and the phenotypes resulting from these mutations were recorded. There is a problem with these mutations: many of them have a pleiotropic effect, and while the mutation was identified by one phenotype (and a name was given accordingly), later studies showed that the same genetic defect also affects other features. The results of this may cause confusion: the name of a mutant that affects the SAM can be unrelated to the SAM.

The names of Arabidopsis genes (as well as of genes in some other angiosperms) are commonly written in capital letters and in italics. An example is *SHOOTMERISTEMLESS* (*STM*). Note that the wild-type *STM* has a *normal* shoot meristem, while the mutants have defective shoots. Furthermore, the recessive loss-of-function mutation (*stm*) was revealed by the phenotype of the Arabidopsis embryo: it had no shoot apex. However, STM also has a prime role in the differentiation and growth of the SAM after germination. Quite a number of Arabidopsis mutants are pleiotropic, meaning that the product of the unmutated genes are required for several activities and can be activated to express their product by more than one transcription factor (in some cases, by the interaction of several factors). One should be aware of another confusing situation. A gene that was identified in one plant (e.g. Arabidopsis) and given a specific name by investigators that used this plant may also be identified in another plant (e.g. maize) and given another name. The two genes having very different names can thus be homologous and have (almost) the same deoxynucleotide sequences.

I would like to introduce another term that should be kept in mind: *homeodomain genes* and their homeodomain-containing protein products. The homeodomain-containing proteins are transcription factors that cause other genes to become active only in cells where the homeodomain proteins are located. When expressed in cells where they are normally shut down, this causes out-of-place differentiation (like legs formed in the head of *Drosophila* flies). In plants, an important family of homeodomain genes was first discovered in maize and termed *knotted*. A knotted type of homeodomain protein was thus revealed. The name was derived from unusual outgrowths from the leaf blades of maize plants that overexpressed *knotted1* in the blades. Homologous *knotted1* (*KN1*) proteins were identified in Arabidopsis, where the respective genes were termed as "*KN1*-like genes". The *STM* gene belongs to these KN1-like genes.

Listing all the genes that were identified as being involved in the differentiation of the SAM and its derivatives is beyond the scope of this book. Therefore, a selected list of such genes and their main roles is provided in Table 4.

Dealing with a plant in which the nuclear genome is fully sequenced has several research advantages. One of these is that after the mRNA of a given gene is available (permitting the synthesis of the respective cDNA), the investigator can swiftly find the promoter of this gene. This promoter can then be linked to a reporter gene in a chimeric gene construct, which can be introduced into the plants by genetic transformation. Reporter genes that are easily detected in plant tissue are available, and the genetic transformation of Arabidopsis is now a standard and efficient procedure. In short, Arabidopsis is now the plant of choice for studying the site and level of gene expression.

Efficient methodologies have become available in recent years to detect specific mRNAs (e.g. by site-specific hybridization of histological preparations with radioactive probes) at the individual cell level. Procedures were also developed to evaluate the existence of specific proteins by histoimmunological techniques. Consequently, the expression of genes can be followed in the cells of the SAM. Thus, the expression of genes that were previously characterized by their phenotypic effects can now be revealed spatially and temporarily. For example, the *STM* transcript was found already in one or two cells of the late globular stage of the embryo. The *STM* expression then spreads to additional cells, but is excluded from cells that formed the cotyledons. Furthermore, the spatial spread of *STM* expression is controlled by certain other genes. I mentioned *KN1*-like genes

Table 4. Genes Involved in Meristem Establishment and Maintenance (Not Including Genes Involved in Hormone Synthesis)

Full Name	Abbr.	Protein Type	Role
Names of genes are from At* unless indicated to be from other species			
AGAMOUS (PLENA Am)	*AG*	MADSTE	AC function floral identify gene. In addition, active in floral meristems to repress WUS. When mutated, floral meristems are indeterminate and flowers are formed within flowers.
ASYMMETRIC LEAVES1 (*PHANTASTICA*, Am, *ROUGH SHEETH2*, Zm)	*AS1*	MYB TF	Restricts KNOX proteins such as BP and KNAT2 from leaves of many plans, where KNOX level and is kept at bay.
ASYMMETRIC LEAVES2	*AS2*	LOB TF	Forms a complex with AS1.
CLAVATA group	*CLV*		
CLAVATA1 (*THICK TASSEL DWARF1* Zm)	*CLV1*	LRR Receptor Kinase	Part of a membrane receptor and signal transduction complex with CLV2.
CLAVATA3	*CLV3*	CLE peptide	Small polypeptide ligand for the CLV1/CLV2 complex with restricts WUS activity.
CUP SHAPED COTYLEDON 1/2/3 (*NO APICAL MERISTEM* — Ph *CUPULIFORMIS* — Am)	*CUC1/2/3*	NAC TF	Specify organ boundaries and help to define STM embryonic expression. CUC1&2 are negatively regulated by miR164.
HAIRY MERISTEM Ph	*HAM*	GRAS TF	Expressed "outside" of the meristem proper and probably required for cellular responses to WUS and STM.

(*Continued*)

Table 4. (*Continued*)

Full Name	Abbr.	Protein Type	Role
Names of genes are from At* unless indicated to be from other species			
KNOTTED-LIKE group-	*KNOX*		
SHOOT MERISTEMLESS (*KNOTTED* Zm)	*STM KN*	Homeodomain TF	Required for establishment and maintenance of the SAM. Promote cytokinin biosynthesis and negate GA. Expressed in the shoot meristem, but expression disappears when leaf primordial are specified. Inhibit leaf differentiation, and thus are not expressed in simple leaves but are expressed in compound leaves. Proteins are able to move through plasmodesmata
BREVIPEDCELLUS/KNAT1 *KNAT2/6*	*BP/KNAT1* *KNAT2/6*	Homeodomain TF	Have redundant role with STM. BP regulates pedical elongation and lignin synthesis.
LEAFY (*FLORICULA* Am, *FALSIFLORA Sl*, *BERRANT LEAF AND FLOWER* Ph, *UNIFOLIATA* Ps)	*LFY*	LFY TF	Specify the fate floral meristems as flowers and not as leaves.
PHABULOSA, *PHAVOLUTA*, *REVOLUTA*, *CORONA*	*PHB PHV* *REV COR*	Class III HD-ZIP TF	Maintain the SAM, specify the adaxial side of lateral organs and promote xylem formation in vascular strands. Negatively regulated by miR165/6.
MONOPTERUS	*MP*	ARF TF	An auxin-response factor that facilities root meristem pole and flowers formation.

(*Continued*)

Table 4. (*Continued*)

Full Name	Abbr.	Protein Type	Role
Names of genes are from At* unless indicated to be from other species			
STIMPY/WOX9	*STIP*	WOX TF	Promote cell divisions in the vegetative SAM but can be replaced by sucrose.
WUSCHEL (ROSULATA Am*)*	*WUS*	WOX TF	Specify a sub-domain of the central zone, just below the CLV1 domain. Required for the SAM maintenance by a regulatory loop that involves the remote CLV3 expression.
ZWILLE/PINHEAD *ARGONATUE1*	ZUPNA *AGO1*	AGO member of the RISC	Function as a "slicer" in the RISC complex. Acts together with AGO1 to establish and maintain the apical meristem through proper regulation of miRNAs and their target genes.

*At — *Arabidopsis thaliana*, Am — *Antirrhinum majus*, Ph — *Petunia hybrida*, S1 — *Solanum lycopersicum*, Ps — *Pisum sativum*, Zm — Zea maize, TF — transcription factor.

This table was complied on the basis of several publication as Blazquez *et al.* (2006); Callos *et al.* (1994); Kerstetter and Hake (1997); Carles and Fletcher (2003); Carraro *et al.* (2006); Castellano and Sablowski (2005); Hudson (2005); Jonsson *et al.* (2005); Reddy *et al.* (2004a); Reddy and Meyerowitz (2005); Timmermans *et al.* (1999); Traas and Bohn-Courseau (2005); Tsiantis *et al.* (1999); and Williams and Fletcher (2005).

I am grateful to Dr. Yuval Eshed, Plant Sciences, The Weizmann Institute of Scinence, Rehovot, Israel, for brining this table to its final form.

to which the *STM* gene belongs, but there is also a rice homologous gene termed *OSH1*. The expression of this gene is also excluded from the scutellum of the rice embryo. Considerable progress was made during the 1990s in the identification of genes which play roles in the differentiation that takes place at the SAM and at organs (e.g. leaf initials, stem components) constantly derived from the SAM. These genes were mostly identified in Arabidopsis and in maize. The list in Table 4 summarizes the information on some of these genes, and publications by Doebley *et al.* (1997), Kerstetter and Hake (1997), Evans and Barton (1997), Lenhard and Laux (1999), Haecker and Laux (2001), Clark (2001) and Jürgens (2001) provide a good coverage of this progress. An outstanding review on the information and maintenance of the SAM was provided by Bowman and Eshed (2000). The latter authors took into consideration the main genes that were identified at that time and that are involved in the maintenance of the functional domains of the SAM.

Bowman and Eshed (2000) stressed the role of the *stem cells* (SCs) that are situated at the summit of the SAM in a domain termed *central* the *zone* (CZ). The pluripotent CSs serve three main functions:

- They are the source of lateral organs (e.g. leaves) that are initiated at the *peripheral zone* (PZ);
- They contribute the cells that constitute the *rib zone* (RZ) and to the majority of the shoot (stem) tissue;
- They replenish themselves to compensate for cells that went to fulfill the two former functions.

One should recall that we are looking at a dynamic process that on one hand causes continuous acropetal growth of the shoot and, on the other hand, produces intermittently lateral outgrowth at precise phyllotactic locations. While the manifestation of this dynamic process occurs only after germination, there is evidence that the foundations for this process are already gradually laid down during embryogenesis. A major aspect of the review of Bowman and Eshed was the maintenance of the CZ at a constant size, throughout the growth of the indeterminate shoot. This maintenance is primarily regulated by the CLAVATA/WUSCHEL "loop". Since important information has been added to this subject since 2000, this loop is handled in a separate section.

Maintenance of the Stem Cell Niche by the CLAVATA/WUSCHEL Loop

The lever that ultimately led to an understanding of the main regulatory mechanism maintaining the stem cell niche in the SAM was a phenomenon that was known to botanists for centuries: fasciated stems. In such stems, there are many fused shoots that look like a bundle of shoots. Hence, the term "fasciated" stems from the Latin *fascia* ("bundle"), which was a symbol in the Roman Empire; and in the 20th century, the Latin root was adopted in the term "fascism", to describe a totalitarian regime (first by Benito Mussolini in 1922, and then in Germany and Spain).

The awareness that specific mutations in plant genes may result in a fasciated phenotype in the shoot was reported only about a dozen years ago by S.E. Clark and others of the laboratory of E.M. Meyerowitz at Caltech (e.g. Clark *et al.*, 1993, 1996). After several more years of investigations, the Meyerowitz laboratory came up with a clearer picture of the fasciation (Fletcher *et al.*, 1999) that resulted mainly from genetic studies with the *CLAVATA* genes, *CLV1*, *CLV2* and *CLV3*. It should first be noted that the fasciation (caused by a mutation in *CLV3*) is not limited to the vegetative shoot — quite the contrary, the fasciation of the inflorescence is even more impressive morphologically, as the flowers emerge from a "ring" rather than from a point. Moreover, flowers of the *clv3* mutants have extra organs, such as extra stamens and extra carpels. The latter study focused on *CLV3*.

Several mutants of this gene were isolated after insertional (T-DNA) mutagenesis. The amino acid sequence of the protein encoded by *CLV3* was determined: a small protein with 96 amino acids was revealed. This small protein had no similarity to other, then known, proteins. It had a putative signal to direct it into the secretory pathway, but no sequence that would cause its retention inside the cell. The investigators thus assumed (correctly) that the *CLV3* protein may be able to move into the extracellular spaces and serve as a ligand for the *CLV1* protein, which was previously identified by the Meyerowitz laboratory (Williams *et al.*, 1997) as having an extracellular receptor kinase.

Interestingly, no mention was made in the Fletcher *et al.* (1999) publication of the interaction of the *CLV* genes with *WUSCHEL*. *WUSCHEL* was already known as a gene being involved in SAM maintenance (e.g. Laux *et al.*, 1996). Moreover, the German investigators, then all in Tübingen (e.g. A. Haecker, G. Jürgens, T. Laux, M. Lenhard), already suggested an interaction between WUSCHEL and *CLAVATA* (Mayer *et al.*, 1998). In further studies, these investigators substantiated this suggestion by a series of studies on *WUSCHEL* (see Lenhard and Laux, 1999; Schoof *et al.*, 2000; Haecker and Laux, 2001; Weigel and Jürgens, 2002). Meanwhile, T. Laux moved from Tübingen to Freiburg, and J.C. Fletcher moved north from Pasadena (CA) to Berkeley/Albany, where she continued with her SAM studies and, in a very useful review, merged the studies on *CLAVATA* and *WUSCHEL* (Carles and Fletcher, 2003; see also Lenhard and Laux, 2003).

It became clear that *CLV1*, *CLV2* and *CLV3* act in the same pathway to limit the expansion of the undifferentiated stem cells at the summit of the SAM and likewise in the floral meristem. Also, *CLV1* and *CLV3* gene products have a quantitative interdependence, meaning that a complex can be formed between these two products. As for the proteins *CLV1* and *CLV2*, they are identified as a receptor-like kinase (RLK) and a receptor-like protein, respectively. The genes that encode these proteins are members of a large gene family. As shown in Fig. 29, the extracellular region of *CLV1* contains 21 leucine-rich repeats (LRRs), a transmembrane region and an intracellular serine/threonine kinase domain. The *CLV2* contains 20 LRRs and a transmembrane region, but lacks the intracellular domain. Good evidence indicates that the 96-amino-acid–long *CLV3* protein serves as a ligand and is bound between *CLV1* and *CLV2*. The two latter proteins are also linked to each other by an S–S bond. The expression of the *CLV3* gene (i.e. the small ligand protein) is probably restricted to the two tunica layers L_1 and L_2 of the CZ. The mRNA expressed by *CLV1* is found in the L_3, probably as an envelope around the location of the *WUS* expression. There seems to be a spatial difference between the expression of *CLV1* and *CLV2*. The expression of the latter gene can be found in several plant tissues in addition to the SAM (e.g. in the floral meristem). The widespread potential expression of *CLV2* is consistent with the pleiotropic phenotype of *clv2* mutants. Now comes the possibility of a confusion for those who are not experts in this field: there is a *CLV2*-like gene in maize, but this gene is termed *FASCIATED EAR2* (*FEA2*). The mutant *fea2* of maize causes fasciated inflorescence meristem phenotypes. This suggests that the *CLV/WUS* interaction evolved before the separation of monocots from dicots.

The heterodimer of the two proteins *CLV1* and *CLV2* is probably functional only when they interact correctly. Hence, when the *clv2* mutant expresses a malformed protein, there is no functional interaction. This assumption explains why a *clv2* mutant can be dominant: there is a binding between the *CLV1* protein and the *clv2* protein, but the resulting complex is not functional; while the wild-type *CLV1* protein is "wasted" in this nonfunctional interaction. Moreover, the *CLV3* ligand, which may be the limiting factor in the *CLV* complex, can also be wasted in this nonfunctional complex. The fully functional *CLV* complex has a molecular mass of 450 kDa. Such a complex was not found in the *clv2* mutants. The overall picture that emerged indicated that the *CLV3* protein is secreted from the overlying *L1* and *L2* cell layers and moves intercellularly between the cells into the underlying *L3*, where it joins as ligand to the extracellular region of the *CLV1/CLV2* heterodimers.

At this point, the *WUS* gene may enter the scene. The wild-type *WUSCHEL* ("tousled hair" in German) gene produces a noncell autonomous signal that activates the cell division in the stem cells. The *WUS* is likely to act downstream of *CLV*, thus *wus* mutants are epistatic over *CLV*, meaning that it does not make a difference whether *CLV* is wild-type or defective. In the presence of defective (mutant) *wus*, the stem cell niche of the SAM will shrink or disappear. When *CLV* genes are mutated, the expression niche (domain) of *WUS* is enlarged, while ectopic expression of *CLV3* causes the

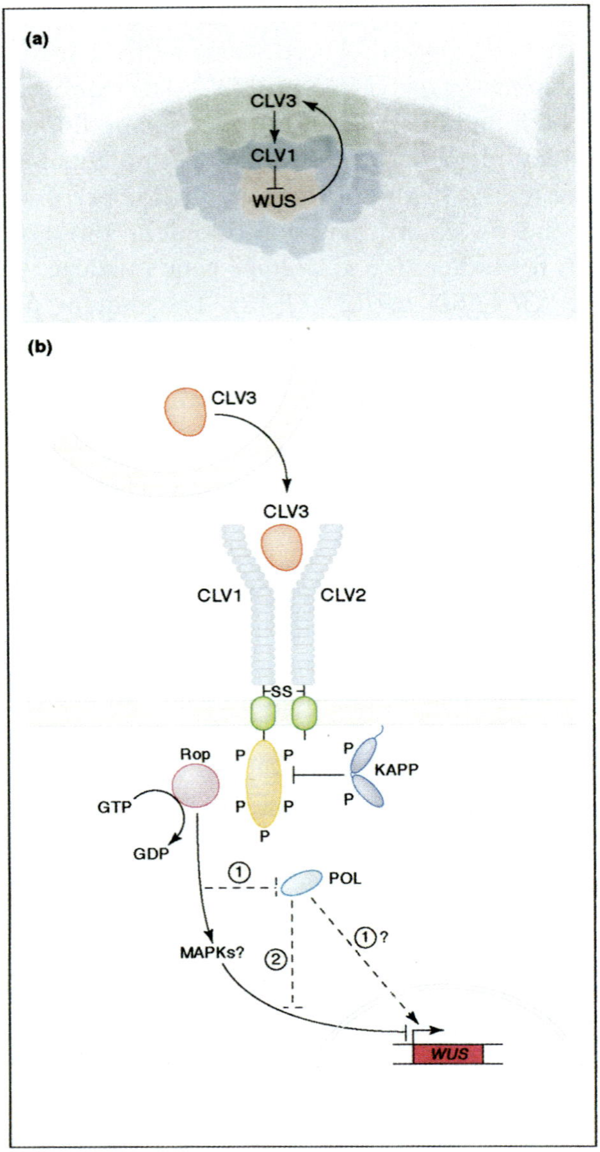

Fig. 29. Coordination of cell proliferation and cell fate decisions across the shoot apical meristem (SAM). **(a)** Stem cell regulation in the SAM via the CLAVATA-WUSCHEL feedback loop. *CLAVATA3* (*CLV3*) expression is restricted mainly to the L1 and L2 of the central zone, whereas *CLV1* mRNA can only be detected in the L3

repression of *WUS* expression. In wild-type plants that have unmutated *CLV* and *WUS* genes, the *CLV3* protein spreads centrifugally and basipetally. When this protein meets the extracellular *CLV1/CLV2* complex, it binds to it and a signal transduction, in which cytoplasmic components are involved, causes the repression of *WUS* expression. This causes the establishment of a loop: *WUS* transcription sends a signal to the tunica cells in the CZ and promotes the cell division in this CZ. The cells in which division was promoted now express the *CLV3* protein, and the latter is active in a transduction that enters the nucleus, where *WUS* is transcribed and represses this transcription.

As noted above, the level of the *CLV3* protein seems to be the limiting factor. Because the *WUS*-expressing cells are enveloped by *CLV1/CLV2*-containing cells, these cells may constitute a trap for free *CLV3* proteins, and thus place a hurdle between the source of the *CLV3* and the *WUS*-expressing cells. This hurdle is dumping the

---

of the central zone and rib zone. *WUSCHEL* (*WUS*) is expressed in few central cells of the rib zone. The CLV-WUS feedback loop consists of a WUS-mediated signal from the organizing center (OC) that specifies stem cell identity in the outermost layers, which signal back via the CLV pathway to limit the size of the *WUS*-expressing OC. **(b)** CLV transduction pathway at the cellular level. The CLV1 and CLV2 leucine-rich repeat (LRR) receptor proteins might homo- or heterodimerize through the formation of disulfide bridges between the conserved cysteines pairs (SS) flanking the LRR domain. The binding of CLV3 to CLV1 and/or CLV2 could promote the assembly of the 450 kDa complex, which also contains a protein phosphatase (KAPP) and a Rho-like GTPase (Rop). KAPP is a negative regulator of CLV signaling, capable of dephosphorylating CLV1; Rop transduces the signal from the CLV complex to downstream targets. By analogy with animal systems, the binding of Rop to CLV1 might be mediated by a linker protein, the nature of which remains to be determined. The signaling cascade from the plasma-membrane-bound CLV complex leads to the downregulation of WUS transcription in the nucleus. This intracellular signaling cascade might involve mitogen-activated protein kinases (MAPKs), based on the example of other LRR receptor-like kinase signaling pathways. The protein phosphatase 2C POLTERGEIST (POL) acts either: 1) as a positive regulator of *WUS* transcription that is antagonized by CLV signaling; or 2) as a direct inhibitor of CLV signal transduction, analogous to KAPP but acting at a different point in the pathway. (From Carles and Fletcher, 2003.)

repression of *WUS*. It appears that this controlled repression still permits, in wild-type SAMs, sufficient *WUS* expression to maintain a normal-size niche of stem cells. The equilibrium at the SAM between the sizes of the niches of *WUS*- and *CLV*-expressing cells and the number of stem cells is typical to the various plants, and seems to vary only little during the vegetative growth of the shoot. This feedback loop was termed: *meristem homeostasis*. By 2003, there was already evidence that, overall, additional proteins are involved in the transduction of the *CLV1/CLV2/CLV3* signal to the nucleus, where the transcription of *WUS* is controlled by factors that affect its promoter. This is shown in a highly schematic way in Fig. 30.

While the *CLV* and *WUS* genes are major players in maintaining meristem homeostasis, they are not the only players in this maintenance. A few other genes are noteworthy, especially those studied in Arabidopsis. The two *FASCIATED* (*FAS*) genes, *FAS1* and *FAS2*, encode proteins that are components of a chromatin assembly factor, and the phenotypes of the respective mutants are linked to the distortion of the *WUS* expression pattern (see more on this below, where chromatin remodeling is discussed).

Another gene is the *DORNRÖSCHEN* (*DRN*) gene. While the molecular genetic role of *DRN* is still under study, it was found that when *drn* mutants (that have a shrunken SAM) are induced to express *DRN* ectopically (i.e. by genetic transformation with a hybrid plasmid in which *DRN* is activated with a strong universal promoter), there is a dramatic increase of the SAM. In wild-type Arabidopsis, *DRN* is expressed in the CZ of the SAM and in primordia of lateral organs derived from the SAM. Moreover, *DRN* seems to impose its effect independently of another gene, *SHOOTMERISTEMLESS* (*STM*), which is mentioned in the following paragraph.

The *KNOTTED1* (*KN1*) gene was isolated from maize by Sarah Hake and associates many years ago (Vollbrecht *et al.*, 1991). Since then, the homeodomain transcription factors encoded by this gene and *KNOX* (*KNOTTED1*-like homeobox) genes were studied in the laboratory of S. Hake as well as in other laboratories (see review by Hake *et al.*, 2004 and references below). The homeodomain-containing transcription factors encoded by *KNOX* genes exist in all analyzed plants, including Arabidopsis. In general, it was found that these factors are required for the maintenance of meristems and for the proper patterning of lateral organ initiation, as well as in the final patterning of leaves that are not simple leaves, but rather dissected leaves.

For our deliberations on the patterning of the SAM in Arabidopsis we should focus our attention on the *STM* gene, which is the Arabidopsis ortholog of *KN1*. Loss-of-function *stm* mutants of Arabidopsis fail to establish and/or maintain a normal SAM: the SAMs do not produce lateral organs. Such mutants do not maintain the indeterminant cell fate and do not prevent the shift to differentiation. Contrary to *WUS* and *CLV3*, the expression of STM is widespread in the SAM. Arabidopsis has additional *KNOX* genes (*KNAT1*, *KNAT2* and *KNAT6*) that seem to function redundantly with STM. Experimental results also indicated that the latter *KNOX* genes play a role in restricting the expression of *ASYMMETRIC LEAVES1* (*AS1*) and *AS2* genes to lateral organ primordia (e.g. leaves); hence, preventing leaf initiation from inappropriate sites on the flanks of the SAM. The two latter maize genes encode somewhat different proteins. *AS1* encodes a Myb-domain transcription factor, while *AS2* encodes a putative transcription factor that contains a leucine-zipper domain. Arabidopsis and *Antirrhinum* have genes that are homologous to *AS2*, but in these plants the genes were given a different name: *PHANTASTICA* (*PHAN*) (Timmermans *et al.*, 1999; Tsiantis *et al.*, 1999). While ectopic expression of *STM* will induce the cell division process, this expression will not activate the *CLV3* gene. Hence, the *CLV/WUS* and the *STM* genes seem to have independent functions.

A scheme suggested by Bäurle and Laux (2003) that also takes the effects of plant hormones into consideration is provided in Fig. 30. These authors summarized that *STM* restricts (lateral) organ initiation to defined sites of the SAM flanks by repressing *AS1* and *AS2*, which in turn repress the expression of *KNOX* genes. Consequently, the

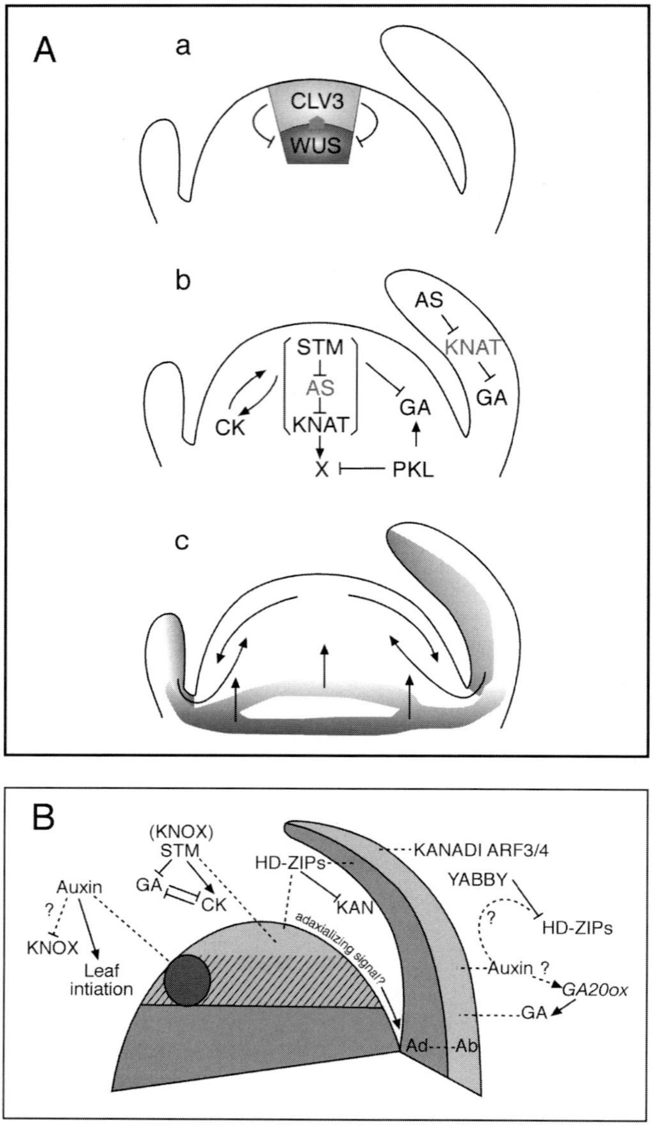

Fig. 30. **(A)** Regulatory pathways in the shoot meristem. **(a)** Local regulation of stem-cell maintenance by the *WUS-CLV* feedback loop. The *WUS*-expressing OC (blue) confers stem-cell identity upon the overlying cells (pink), which in turn restrict the *WUS* expression domain via the *CLV* signaling pathway. **(b)** Factors regulating meristem maintenance and organ initiation. In the center of the SAM, *STM* represses

prevention of meristem differentiation by *STM* is indirect: it allows *KNOX* gene expression. The sequence of events that leads to the lateral organ formation in Arabidopsis, as suggested by Bäurle and Laux (2003), is the following: A group of 15–30 cells that are derived from all the three SAM layers becomes assigned to an incipient organ primordium. In these cells, there is a dowregulation of *STM* expression, while this expression is retained in other parts of the SAM. This downregulation permits the expression of *AS1* and *AS2* in the primordium. The latter expression represses meristematic cell fate by downregulating the *KNOX* genes (e.g. *KNOX1*, *KNOX2* and *KNOX6*). This appears to be a plausible scenario, but does not answer an important question: what determines the exact position of these 15–30 cells? By human logic, the incipient primordium should be located at a given distance from the summit of the SAM and at a

---

the *AS* genes, thereby allowing expression of *KNAT* genes. *STM* and *KNAT* repress GA biosynthesis in the SAM. In young leaf primordia, *STM* expression is downregulated; *AS* gene products are present and repress *KNAT* gene expression while GA biosynthesis is upregulated. Positive interactions between Cytokinin (CK) and *KNOX* genes promote SAM activity. It is suggested that *PKL* promotes the transition from a meristematic to a differentiated cell state by repressing *KNOX* target genes and through GA biosynthesis. **(c)** The shoot apex affects adaxial-abaxial polarity of lateral organs. Vice versa, cells in the adaxial part of the young leaves (green) and in the deeper layers of the stem and the young leaves (symbolized by a brown ring of *HAM* expression) promote shoot meristem maintenance via yet unidentified signals. (From Bäurle and Laux, 2003.) **(B)** Interactions between hormones and transcription factors that mediate SAM function and adaxial-abaxial patterning in the leaf. Arrows indicate positive and T-bars negative regulation. In the Arabidopsis SAM, the KNOX protein STM suppresses GA biosynthesis and promotes CK biosynthesis. At the flanks of the meristem, in the PZ (hatched area), the accumulation of auxin prompts the initiation of a leaf primordium, possibly by downregulating *KNOX* expression. In the leaf (right-hand side), adaxial (Ad) and abaxial (Ab) identities are specified by the mutual antagonism between HD-ZIP and KANADI/YABBY transcription factors, apparently under the influence of an adaxialinzing signal that emanates from the meristem. The repression of HD-ZIP expression also involved ARFs and possibly asymmetric auxin accumulation. Auxin in the leaf might also promote GA levels by positively regulating the GA biosynthetic gene, *AtGA20ox1*. (From Kepinski, 2006.)

given angle from the previous organ initial. However, does the plant use the same logic?

Advanced Studies and the Use of *in vivo* Manipulations

Microsurgical and laser ablation of tomato SAMs

Microsurgical experiments with plant apical meristems were performed about 80 years ago (see Steeves and Sussex, 1989). These were rather rough surgical operations, such as needle pricking, and the interpretations of the experimental results were not unequivocal. In general, the experiments indicated that the destruction of the summit of the SAM caused one or more new apical meristems to substitute for the destructed meristem. The ablation approach was taken up by Cris Kuhlemeier and associates (Reinhardt *et al.*, 2003) of the University of Berne (Switzerland).

Reinhardt *et al.* (2003) used tomato shoots, but took into account the accumulated information of the molecular genetics of SAM patterning in Arabidopsis and other plants. The latter information was especially useful for tracing the expression of specific genes before and after the different ablation treatments. The investigators used tomato shoots that were cultured *in vitro*, and the laser beam was of an infrared wavelength (2.94 μm) with a focal diameter of 40 μm at the ablated SAM. Pulses of 60 seconds were applied, and the depth of the ablation could be regulated by the number of pulses provided at a given site. Hence, either only the upper layer (L_1) or more layers (L_1–L_3) could be ablated. The ablation could also be aimed at the periphery rather than at the summit of the SAM. A shallow ablation had very little effect on the further differentiation in the SAM, but when the ablation was deeper and included all the *WUS*-expressing cells, an ectopic expression of *WUS* cells was established with an occasional emergence of an additional meristem center.

These investigators combined their ablation procedure with molecular methods: the location of *WUS* expression was determined by *in situ* hybridization of the tomato *WUS* (Le *WUS*) with an appropriate riboprobe. It should be noted that the ectopic *WUS* expression occurred only when the ablation was performed at the CZ, but not when the ablation was performed at the PZ, at the site of an incipient leaf formation. In the latter case, the ablation only affected the position of the next new leaf: the laser ablation suppressed the formation of the expected new leaf, or the leaf primordium was displaced. However, this PZ ablation did not cause ectopic *WUS* expression.

Another noteworthy conclusion from these ablation experiments was that the destruction of the stem cells at the CZ did not have an immediate effect on the regular formation of new leaves. Also, it was observed that the ablation of the cells in the L_1 layer caused a change in the direction of cell division. In the sustaining cell layer, the division became periclinal rather than anticlinal. It seems that the L_1 layer normally controls the direction of the cell division of lower layers. Hence, this L_1 layer probably has a prime role in the initiation of new lateral organs.

In further experiments (Reinhardt *et al.*, 2005), the team of C. Kuhlemeier augmented the laser ablation with fine microsurgical operation at the tomato shoot apex. These experiments led to the suggestion that the two youngest primordia influence the size and position of new lateral organs, meaning they determine the phyllotaxis. These experiments also indicated that the L_1 cell layer of the SAM is essential for the dorsoventral patterning of the leaf primordium.

The combination of advanced optical methods with interference in cell division and DNA replication can be used for the study of differentiation of shoot apices

A team of investigators in Versailles, France (Grandjean *et al.*, 2004) employed intact young Arabidopsis seedlings to follow the differentiation of living shoot apices, especially to trace the initiation of lateral primordia. They visualized the activation of several genes that are presumed to be involved in primordia initiations by using their promoters to activate a reporter gene (GFP) that could be observed microscopically. The promoters were constructed in a manner that enabled their activation by ethanol (a solution of 10% ethanol was used in the medium in which the seedlings were cultured). The optical analysis was performed with a confocal laser scanning microscope. The fluorescence and the (red) vital staining were observed with an inverted microscope; hence, for such observations the seedlings were observed with their meristem down. Before and after observation, the seedlings were maintained in their normal direction.

The vital staining was useful to follow the rate of cytokinesis of the cells in several regions of the SAM. As expected, the cells at the center divided slower than more peripheral cells. There was no synchrony of cell division: some cells divided faster than others; even neighboring cells had different cell cycle rates. Obviously, such differential cell cycle rates cannot be determined by analyzing fixed SAMs. The activation of expression was then followed by the use of reporter (GFP) genes attached to promoters, such as those of *ANT* (involved in primordium outgrowth) or *FLY* (which, in combination with *WUS*, activates AG), that are active in the control of primordium initiation. By that, the investigators could estimate the number of cells that were committed to initiate the new meristem. It seemed that the recruitment of cells to join the new meristem is a two-phase process. Apparently, the recruitment terminated when between 30 and 60 cells were recruited. The application of Oryzalin (a drug that depolymerizes microtubules, causing arrest of cytokinesis) caused quick arrest of cell division, while nuclear division and cell enlargment were not arrested. This Oryzalin treatment did not prevent differentiation. On the other hand, the application of drugs that inhibit DNA replication (aphidicolin and hydroxyurea) inhibited cell growth and differentiation. This suggested that DNA synthesis is important (or even essential) for growth and patterning at the SAM.

Real-time lineage analysis of cell division in living shoot apices of Arabidopsis reveals details of SAM patterning

The laboratory of Elliot M. Meyerowitz at Caltech (Pasadena, CA) can be considered as a "fountain" not only for information regarding plant patterning, but also as a renewal source of talented investigators in this field (e.g. J. Fletcher, S.E. Clark, J.L. Bowman, G.V. Reddy and several others). In a study that was probably conducted in parallel to the study of Grandjean *et al.* (2004), the Meyerowitz team (Reddy *et al.*, 2004a) used similar methodologies to analyze cell division patterns in living and actively growing SAMs of Arabidopsis. In such SAMs, all the cells that finally structure the stem have as their source a very small number of stem cells that constitute a "fountain" for the life of the shoot.

The uniqueness of the study of Reddy *et al.* (2004a) was that they used time-lapse imaging data to reconstruct cell division events over time and employed age-registration algorithms to visualize morphogenesis in relation to cell behavior. Estimating the rate of cell division by evaluating the percent of cells that are in the S-phase or at the metaphase in fixed plant preparations can be misleading. In cases where the S-phase and/or the metaphase are either extended or shortened, the estimation will be wrong. The live-imaging technique can provide a relatively accurate spatial and temporal map of cell division patterns at the SAM. The study of Reddy *et al.* (2004a) was also performed with flower primordia (FP), from the stage in which four initial L_1 layer cells are committed to form a flower until a floral primordium is shaped.

The main procedures used by Reddy *et al.* (2004a) were as follows: Seeds were germinated and plants of 10 days after germination were analyzed; other floral primordia were removed to expose the SAM. To mark the boundaries between the cells, transgenic plants were produced that contained transgenes with appropriate promoters and reporter genes (the latter were yellow fluorescent protein or GFP). After screening numerous transgenic plants, those best fitted to imaging were selected. Laser beams with the appropriate wavelengths were used to activate the fluorescence, and confocal optics, in conjunction with a multimodality image-registration program, were employed to analyze the imaging results. Likewise, to mark the entrance of cells into the mitotic cycle, transgenic plants that had a hybrid transgene with a cyclin B promoter and a GFP reporter (expressed during G2-phase to metaphase) were analyzed. The expression of the respective cyclin B1; 1; GFP construct was counterlabeled with a vital dye (FM4-64) that clearly showed cell boundaries. The whole process could be followed dynamically by a time-based movie. The main conclusions from this study were the following:

- Cell division at SAM is not uniform in the SAM and (as reported by Grandjean *et al.*, 2004) it is also not uniform among neighboring cells. There are temporal variations in mitotic activity that are not correlated to plastochron length or diurnal rhythms; the cell cycle length can vary considerably: from between 12 to 18 hours up to between 90 to 96 hours.
- Generally (as expected), cells in the peripheral zone (PZ) divided faster than cells in the central zone (CZ) of the SAM: the division rate is more uniform in the PZ than in the CZ.
- In the cells that will initiate a floral bud, four L_1-layer cells undergo a typical change in cell division orientation and mitotic activity. The newly oriented cell division is followed by a burst of cell expansion and cell division, until a clear three-dimensional floral bud is established.
- A clear lack of cell expansion occurs in a narrow band of cells that form the boundary between the SAM and an emerging floral bud.

Programmed silencing of CLV3 gene led to answer questions regarding the control of the CZ size in the SAM

In a more recent study, Reddy and Meyerowitz (2005) found a way to silence the *CLV3* gene in a controlled manner. The idea was to cause this silencing by a specific microRNA that would cut the mRNA of *CLV3* and thus suppress the formation of its encoded 96-amino-acid protein. By that, a temporary loss of function of the *CLV3* could be achieved. For that, a region of the *CLV3* transcript was used to construct a foldback sequence that was processed in the plant to a microRNA, causing the required cleavage in the *CLV3* mRNA. By crossing two transgenic Arabidopsis plants — in one of which there was a construct that could be induced by dexamethasone (DEX, 10 μm) with another plant that contained a coding sequence for the microRNA, so that the transcription of the latter sequence could be induced by the activation of the former transgene — they had the required system; only under treatment with Dex, the *CLV3* was silenced.

After selection, among the double-transgenic progeny, four plants were used for further study. Indeed, in these plants, the *CLV3* was silenced by Dex treatment as revealed by the reverse-transcription polymerase chain reaction (FT-PCR). The location in the SAM, where *CLV3* is transcribed, could be evaluated by another transgenic plant. The latter plant resulted from crossing the double-transgenic plant with a transgenic plant that contained a transgene with a *CLV3* promoter in front of a GFP reporter sequence. This enabled the investigators to follow the changes in the CZ caused by downregulating the level of the *CLV3* product. When mock-treated, a regular small size of CZ was marked by the GFP fluorescence; but when the plants were treated (e.g. for 18 days) with Dex, there was a great increase in the CZ, meaning a much enlarged SAM dome. Also, the analysis of Dex-treated plants (e.g. for 7 days) showed that the volume of *WUS*-expressing cells was increased, even though the level of *WUS*-expressing domain appeared "patchy".

Furthermore, in order to better see the cell boundaries, a fourth kind of transgenic plant was obtained. Such plants resulted from a cross between a triple-transgenic plant and a plant that had a transgene causing yellow fluorescence (by YFP) of plasma membranes. With those tools in hand (e.g. the quadruple transgenic plants), the dynamics of the CZ and SAM changes in size could be followed in living shoot apices with or without Dex treatment.

For those not familiar with the technical aspects of securing transgenic plants, I would like to note that obtaining a quadruple transgenic plant that functions as expected is an elaborate procedure. Although genetic transformation of Arabidopsis is an established process, adding another transgene to a transgenic plant can cause problems (e.g. cosuppression) and only a small fraction of the progeny consists of useful plants (see Galun and Breiman, 1997).

The changes in the size of the *CLV3* expression domains and the CZ size, imposed by the treatments of the quadruple transgenic plants, did not affect the process of flower primordia formation. These primordia were always situated outside of the perimeter of the expanding *CLV3* expression (i.e. in the PZ). The results also indicated that the increased SAM size is not only due to a reduction in the rate of PZ cells that enter the differentiation pattern. Furthermore, there were good indications for a scenario that the gradual expansion of the CZ after treatment, which reduced the *CLV3* product, resulted from respecification of PZ cells to CZ cells. In untreated SAMs, there is probably a dynamic balance between PZ and CZ cells. These conclusions were supported by observing the dynamics of cell division in the L_1 cell layer that resulted from Dex treatment (i.e. reduction of *CLV3* product and increase in CZ size). The authors thus concluded that *CLV3* signaling in meristems mediates both cell fate specification and growth control through the inhibition of cell division rate.

Clearly, further studies in which the synthesis of the *WUS* product can be regulated temporarily will add information on the role of the *CLV/WUS* loop in the patterning of SAM.

Initiation of Leaves, Phyllotaxis and Plant Hormones

At the beginning of his review on leaf initiation and phyllotaxy, Fleming (2005) cited a passage (in German) from Hugo von Hoffmannsthal (an Austrian poet and dramatist, 1874–1929): "If I knew how a leaf from a branch grew, precisely, forever silent would I remain, for my knowledge would suffice." While opening a discussion with a citation from a person with a creative mind appeals to me, I strongly disagree with the content of this citation. I hold the opinion that the acquisition of knowledge causes a craving for additional knowledge. One can bring this subject to absurdity: a person who lacks any knowledge has no desire to receive knowledge. This brings me again to the "Law of Memory Storage" by Amiel Ben David Halevy (my pseudonym) that says: "When memory is added to a human's mind, it causes an ability to increase the memory-storage." Well, probably up to a certain, very high limit. Here, one may also go to an absurd situation: no memory can be added into a brain that lacks previous memory.

In this section, we shall see how some fundamental knowledge on the SAM and the initiation of new leaves were followed by streams of information from different directions (such as information about auxin, gibberellins, cytokinins and brassinosteroids) that assembled and led to a leap of knowledge that emerged during about 1 year.

The involvement of the interaction between genes expressed in the CZ of the shoot apex and in the leaf initials with the two kinds of plant hormones, cytokinins (CK) and gibberellin (GA), was noted already in a previous section of this chapter (e.g. see Fig. 30). It was revealed that in the center of the SAM, the expression of *STM* causes the repression of AS genes. This repression derepresses the *KNAT* gene; as *KNAT* represses the synthesis of GA, the net result of *STM* activity is the reduction of GA in the center of the SAM. In young leaf primordia, the *STM* is downregulated and the *AS* gene products repress *KNAT*; this repression permits the upregulation of GA synthesis. Also, the positive interactions between CK and *KNOX* gene products promote the *STM* activity in the SAM. It was suggested that another gene in the SAM, *PKL*, is involved in balancing the level of GA biosynthesis. While these interactions seem logical, they require substantiation.

A team led by Robert Martienssen of the Cold Spring Harbor Laboratory (NY) (Byrne *et al.*, 2003) identified and characterized a gene that is involved in lateral outgrowth from the SAM, and is thus involved in the phyllotaxis of Arabidopsis. These investigators found evidence supporting the claim that the Arabidopsis gene *BELL-RINGER* (*BLR*), which encodes a homeobox protein, is required to maintain the normal spiral phyllotaxis of Arabidopsis. In the absence of this gene's activity, the regular initiation of lateral organ initiation is disturbed, and these organs are initiated more frequently than in the presence of *BLR* activity. In the absence of the activities of *STM* and *AS*, *BLR* contributes towards the maintenance of stem cell fate in the SAM. The *BLR* protein is related to the BELL1 subclass of homeodomain transcription factors of 12 members that exist in Arabidopsis. It appears that the *BLR* protein can interact with the *KNOX*-class homeodomain transcription factors by protein–protein interaction, and that the respective complex is then involved in the regulation of phyllotaxis patterning. Again, an appealing possibility that requires substantiation.

While the abovenoted suggestions for the regulation of leaf outgrowth and normal phyllotaxis require additional experimental support before they can be regarded as established regulatory processes, the evidence obtained on the roles of plant hormones in these patternings is nevertheless rather impressive. Before reading the following sections, it is recommended that the readers should look again at Chapter 2, where basic knowledge on auxins, GA and CK is provided.

Auxins and lateral organ initiation

After the Arabidopsis *PINOID* gene was cloned and found to encode a protein serine/threonine kinase, Benjamins *et al.* (2001) of Leiden University (in the Netherlands) looked into the role of *PINOID* in the regulation of organ development. It was recorded that this gene is induced by auxin, and that the protein kinase is normally present in primordia of leaves and floral buds as well as in certain other tissues (e.g. vascular tissue, cotyledon initials). When *PINOID* was overexpressed by a transgene, with a strong promoter that activates the transcription of the PINOID mRNA, the respective transgenic plants had an altered sensitivity to auxin transport that could lead to a collapse of the primary root meristem. Such phenotypes could be rescued by naphthylphtalamic acid (NPA), an inhibitor of polar auxin transport. It was thus suggested that the PINOID protein exerts its effect by functioning as a positive regulator of polar auxin transport.

The interaction of a gene that is involved in auxin flow in the apex of the Arabidopsis embryo (*PIN-FORMED1*) and several other genes that are expressed in this apex (e.g. *MONOPTEROS*, *CUC1*, *CUC2*, *STM*) was analyzed by Aida *et al.* (2002). These authors suggested a model for the role of auxin distribution in the patterning of the SAM during embryogenesis.

A team of investigators from the University of Berne, mentioned above (Steiger, Reinhardt and Kuhlemeier, 2002), evaluated the function of the auxin efflux carriers in leaf positioning. Shoot apices of tomato plants were used in this study, and these were grown in a culture medium in which drugs that affect auxin flow could be added. As there were previous data supporting the possibility that the inhibition

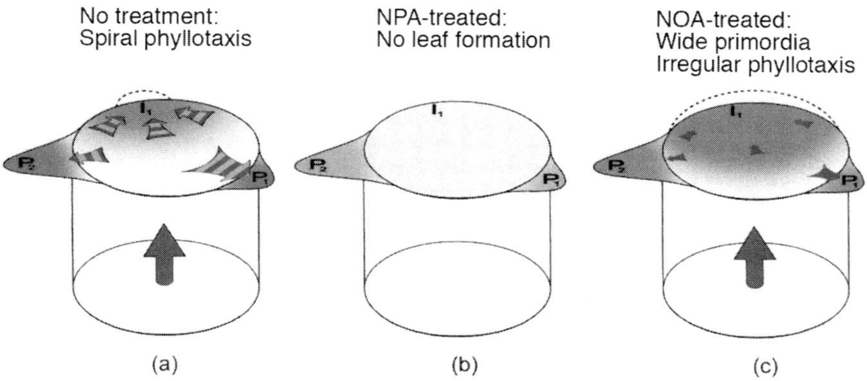

Fig. 31. Model for the regulation of phyllotaxis by auxin gradients. Shown are representations of a meristem decapitated at the site of leaf initiation. **(a)** Auxin reaches the meristem via acropetal transport. P_1 and P_2 act as sinks for auxin, leading to redistribution and to establishment of an auxin gradient with a maximum at I_1. Acropetal auxin transport into the meristem is controlled by the efflux carrier, whereas redistribution of auxin within the meristem involves both, the influx and the efflux carrier. **(b)** Inhibition of the efflux carrier by NPA stops the acropetal auxin transport as well as auxin export from the meristem to the primordial, leading to decreased auxin contents and the loss of an auxin gradient in the meristem. Therefore, primordium formation ceases. **(c)** Inhibition of the influx carrier by NOA reduces directed auxin transport, resulting in a shallower auxin gradient and therefore, primordia develop as wider structures. I_1, site of incipient primordium formation; P_1 and P_2, pre-existing primordia with P_1 being the younger; dashed lines, shape of the future primordium at I_1. (From Stieger *et al.*, 2002.)

of auxin efflux causes the cessation of leaf initiation, the authors focused on auxin influx carriers. Actually, the authors took a pharmacological approach and used the auxin influx inhibitor naphthoxy-1 acetic acid (NOA) in their experiments. Their conclusion was that the influx of auxin has a major role in regulating the sites of new leaf initiation from the shoot apex, i.e. in controlling the normal phyllotaxis. The model suggested by Steiger *et al.* (2002) is presented in Fig. 31. While Steiger *et al.* (2002) used a pharmacological approach, Michael J. Scanlon of the University of Georgia (USA) combined this approach with the evaluation of the expression of the *KNOX* gene in the SAM. Rather than tomato shoots, Scanlon (2003) cultured maize shoot apices *in vitro* and also exposed them to NPA. The findings of this study indicated that the lower sheath margins emerge from the maize leaf base late in leaf development, and that the separation of these leaf margin domains is correlated with auxin transport and the downregulation of *KNOX* proteins.

Based on the information mentioned above and on additional information, Castellano and Sablowski (2005) suggested a model for the initiation of new leaf primordia that involves auxin flow. According to this model, the auxin in and around the SAM is trafficked into existing leaf outgrowths. This flow causes the removal of auxin from other parts of the shoot apex. The site of the new leaf primordium will then be where auxin was least depleted.

Didier Reinhardt moved from Kuhlemeier's laboratory in Berne to the University of Fribourg (only 30 km southwest of Berne), and updated the involvement of auxin flow with phyllotaxis (Reinhardt, 2005). Possibly inspired by the landscape of Fribourg, Berne and the Berner Oberland, he titled his review as "Phyllotaxis — a new chapter in an old tale about beauty and magic numbers." His contribution was that he attempted to explain most aspects of phyllotaxis by the expression pattern of the subcellular localization of *PIN1* (that is required for auxin efflux). Accordingly, the PIN1 is localized at one side of each of the L_2 and L_3 cells in a young leaf primordium, and thus moves auxin to a center in this primordium. On the other hand, in the epidermal L_1 layer, the PIN is situated intracellularly in a manner that causes auxin flow from cell to cell in this layer. Auxin is transported from inner cells into the L_1 cells by the influx carrier AUX1. However, from where the auxin came (where the "fountain" of auxin is in the SAM) is not yet known. It seems, though, that there is an agreement among those investigating the localization of future new leaf outgrowths that polar auxin transport in the plant's apex generates local peaks in auxin level and consequently determines the site of the new organ.

Support for the role of auxin flow and auxin levels, as well as the integration of the roles of several genes that are expressed during initiation of lateral organs at the shoot tip of Arabidopsis, came from a study of the Meyerowitz team (Heisler *et al.*, 2005). In this study, Arabidopsis shoot tips were cultured in a manner seen in previous studies of this team (e.g. Reddy *et al.*, 2004a). Using confocal imaging of living shoots during the initiation of new lateral organs, the authors monitored the changing levels of expression of several genes (*PIN1*, *STM*, *CUC2*, *REV*, *FIL*, and *FLY*) in the shoot apex. The lateral organs that were the subjects of this study were floral buds. The results suggested that cycles of auxin buildup and depletion (at specific sites of the SAM) accompany, and possibly direct, different stages of the primordium development. Imaging of several GFP–protein fusions showed that these dynamics also correlate with the specification of primordial boundary domains, organ polarity axes and the sites of the new lateral organ (floral meristem) initiation. A summary of the expression changes of several genes and of presumed auxin flow at the shoot apex is presented in Fig. 32.

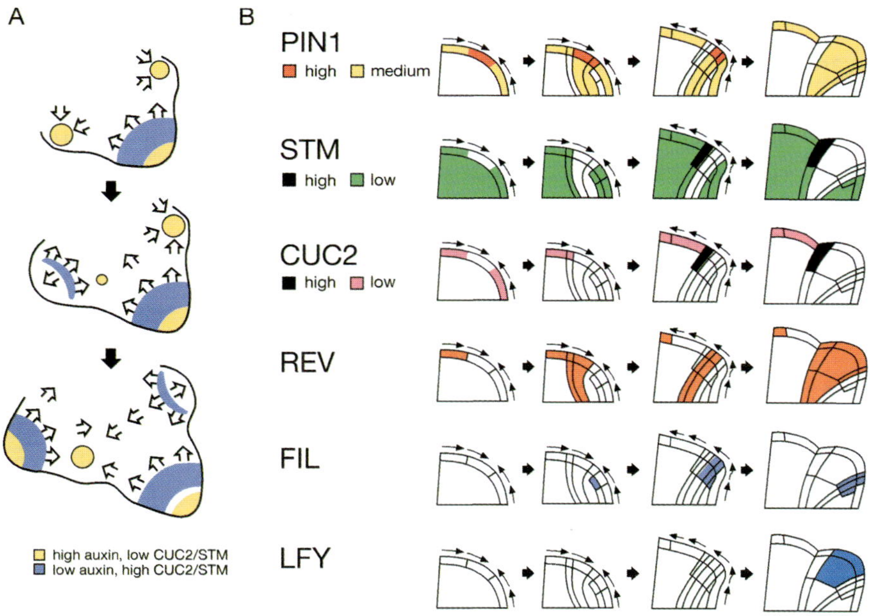

Fig. 32. SAM primordium development — a conceptual summary. **(A)** PIN1 polarity (*arrows*), presumed auxin concentration dynamics, and CUC2/STM expression levels. High auxin, low STM/CUC2 (yellow); low auxin, high STM/CUC2 (blue). **(B)** Summary of observed expression patterns during early stage of primordial development. Arrows indicated direction of auxin transport in the epidermis. (From Heisler *et al.*, 2005.)

Cytokinins, gibberellins and lateral organ initiation

The known information on the signal transduction of cytokinin (CK) was summarized in Chapter 2 of this book, where this transduction in Arabidopsis was schematically described. Briefly, there is a complex that has a CK receptor outside of the cell and an intracellular histidine kinase (CRE1/WOL/AHK4). From the cytoplasm, the signal is transferred into the nucleus, from where the signal is further mediated by Arabidopsis-response regulators (ARRs) to cause CK responses. The latter act in negative regulation of CK responses. In Arabidopsis, there is a family of 10 *ARR* genes that are partially redundant (To *et al.*, 2004). In maize, there is a gene, *ABPH1*, which has similarities to ARRs (Giulini *et al.*, 2004). Like ARRs in Arabidopsis, *ABPH1* is induced by CK in maize.

Giulini *et al.* (2004) of the Cold Spring Harbor Laboratory (NY) searched for the role of CK in the control of phyllotaxis in maize. The normal phyllotaxy of maize is distichous, meaning the leaves are alternative with a 180° angle. However, in *abph1* mutants, the leaves appear in a decussate pattern (paired at an angle of 180°). Significantly, the decussate phyllotaxy is maintained in the loss-of-function mutant throughout the life cycle of the plant, and even the ears are initiated as opposite pairs. Several tested *abph1* mutants did not produce the normal *ABPH1* transcript.

By *in situ* hybridization, weak *APBH1* expression was detected at the sites where SAM evolves during the transition stage of embryogenesis. At a later stage, the expression was increased in the SAM and persisted there at the first leaf stage. The *abph1* apex did not show such an expression. In maize seedlings of 2 weeks (~10–15 leaves), *ABPH1* expression was localized to a small domain of the SAM at the site of incipient leaf primordia. However, the expression did not cover the entire incipient leaf primordia. The *ABPH1* was also expressed in the female inflorescence (ear). Exposure of seedling shoots to a solution that contained cytokinin induced a threefold-to-eightfold increase of *ABPH1* expression in the shoot tips (but not in the young leaves). The investigators also found that when wild-type (*ABPH1*) embryos were exposed to cytokinin, a phenocopy of the mutant (*abph1*) was obtained that had an extended SAM. It was thus suggested that the ABPH1 protein restricts in the embryo (possibly also later) the cytokinin-induced SAM proliferation, keeping the latter in its required size. How this cytokinin regulation interacts with the auxin-regulated control of lateral organ initiation is an open question to be answered by additional research.

Indeed, additional studies were reported in two consecutive articles in *Current Biology* (Jasinski *et al.*, 2005; Yanai *et al.*, 2005) that were reviewed by Hudson (2005) in the same issue. Hudson first reviewed the information on the involvement of transcription factors in the regulation of plant hormones (e.g. cytokinin and gibberellin hormones). As for the roles played by the *KNOX1* gene, *STM*, in hormone regulation in the shoot apex of Arabidopsis, the following scenario emerged: The *STM* expression in the SAM has two effects. It represses GA synthesis and promotes CK synthesis. The CK induces the formation of an enzyme, GA2-oxidase, which can degrade GA. By the formation of GA2-oxidase at the base of an incipient leaf, the GA in the distal part of the leaf is barred from entering the SAM. The direct effect of *STM* in the SAM towards GA synthesis is that the *STM* protein represses the enzyme GA20ox. The latter enzyme is a key component in the metabolism that leads to active GA. Taken together, *STM* acts to assure low GA levels and high CK levels at the SAM, and maintains its undifferentiated cell proliferation (this is the "SAM-cell identity").

This general picture was supported by the studies of Jasinski *et al.* (2005) and Yanai *et al.* (2005). The team of Naomi Ori (Yanai *et al.*, 2005) of the Faculty of Agriculture of the Hebrew University (Rehovot, Israel) described experimental procedures that implied that *KNOX1* transcription factors (e.g. *STM*), which are essential for meristem functions, act in part through the induction of local biosynthesis of CK. The *KNOX1* proteins were thus identified as central regulators of hormone levels in the meristem. The main experimental findings of Yanai *et al.* (2005), which substantiated the claim of the central role of *KNOX1*, are the following: The *STM* expression could be induced in transgenic Arabidopsis plants into which the transgene *35S:STM-GR* was inserted. The induction of *STM* took place when a GR-binding factor was applied. This factor, DEX (dexamethasone), was already mentioned above.

When such an induction was activated, there was a quick and strong elevation of a gene (*AtIPT7*) that codes for isopentenyl transferases (IPTs), and thus is essential in the biosynthesis of CK. Hence, there was evidence that the KNOX1 gene *STM* facilitates CK synthesis. The levels of two cytokinins were the most affected by the *STM* induction: *trans*-zeatin riboside 5′ monophosphate (tZMP) and *trans*-zeatin riboside (tZR); the synthesis of both requires isopentenyl transferase. The levels of several other cytokinins were not affected by the increase of *STM* expression. In addition, the authors found that the activation of *STM* also increased the level of the expression of the primary cytokinin response gene, *ARR5*.

The mutants of *stm* cause severe malformation, and no normal SAM is formed. When cytokinins such as zeatin or 6-benzylaminopurine (BA) were added to such mutants, many of the treated mutants were partially recovered. The recovery was specific to *stm* mutants; *cuc* mutants were not recovered by CK treatment. Finally, a good recovery was obtained when loss-of-function *stm* mutant plants were genetically transformed to contain a transgene, having a *STM* promoter fused to the coding sequence of the *AtIPT7* gene. The experimental results of Yanai *et al.* (2005) indicated that *STM* acts upstream of CK, at the level of CK biosynthesis, to maintain a functional and active SAM. Is there also a positive feedback loop: CK promoting *STM* expression? This publication did not furnish an answer to this question. The accompanying publication in *Current Biology*, by Jasinski *et al.* (2005), a team of nine investigators from the University of Oxford and Rothamsted Research in England, presented evidence that is in line with the conclusions of Yanai *et al.* (2005); but the former team also analyzed the synthesis and degradation of gibberellins (GA). Jasinski *et al.* (2005) also provided evidence for the suggestion that CK activity is both necessary and sufficient for the stimulation of GA catabolic gene expression (leading to GA2-oxidase). On the other hand, *KNOX* proteins repressed the transcription of genes encoding the GA20-oxidases (enzymes essential for the biosynthesis of active GAs). The *KNOX* (*STM*) proteins were thus claimed to have a dual role: they elevate CK, which stimulates GA degradation, and they repress the transcription of a gene involved in GA biosynthesis (as mentioned above). Therefore, the KNOX proteins emerged as general orchestrators of growth-regulator homeostasis of the SAM of Arabidopsis, positively by elevating CK and negatively by keeping low GA levels at the apical meristem.

More recent research furnished additional information on the meristem's function and the CK in the SAM. A team of eight investigators from Tübingen, Germany, and the University of North Carolina looked at the roles of *WUS* and the ARABIDOPSIS *RESPONSE REGULATOR* genes (*ARRs*) in regulating the normal function of Arabidopsis SAM. As noted above, *WUS* is a positive regulator of stem cells at the SAM (and it is balanced by the CLV ligand–receptor system). *WUS* directly represses *ARRs*, and the latter act in the negative feedback loop of CK signaling. *ARR* genes probably mediate a negative influence on the meristem size.

The authors found that the *WUS* protein can bind to a sequence of the *ARR7* gene that is about 1000 base pairs upstream of the start codon of this gene (i.e. it binds to its promoter). This binding probably represses the transcription of the *ARR7* mRNA. On the other hand, the authors provided evidence for the effect of *ARR7* on WUS: when *ARR7* was overexpressed in transgenic plants that contained the transgene 35S:*ARR7*, they had a lower level of *WUS* mRNA. In spite of the reduced *WUS* mRNA, these plants had no aberrant phenotype, probably because the low *WUS* was sufficient for an apparently normal SAM. Notably, the repression of *ARR* genes could be temporarily regulated in plants harboring a plasmid in which the transcription of *WUS* can be induced by Dex. Dex treatment (and the resulting elevated *WUS* protein) caused a reduction of *ARR* transcripts already 4 hours after the start of Dex application.

Chromatin and the Function of SAM

By human logic, an efficient way to shut off the expression of genes that are not involved in the patterning and metabolism of an organ would be to bar the expression of these genes at the chromatin level. If genes are made inaccessible to transcription by PolII, there could be an upstream means, epistatic over *cis*-regulatory domains, to silence these genes. There are two problems with this logic. One is that the logic of angiosperms may differ from human logic. The other problem is that unlike the situation in prokaryotes, where several genes that code for enzymes in a certain metabolic pathway are clustered, the genes involved in plant patterning do not seem to come in clusters. Silencing a cluster by the respective chromatin in which this cluster resides would render the silencing rather efficient. However, silencing individual and scattered genes by the respective modification of their encompassing chromatin does not appear to be efficient.

After this theoretical consideration, the question is whether there is evidence for the involvement of chromatin in the control of the expression of certain genes at the SAM of angiosperms. The emerging answer to this question is that there is evidence, but up to the present, it is still meagre.

One piece of evidence came from a team of Japanese (Kyoto University) and US (Cold Spring Harbor Laboratory) investigators (Kaya *et al.*, 2001). These investigators looked at the *FASCIATA* (*FAS*) genes of Arabidopsis. To appreciate their study, let us first look at a factor that was found in human cells: chromatin assembly factor 1 (CAF1). This factor consists of three subunits (p150, p60, p48) and putatively supports the assembly of nucleosomes onto replicating DNA. CAF1 is also associated with newly synthesized histones H3 and H4, and is localized at the replication loci in replicating human cells. It may thus function to ensure stable propagation of the epigenetic state and the maintenance of genome integrity. This state is probably maintained by rapid reformation of the chromatin structure (i.e. DNA encircling a histone complex, forming the nucleosomes) after the passage of a DNA replication fork.

Is there a connection between *CAF1*, *FAS* and maintenance of the normal function of the plant's apical meristems? Amazingly, there is, as detected by Kaya *et al.* (2001). When the amino acid sequence encoded by *FAS1* was analyzed, it was found that it has significant similarity to the amino acid sequence of the larger subunit (p150) encoded by the human *CAF1*. *FAS2* encodes an amino acid sequence that has similarity to the p60 subunit encoded by CAF1. Another family of Arabidopsis genes, $ATMSI_1$–$ATMSI_4$, encodes an amino acid sequence that is similar to the third (p48) subunit of CAF1. This strongly suggested that *FAS1*, *FAS2* and *ATMSI*-encoded proteins are components of trichromatin assembly factor in Arabidopsis. This suggestion is in line with the phenotypes of the *fas* mutants: the shoot of these mutants is broader than wild-type, and the former have a fasciated phenotype with distorted organization of their SAM — the normal organization into functional zones and cell layers is distorted, and the shoots show various degrees of failure to develop leaf primordia.

By *in vitro* analyses, it was revealed that *FAS1*, *FAS2* and *ATMSI* formed a complex that had replication-dependent nucleosome assembly activity, just as was found previously for the p150, p60, and p48 subunits of the human CAF1. This further supports the suggestion that the *FAS1* and *FAS2* gene products are subunits of the Arabidopsis counterpart of CAF1. As for the relation of chromatin organization and *WUS*, there was a significant observation by Kaya *et al.* (2001): in the SAM of *fas* mutants, the expression domain of *WUS* expands significantly relative to wild-type SAMs. The expansion is both lateral and acropetal — in *fas* mutants, mRNA of *WUS* was detected not only in L_3 cells, but also in L_2 and occasionally in L_1 cells. Hence, the domain of *WUS* expression is clearly changed in *fas* mutants.

During the years 2001 to 2004, the involvement of chromatin remodeling in the spatial and temporal expression of genes in the SAM was studied by several research groups. These studies were reviewed by Guyomarch *et al.* (2005) of Orsay, France. The studies were summarized in two tables. One table listed chromatin assembly factors and ATP-dependent chromatin remodeling complexes, as well as histone modification, involved in the regulation of meristem function in plants. The other table summarized the information on heterochromatin formation factors, as well as the polycomb-like and trithorax-like complexes involved in meristem function.

While about 40 factors/genes of Arabidopsis were listed in these tables, and their similarities to factors/genes in yeast and/or *Drosophila* were indicated, substantial and detailed information was available for only a very small number of factors/genes in Arabidopsis. Of these, four are genes involved in chromatin assembly: *FAS1*, *FAS2*, *ATMSI1* and *TSL*. The three former were mentioned above, and TSL is a kinase that interacts with and phosphorylates the Arabidopsis homolog of the chromatin assembly complex Asf1. A fifth factor worth mentioning is ATBRM. It seems to belong to the ATP-dependent chromatin-remodeling complex and serves as a repressor of certain genes such as *SOC1*, but not *FLC*. Another factor/gene with yet unknown functionality is BRU1, TSK. This is a protein that has a leucine-rich region (LRR) and appears to regulate *WUS* expression in the SAM, *AGAMONS* (*AG*) expression in the floral meristem and *SCR* expression in the root apical meristem.

It is expected that more detailed information on the roles of chromatin remodeling in maintaining the balance of cellular activities in the SAM will emerge from future studies. Meanwhile, several general theories are being suggested. One of these concerns the difference in patterning between animals and angiosperms. In this theory, the starting point is that, in animals, after the mature embryo is formed, there is very little further differentiation. On the other hand, in angiosperms, the bulk of differentiation starts after the mature embryo is established. Consequently, according to this theory, in animals there would be an advantage if all those genes involved in differentiation would be silenced epigenetically (by chromatin remodeling). In angiosperms, no such "bulk" silencing is required. The silencing should better happen at defined developmental stages and involve a choice of patterning genes. This may mean that the epigenetic silencing in angiosperms should be more fine-tuned and reversible than in animals. Experimental data that either refute or verify this theory are not yet available.

A short but educative review on the regulations involved in the maintenance of the shoot apical meristems (focusing on the regulation of stem cell maintenance) of angiosperms was presented by Williams and Fletcher (2005). This review summarizes the knowledge on SAM maintenance by the CLV/WUS feedback loop, on the induction of the shoot's stem cell identity and on the regulation of

WUS expression imposed by chromatin. Most of the topics in this review were already presented above in this chapter. However, some recent additional information is noteworthy. The BRU1 protein (encoded by the *BRUSHY1* gene of Arabidopsis) that was found to be involved in the postreplicative stabilization of chromatin structure has indeed been recently found to be associated with the regulation of the size of the *WUS* expression domain (Takeda *et al.*, 2004). Furthermore, it was found (Kwon *et al.*, 2005) that *WUS* is the direct target of the chromatin remodeling factor, *SPLAYED* (*SYD*).

SYD is a member of a family of ATPases that facilitates the transcription by creating a DNA template which is accessible to transcription. Adult *syd* mutants undergo a premature termination of their SAM that is correlated with the reduction of the mRNA of *WUS* (and also the mRNA of *CLV3*). It was found that SYD is specifically recruited to the proximal promoter region of *WUS*. This implies that SYD is a *direct* and *specific* positive upstream regulator of *WUS*, and is therefore required for the maintenance of the proper level of *WUS* expression at the SAM.

Finally, here is a remark that is not specifically related to plant patterning. Williams and Fletcher (2005) furnished an authoritarian and clear update on the regulatory mechanisms of the SAM; however, I have one reservation. The opening sentence of this review includes two words that do not appeal to me: "higher plants". This term was also used by other authors writing on plant patterning, but I regard it as a bad term. First, the question can be asked: "higher than what?". Then, it is not clear what the authors really mean: should it include all phanerogams, gymnosperms and angiosperms or only angiosperms? Why not use good and old botanical terms? The tendency to shunt from botany is rather fashionable. Departments of Botany changed their name to Department of Plant Science. Similarly, Department of Biology changed to Department of Biological Sciences. The last name is obviously wrong because biology already has "science" in it: it is the science of living organisms. Using quasi-sophisticated names does not increase knowledge.

Chapter 9

PATTERNING OF ROOTS

The human mind favors definition of entities. Such definitions are not only helpful, but in some fields of knowledge they are the essential basis for any progress. Geometry is an example of such a field in which definitions are the foundations for handling the subject. The tendency to arrange entities in defined "drawers" was also adopted by biologists, botanists included. Hence, before dealing with *roots* of angiosperms, investigators found it necessary to give a clear answer to the question: "What are roots?" They would prefer a short sentence to define the roots of angiosperms. However, nature is not a chest of drawers. Hence, to define the root as the component of a plant that is below the soil surface would be wrong in many cases. First, because there are roots that are not in the soil, such as the roots of epiphytes (e.g. many orchid genera) and aquatic angiosperms. Also, there are plant components that are in the soil but are not roots (e.g. the stolons of potato and the fruits of peanut and other subterraneous legumes).

A somewhat more appropriate definition of roots could be all the plant components that are derived from that part of the seedling below the hypocotyl. However, again, there are several problems with such a definition, because in many plants roots can develop from branches that are far above the soil or from nodes that are above the soil (e.g. in maize). On the other hand, in this book we shall deal with roots of young seedlings and focus on dicots, with only some reference to cereal crops. In fact, most of our deliberations will be on the primary root of Arabidopsis seedlings, because the bulk of structural, genetic and molecular studies during the past 15 years has been conducted with the latter roots. We shall thus handle roots without having an exact definition.

Even while dealing with the primary root of Arabidopsis seedlings, the problems of definitions come up. Two recent publications dealt with such definitions: Baluska *et al.* (2005b) and Friml *et al.* (2006). These publications were written by 13 and 20 authors, respectively. The publication of Baluska *et al.* (2005b) focused on the question: what is "apical" and what is "basal" in root development. Without going into details, one of the confusions stems from the term "base". Seedlings have two bases. One is the basis of the above-soil part of the seedling, while the base of the root is the *uppermost* part of the root. Hence, both bases fuse into one location. These seedlings have two apices, one at the top of the stem and the other is the apex of the young primary root that points downwards. The confusion is increased when we go back to the immature embryo. This embryo is attached to the mother plant by the *suspensor*. While this suspensor is still attached, the point of attachment can be considered the base of the embryo. However, this end of the embryo is later converted into the root apex. Also, the terms "acropetal" and "basipetal" can be confusing: is a basipetal movement of auxin a movement from the root apex towards the base of the root? The authors thus suggested that terms such as "apical", "basal", "acropetal" and "basipetal" shall be used only after the establishment of the embryonic root/shoot polarity. I shall attempt to avoid ambiguousness in the use of root terms.

The publication of Friml *et al.* (2006) is actually a reaction to the publication of Baluska *et al.* (2005b). The former did not fully agree with the latter. The former claimed that anatomically-based terminology, when used for the description of *cellular* polarity, has important deficiencies. This is exemplified by describing the flow of auxin (from the shoot tip to the lower parts of the root). By anatomical terms, this flow is *basipetal* from the shoot tip, but when the flow continues (in the same direction), it becomes *acropetal* in the root. Furthermore, in recent years it became evident that many of the gene activities starting in the embryo are carried on to the young seedlings. Therefore, there is no sense in a complete separation of embryos from seedlings. Hence, using different terminology does not make sense. The authors therefore recommended referring to the first cell division of the zygote, when the upper cell (the cell that is away from the suspensor) is the "apical" cell and the cell wall that is opposite of the first cell plate of the divided zygote is consequently the apical side of the cell.

This polarity is maintained in the embryo as well as in the future seedling. Consequently, Friml *et al.* (2006) made the logical suggestion to keep the anatomical terms (*apex*, *base*, *acropetal*, *basipetal*) for postembryonic development, but to use the zygote-derived apical–basal body axis as a reference for the description of the polarity of individual cells in cell biology.

The root serves the angiosperm plants for two main purposes. It anchors the plant to a substrate, which is commonly soil, and it supplies the plant with water and mineral nutrients. We shall see that the availability of water and nutrients, such as phosphates, in the soil substrate has an impact on the patterning of the root system.

Roots also fulfill two additional tasks that I shall only mention briefly in this book without elaborating on them, because each of these constitutes an extensive field of endeavor that is outside of the scope of this book. The first of these tasks is shared by the roots of several angiosperm families, but clearly not by the roots of all plant families. This task is to supply the plants with fixed nitrogen produced by nitrogen-fixing bacteria. The best known of these systems is the symbiosis of nitrogen-fixing bacteria (*rhizobia*) with legume (Fabaceae) roots and with roots of another specific nonlegume genus, *Parasponia* (Ulmaceae). Another symbiosis that transfers fixed nitrogen to the roots is between bacteria of a Gram-positive genus *Frankia* and roots of about nine dicot families, usually roots of trees (e.g. of the genera *Alnus*, *Myrica*, *Casuarina* and *Ceanothus*). Roots (e.g. of maize) may have a close association with certain bacteria that are capable of nitrogen fixation (*Azospirillum*); however, the direct transfer of fixed nitrogen by the latter bacteria into the roots is questionable.

The other, additional task of roots is the utilization of a range of strategies to increase the availability of phosphates to the plant. It should be noted that because of the high sorption of phosphate to soil particles, phosphorus (P) is the macronutrient that is least available to plant roots. Phosphate ions usually bind tightly to the surface of soil clay minerals. One of these strategies is the establishment of *mycorrhizal associations*. Such associations exist in many angiosperms, but probably not in all of them. There are two types of these associations, the *endomycorrhizae* (also known as *vesicular–arbuscular mycorrhizae* or

VAM) and the *exomycorrhizae*. The former association is much better known to scientists than the exomycorrhizae. The latter are known to the public because edible mushrooms are fruiting bodies of exomycorrhizae. Both types of mycorrhizae are instrumental in the absorption of P from the soil into the roots and, from there, to the whole plant. There are additional strategies to increase the availability of P to the roots, such as the release of organic acids by the roots; however, again these strategies are outside the scope of this book, and information on these strategies can be obtained by reading the chapter of Kochian (2000).

In the present chapter, I focus on the patterning of roots and deal with the following subjects:

- The structure of the primary root is outlined, with the Arabidopsis root serving as an example for roots of dicot seedlings;
- The structure of the primary root of Arabidopsis is compared with that of monocot seedlings (maize and rice);
- The organization of the root apical meristem (RAM) and the main genes that participate in the maintenance of the RAM are discussed;
- The epidermis of the young root and the differentiation of root hairs are handled;
- The origin of lateral roots, the genes involved in lateral root formation and the impact of environmental effectors on root architecture are reviewed;
- The awareness of the major role of plant hormones in plant patterning has increased in recent years, and the involvement of these hormones in root patterning is clarified by recent investigations; hence, I provide the relevant information on this subject;
- Approaches to identify and characterize genes involved in root patterning are briefly covered.

The Root of Young Arabidopsis Seedlings: The Meristem and the Differentiation of Projected Cell Files

In the late 1980s, it was recognized by many investigators that Arabidopsis is an excellent model plant for structural and molecular genetic investigations. Meyerowitz (1989) defined it as "a useful weed". In fairness to George Reddy, it should be remembered that he brought the gospel of Arabidopsis with him when he emigrated from Hungary to the United States. Consequently, when detailed studies on angiosperm roots were initiated, the root of Arabidopsis became a favorite research object. One of the teams that performed the early studies on the cellular organization of the *Arabidopsis thaliana* root was Dolan *et al.* (1993), a team composed of investigators from the University of Pennsylvania (D. Dolan and S. Poething), the University of Utrecht in the Netherlands (B. Scheres) and the John Innes Institute in Norwich, UK (P. Linstead and K. Roberts). Members of this team and additional investigators published further important papers on cell patterning in Arabidopsis roots and on the functions of root apical meristem (e.g. Dolan *et al.*, 1994; van den Berg *et al.*, 1995, 1997). The outcome of these studies is summarized in the following paragraphs.

The root meristem is derived from the cells of the hypophysis and adjacent cells of the embryo. In the meristem of the root of young Arabidopsis seedlings, the cell number and location of the different cell types is relatively invariant (see Fig. 33). Hence, there are eight cell files of cortex cells and eight cell files of endodermis files. The number of epidermal files and pericycle files can be greater and variable. There is a quiescent center (QC) of four central cells that are located between the root cap columella (at the tip of the root) and cells that are projected towards the upper side of the root. The plate of the four QC cells is surrounded by cells above them that are the initials of the files of the cortex, the endodermis, the epidermis, the pericycle and the stele cells. Below the QC are dividing cells that contribute to the distal root cap and the cells that form a ring around the central cells. They contribute to the lateral root cap (see Fig. 33B).

This young primary root of Arabidopsis can be divided into zones. The most distal (tip) zone spans about 250 μm and constitutes the root cap. Then, there are the QC and the cells surrounding the QC; this zone is usually termed the root meristem. Proximal to the latter zone is the elongation zone, which also spans about 250 μm. Proximal to the elongation zone towards the shoot apex is the

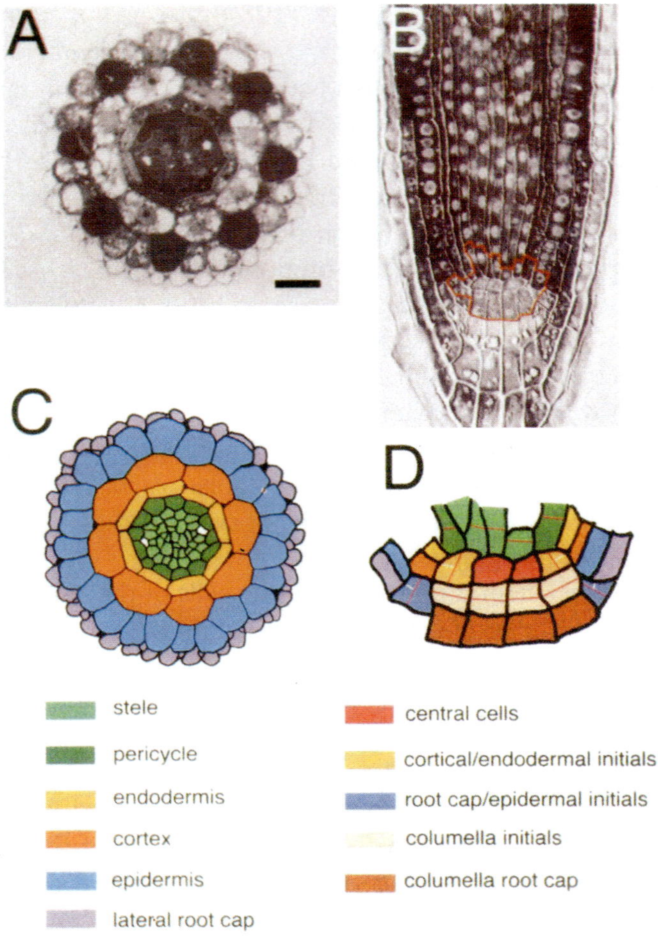

Fig. 33. Cellular organization of the Arabidopsis root. **(A)** Transverse section of an Arabidopsis root from the late meristematic region. A single layer of the lateral root cap cells surrounds the epidermis. At this stage of development, the two differentiating epidermal cell types can be distinguished cytologically; the eight trichoblasts possess densely staining cytoplasm compared with the neighboring atrichoblasts. Bar = 10 μm. **(B)** Median longitudinal section of an Arabidopsis root apex. The cell files converge to the promeristem region (outlined in red). **(C)** Colorized drawing of the transverse section shown in (A). **(D)** Colorized drawing of the promeristem region outlined in (B). The red lines indicate the planes of division that occur in the initials; white lines indicate the secondary divisions that occur in the cortical/endodermal and the lateral root cap/epidermal initials. (From Schiefelbein *et al.*, 1997.)

differentiation zone. The distal end of the differentiation zone can be recognized by initials of root hairs that develop from some of the epidermal file cells, as shall be detailed in a section on root hairs. The central cylinder of the root contains two types of elements: the protoxylem and the protophloem elements. Around this cylinder are the files of the pericycle cells. In the 3-day-old root, at about 1000 μm above the QC, there are about 12 files of pericycle cells. Around the pericycle are the files of (usually eight) endodermis cells. Each of the endodermis cells may face two pericycle cells by facing the cell wall between these pericycle cells. There are commonly eight files of cortex files and these are surrounded by a greater number of epidermal cell files. Epidermal cells that are situated so that they are adjacent to two cortex cells are capable of producing root hairs, while those that are in contact with only one cortex cell will not form root hairs.

The pattern of cell files observed in young primary roots, at a distance of 1000 μm above the QC, will change with maturation when observed further away from the root tip. Also, the maturation concerns further differentiation of the protoxylem and the protophloem. Notably, the protophloem matures much earlier than the protoxylem. Clearly, a great change in the files of cells is visible when we cross the root/shoot boundary and look at the anatomy of the hypocotyl. In the latter, there is a larger number of cell files in the stele and there is an additional outer layer of cortex files. Also, the number of epidermis cell files in the hypocotyl is almost twice the number of these files in the root.

Clearly, the stele contains companion cells and parenchyma cells near the protophloem sieve tube elements. A considerable change in the cytoplasm of stele cells takes place, and connections between cells by plasmodesmata are initiated. The origin of the columella root cap cells is from different dividing cells than the lateral root cap, while the latter share their origin with the epidermis cells. Once the files of epidermal cells are established, they do not differentiate into other cells. On the other hand, at the meristem, cortex initial cells branch into cortex cells and endodermis cells. Notably, nutritional changes, such as the addition of sucrose to the medium of very young Arabidopsis seedling, do not change the anatomy of the root tip.

By [^{3}H] thymidine labeling, the authors (Dolan *et al.*, 1993) were able to determine which cells of the young root are undergoing nuclear division. It appears that nuclear divisions take place in the cells of the young root, with the exception of the four quiescent central cells, although the rate of cell division (as expressed by labeling of the nuclei, which are in the S phase) does vary among the different cell groups. Hence, the highest labeling density was found in the meristematic zone, proximal of the central cells. A much lower density was observed in the cells that give rise to the columella cells. It should be noted that by this labeling, it is not possible to separate labeling that happens before cell division and labeling that will result in endoduplication of DNA.

While the analysis of the anatomy of the young Arabidopsis root is a very useful basis for further genetic and molecular studies, this root does not represent the young roots of other plants. Generally speaking, the young Arabidopsis root has a simpler structure than the structure in young roots of other dicot species. For example, in species of other Cruciferae genera (e.g. *Brassica campestris*), there are two concentric rings of cortical cell files rather than the one ring in Arabidopsis. Clearly, there are many additional anatomic differences, but there is also a uniformity. In all these studied species, the root meristem contains three tiers of initials. The columella is derived from the lowest tier of initials; the lateral root cap cells and the epidermis are derived from the collar of cells located in a ring around the columella initials; the third tier of initials is located above the central cells. From the latter cells, the cortex, the endodermis, the pericycle and the stele are derived.

In a further study, Dolan *et al.* (1994) focused on the patterning of the root epidermis and the formation of root hairs from epidermal cells. In a subsequent section, I shall update the information on root hair differentiation, but here I note some important findings of Dolan *et al.* (1994). Interestingly, a surgical study on the epidermis of Cruciferae roots (radish) was performed much earlier by Bünning (1951) in Germany. Bünning found that separating developing root epidermis caused additional hair-forming cells in the excised tissue at the expense of nonhair cells. Hence, Bünning concluded that in the intact epidermis, hair formation is the default state, but active inhibition of the capability to produce root hairs is caused by the underlying cells (cortex).

I already noted above that, in the young Arabidopsis root, the epidermis cells are derived from the same set of 16 cells which surround the central cells, and that this set of cells is also the source of lateral root cap cells. This was verified in the Dolan *et al.* (1994) study by tracing cell lines by a transgene that contained the GUS gene under a strong promoter (35S), in which the GUS gene had a transposable element (*Ac*) in its coding region. Whenever the *Ac* was excised, the GUS was transcribed and the progenies of cells in which this excision happened could be traced. These authors also found that the number of epidermis cell files does not change much from the root tip up to the differentiation zone; usually there are about 16 such files that face eight files of underlying cortex files. The hair-forming epidermis cells are termed *trichoblasts* and those that do not form hairs are termed *atrichoblasts*. Between the trichoblast files, there are one or two files of atrichoblasts. The hairs grow as a projection from the basal (towards the root tip) end of trichoblasts. The nonhair epidermal cells are longer than the epidermal cells that produce hairs. This difference in length is apparent before the hairs protrude, indicating that the determination of the capability to produce hairs (or to be hairless) takes place rather early in the root meristem.

From the surgical experiments of Bünning mentioned above, Dolan *et al.* (1994) surmised that there should be a morphogen that is transferred from cortex cells to epidermis cells and that this morphogen prevents hair formation. They further assumed that this morphogen could be ethylene. The gene *constitutive triple response* (*CTR1*) is essential for the cascade of ethylene signal transduction. Thus, *ctr1* mutants have a phenotype of plants treated with ethylene. When the authors analyzed Arabidopsis roots of the *ctr1* mutant, they found additional ectopic hairs in the epidermis cells. This indicated that the wild-type gene (CTR1) encodes a negative regulator of hair formation and, in the intact root where ethylene is inducing hair formation, the inhibitor of hairs is regulated by the CTR1 protein. Similar results were presented in a publication by the Dutch investigators

(Scheres *et al.*, 1994) who previously participated in the publication of Dolan *et al.* (1993). The two publications appeared in the same issue of *Development*.

The University of Utrecht people who participated in the Dolan *et al.* (1993) investigation asked a fundamental question (van den Berg *et al.*, 1995): how are the main files of the young Arabidopsis root patterned? There are two possibilities. The destiny may be defined in the meristem and, thus, once the determination is set, the division of the clones will maintain this destiny. The other possibility is that the destiny can be affected by neighboring cells, and the observed clonal maintenance is at least potentially flexible. By using cell ablations, the authors found that there is indeed a flexibility, and one cell file can affect its neighboring cells and contribute substitute cells to replace cells that were ablated. This is positional information. For example, as noted above, cortical initials divide (asymmetrically) to produce cortex and endodermis. When these initials are ablated, the more centrally located pericycle cells will invade the ablated sites and then generate cortex cells. Similarly, if epidermal cells are ablated, they can be replaced by dividing cortical cells that become epidermal cells. From their ablation experiments, van den Berg *et al.* (1995) concluded that positional control is the most important factor in the determination of cell fate. Positional signals can be perpetuated from more mature cells to initial cells, guiding the pattern of meristem cell differentiation. Thus, in the root meristem, inductive processes appear to specify initial cells, whereby the mature daughter cells act as a patterning template. This role of patterning template is a common phenomenon in animal embryology.

In a further ablation study, the same team (van den Berg *et al.*, 1997) obtained an answer to an additional question: what is the role of the nondividing central region cells in the root meristem? The results of this study indicated that these nondividing cells have a role as inhibitors of cell differentiation in the cells that are their close neighbors, and that cell division is maintained in the latter cells due to the presence of the nondividing QC cells. The studies from 1993 to 1997 on the patterning and morphogenesis in young Arabidopsis roots were nicely reviewed by Schiefelbein *et al.* (1997).

Carbon transport (e.g. from the leaves to the root) is commonly performed in plants by the flow of sucrose. At the sink (root), the sucrose can be split either via invertase to glucose and fructose, or via sucrose synthase to fructose and UDP-glucose. Sergeeva *et al.* (2006) found that the elongation of Arabidopsis roots is correlated with the vacuolar invertase activity of root cells encoded by a specific gene rather than with the sucrose synthase activity in these cells.

Root Development in Cereals

As noted above, the Arabidopsis roots served in numerous studies as model roots, but roots of other genera of Cruciferae differ at least in anatomy from Arabidopsis roots. Roots of plants from other dicot families differ substantially from Arabidopsis roots. It should therefore not surprise us that roots of monocot plants are very different in their architecture and anatomy from Arabidopsis roots. A comparative presentation of roots from different plant families is far beyond the scope of this book. However, to illustrate the differences in morphology and anatomy among plant families, I shall provide information on the roots of two cereal plants that were also subjects of anatomical and genetic investigations: maize and rice. It should be noted that these cereals, together with other cereal crops (e.g. wheat, barley), constitute the main source of food production worldwide.

Hochholdinger *et al.* (2004a,b) reviewed the morphology and anatomy of maize and rice, and provided information on genes involved in the patterning of the roots of maize and rice. This section will be based primarily on these reviews, and on the more recent publication of Woll *et al.* (2005).

Contrary to the root of the young Arabidopsis plant, which has the architecture of a main primary root from which lateral roots are branched and further ramify, the roots of maize and rice have a more complex architecture. These plants have several types of roots. Maize and rice also produce an embryonic primary root that becomes visible 2 to 3 days after germination. In maize, there is a variable number of *seminal*

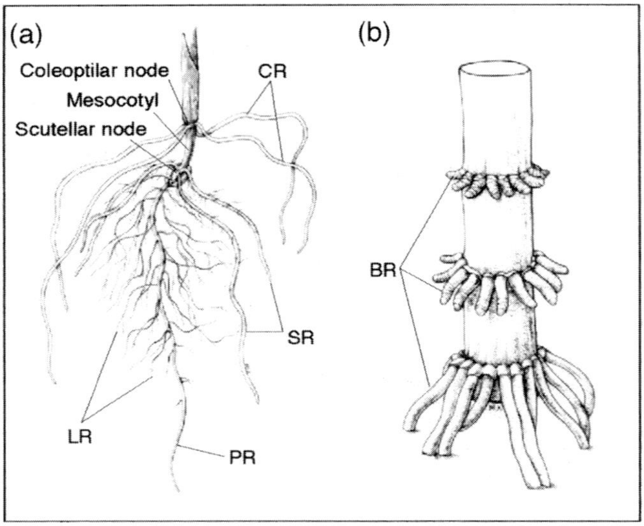

Fig. 34. Maize root system at different developmental stages. **(a)** 14-day-old maize seedling displaying primary (PR), seminal (SR), shoot-borne crown (CR) and lateral (LR) roots. Morphological seedling structures of cereals described in the text are labeled. **(b)** Aboveground root stock of an adult maize plant composed of shoot-borne brace roots (BR). Drawings by Miwa Kojima (Iowa State University). (From Hochholdinger *et al.*, 2004b.)

roots (SR), which are also considered embryonal roots, but they emerge from the scutellar node at about 1 week after germination (see Fig. 34). No such scutellar roots are formed in rice. Later, shoot-borne roots are initiated from underground and aboveground nodes of the stem. These are termed *crown roots* (CR) and *brace roots* (BR), respectively.

The crown roots of maize emerge about 10 days after germination, while the emergence of brace roots occurs only about 6 weeks after germination. Like Arabidopsis roots, maize and rice roots also produce root hairs when they are underground. Brace roots will form hairs only after they penetrate the soil. Cereal roots increase their association with the soil and its flora microorganisms by a structure termed *rhizosheath*, which is composed of root hair-bearing roots with soil particles and associated soil flora (e.g. rhizobacteria). The rhizosheath increases the root surface that is associated with the soil and is probably useful for the uptake of nutrients from the soil. In brief, while the embryo-derived primary root of Arabidopsis is extended and serves this plant until sexual maturity, the primary root of maize is replaced by SR, CR and BR roots as the main functional root system soon after germination.

The morphological and anatomic differences between the roots of cereals (exemplified by maize) and the root of Arabidopsis are listed below.

	In cereals	In Arabidopsis
Embryo-derived roots	Prim. root, SRs (in maize)	Const. the main root
Shoot-borne roots	Extensive	Missing
Root hair pattern	Irregular	Regular on hair cell files
Source of lateral roots	Pericycle endodermis	Pericycle
No. of cortical cell layers	8–15	1
No. of cortical cell files	Variable	8
No. of QC cells	800–1200	4
No. of root initial cells	Few	Several hundred

As was indicated for Arabidopsis, mutations that affect root development were identified in maize and rice. The phenotypes of the monogenic mutations that affect root differentiation are listed in Table 5. In addition to the listed mutants, there are pleiotropic mutations that affect not only root differentiation in maize and rice. Several root mutants of maize and rice cannot be found in Arabidopsis. In some cases, due to a difference in root architecture, Arabidopsis has no shoot-borne roots; hence, there are no mutations in Arabidopsis that affect shoot-borne roots. Interestingly, maize plants can mature into functional plants even if they are homozygous for *rtcs*, and thus lack all the shoot-borne roots (they do have a primary root that has lateral roots). A full molecular and functional comparison between root mutants in Arabidopsis and rice is to be expected because the genomic sequences of these plants are now readily available. Such a comparison may clarify root differentiation phenomena in cereals that were amply studied in Arabidopsis, but are poorly investigated in cereal roots. One such phenomenon is the formation of root hairs.

Table 5. Specific Maize and Rice Mutants Affected in Various Aspects of Root Development

Gene	Species	Phenotype
Shoot-borne roots		
rtcs	*Zea mays*	Complete lack of shoot-borne roots
rt1	*Zea mays*	Fewer shoot-borne roots
crl1	*Oryza sativa*	Fewer crown roots
crl2	*Oryza sativa*	Fewer crown roots and reduced primary root length
Lateral roots		
lrt1	*Zea mays*	No lateral roots at embryonic roots and crown roots at first node
slr1	*Zea mays*	No lateral root elongation at embryonic roots
slr2	*Zea mays*	No lateral root elongation at embryonic roots
rm109	*Oryza sativa*	Lateral root initiation blocked
Primary root		
rm1	*Oryza sativa*	Reduced primary root elongation
rm2	*Oryza sativa*	Reduced primary root elongation
rrl1	*Oryza sativa*	Reduced primary root elongation
rrl2	*Oryza sativa*	Reduced primary root elongation
srt5	*Oryza sativa*	Reduced elongation of all root types
srt6	*Oryza sativa*	Reduced primary root elongation
Root hairs		
rth1	*Zea mays*	Reduced root hair elongation
rth2	*Zea mays*	Reduced root hair elongation
rth3	*Zea mays*	Reduced root hair elongation
rh2	*Oryza sativa*	Missing root hairs

(From Hochholdinger *et al.*, 2004a.)

Also, the roles of plant hormones and the interaction between hormones are much better understood in Arabidopsis than in cereals. On the other hand, there are hormone effects on cereal roots that do not exist in Arabidopsis. An example is the development of shoot-borne roots in submerged rice seedlings. In deep-water rice, the plant hormone ethylene is involved in the emergence of the shoot-borne roots from nodes that are submerged under water.

A recent study of the Hochholdinger team (Woll *et al.*, 2005) focused on one mutation: *rootless with undetectable meristem 1* (*rum1*) in maize. This mutation causes deficiency in the initiation of the embryonic seminal roots and lack of the postembryonic lateral roots emerging from the primary root. The mutation is involved in reduced auxin transport in the primary root, but exogeneous auxin does not induce laterals from this primary root. By microarray analysis of W.T. and mutated pericycles, a great number of genes was found to be preferentially expressed in wild-type pericycle cells, while other genes were preferentially expressed in the mutant pericycle. If such studies would be performed before the actual emergence of branches from the pericycle, they could lead to a better understanding of the molecular processes that lead to branching.

Initiation and Morphogenesis of Root Hairs in Arabidopsis

I mentioned root hairs while describing the various cell files that are projected from the root meristem. However, in this section, I focus on root hair formation. The root hairs in angiosperms have been studied for at least half a century (see Cormack, 1962 for an early review). A detailed anatomical differentiation study on Arabidopsis root hairs was conducted by the team of L. Dolan, P. Linstead, K. Schneider, K. Roberts and others (Dolan *et al.*, 1994). The subject was then updated by a thorough review by Ryan *et al.* (2001) that covered the early studies on specific genes involved in root hair formation. As noted above, the files of projected epidermal cells consist of two types of cells. There are the root hair-forming cells (derived from trichoblasts) and non–root hair-forming cells (derived from atrichoblasts). The two files alternate (see Fig. 35). Trichoblasts are shorter than atrichoblasts and undergo a shorter cell cycle. The formation of hairs from the trichoblasts starts with the budging of the outer cell wall; then, the hair grows until it has a diameter of about 7–10 μm and a length of up to 1 mm. The genetics of root hair

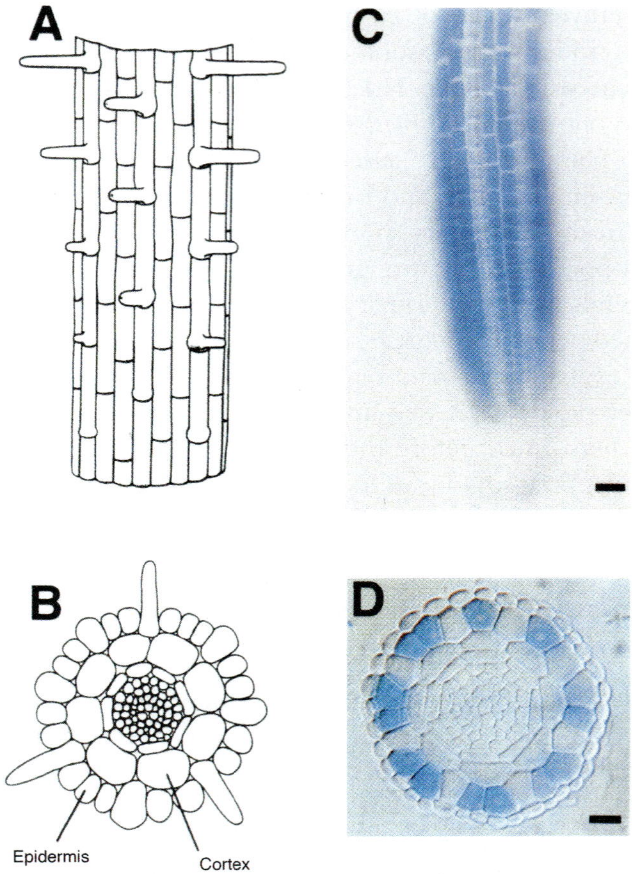

Fig. 35. Cell type determination in the epidermis of the Arabidopsis root. **(A)** Schematic drawing of an Arabidopsis root in surface view in the root hair developmental zone. Files of epidermal cells consist entirely of root hair cells or hairless cells. **(B)** Schematic drawing of a transverse section from the mature portion of an Arabidopsis seedling root. Root hair cells are located in the clefts between adjacent cortical cells. Only three root hairs are shown in this figure because the length of epidermal cells generally prevents all eight root hair projections from being visible in a single transverse section. **(C)** Spatial expression pattern of the *GL2* promoter-β-glucuronidase reporter gene fusion construct in the root apex of a 4-day-old seedling. The *GL2* promoter directs expression preferentially in a cell file-specific manner in the epidermis. Bar = 20 μm. **(D)** Expression pattern of the *GL2* promoter-β-glucuronidase reporter gene fusion construct in a transverse root section from the late meristematic region of a 4-day-old seedling. The *GL2* promoter directs expression preferentially in epidermal cells located over the cortical cells (atrichoblasts). Bar = 10 μm. (From Schiefelbein *et al.*, 1997.)

mutants can be studied easily because even phenotypes that are hairless are usually not lethal.

There are several cytological changes that accompany the initiation of the root hairs. One of these is the migration of the nucleus to a site opposite the newly formed bulge. Hair formation is also involved in specific cytoplasmic streaming. Once the hair is initiated, the cells' nucleus enters the hair. Behind the tip of the root hair, is rich vesicle-containing material (VRZ: vesicle-rich zone) for the extension of the root hair. Root hair cells have an extensive vacuole. The extension of the root hair is similar to other forms of tip growth (fungal hyphae, moss protonema, pollen tubes). The new plasma membrane and cell wall material are supplied via Golgi vesicles to the apical surface. CA^{2+} also seems to be involved in the initiation and growth of root hairs. Growing root hairs have a gradient of Ca^{2+} that is tip-focused. The cytoskeleton of root hairs is composed of microtubules and actin filaments and their associated proteins. Motor proteins such as kinesin and myosins are probably instrumental in polar movement of vesicles towards the hair tip (see Chapter 3 on the cytoskeletons of plant cells). In addition, other cytoskeleton proteins such as profilin and spectrin are located in the growing root hairs, but full understanding of the involvement of the cytoskeleton in the growth of root hairs awaits further investigations.

The growth of root hairs is influenced by the availability of nutrients in the soil (e.g. phosphate). The role of such nutrients in root hair formation shall be mentioned below in the section on the impact of nutrients and environment on the formation of lateral roots and root hairs.

Early Genetic Studies on Root Hair Differentiation

Until mid-2001, eight mutants were characterized by their root hair phenotypes, and these were summarized by Ryan *et al.* (2001). These eight hair mutants are listed in Table 6, where the proposed roles of the proteins encoded by the W.T. genes are also presented. The eight genes are distributed in chromosomes I, III, IV and V (marked in parentheses in the table). While the full description of the phenotypes of these mutants can be obtained from the review of Ryan *et al.* (2001) and the original studies that are cited in this review, here are some specific remarks.

Table 6. Phenotypes of Eight Isolated Root Hair Mutants and the Proposed Role of the Wild-Type Gene

Gene	Mutant Phenotype (chrom. location)	Proposed Function of the W.T. Gene
RHD1 *(ROOT HAIR DEFECTIVE1)*	Normal root hairs with excessive bulging at base (I)	Involved in cell wall modification at bulging site prior to tip growth
RHD2 *(ROOT HAIR DEFECTIVE2)*	Excessive bulging; root hairs fail to initiate (V)	Involved in cell wall modification at bulging site prior to tip growth
RHD3 *(ROOT HAIR DEFECTIVE3)*	Short wavy root hairs (III)	Localizes growth during tip growth
RHD4 *(ROOT HAIR DEFECTIVE4)*	Short root hairs; diameter varies along length of hair (III)	Localizes growth during tip growth
RHD6 *(ROOT HAIR DEFECTIVE6)*	Fewer hairs per root; basal shift in initiation site; multiple root hairs from single trichoblasts (I)	Involved in selection of hair initiation site
TIP1 (TIP DEFECTIVE1)	Excessive bulging; short sometimes branched hairs; multiple hairs from single bulge (V)	Localizes growth during bulging and tip growth by influencing MT dynamics
COW1 *(CAN OF WORMS1)*	Short hairs; diameter varies along length of hair; multiple hairs from single bulge (IV)	Localizes growth during bulging and tip growth by influencing MT dynamics
KJK (KOJAK)	Excessive bulging; hairs burst prior to initiation (III)	Involved in cell wall modification at bulging site prior to tip growth

From Ryan *et al.* (2001).

The mutant *rhd6* has fewer root hairs than the W.T.; moreover, the fewer hairs occasionally emerge from incorrect positions on the outer cell wall of the trichoblasts, often with several hairs emerging from a single trichoblast. The mutant's phenotype can be rescued by the external application of either auxin or ethylene. The mutation could be phenocopied by the application of aminoethoxyvinylglycine (AVG), which is an inhibitor of ethylene effects. This indicates that *RHD6* is involved in a process that includes auxin and ethylene.

The *kjk* mutant fails altogether to produce root hairs. The trichoblasts do produce swellings of bulges, but these tend to burst rather than grow into root hairs. This causes the death of the respective trichoblasts.

The eight genes involved in root hair differentiation can be grouped into three classes. The genes *RHD1*, *RHD2* and *KJK* are active during the bulging. The genes *RHD3* and *RHD4* seem to be required for the polarity of the root hair tip growth; the cytoskeleton is probably involved in this tip growth. The genes *RHD6*, *TIP1* and *COW1* probably have roles in both the early (bulging) and the later hair growth.

On the basis of phenotypic characterization and the analyses of double mutants affecting root hairs, Ryan *et al.* (2001) proposed the following model of gene interactions:

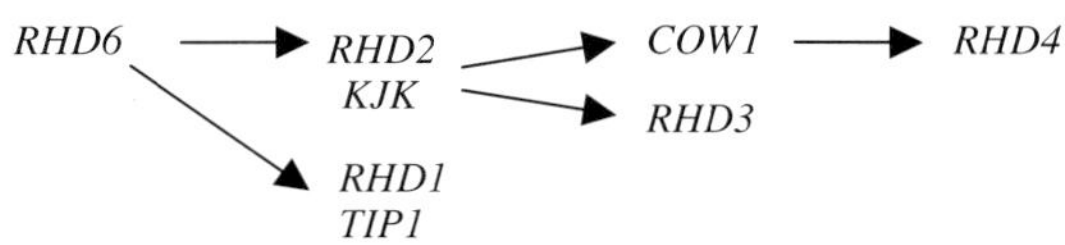

Most of the genetic and molecular genetic studies on root hairs of Arabidopsis in recent years were on the determination of epidermal cells to become either hair-forming cells or cells without hairs; much less attention was paid to the differentiation of the hairs themselves. An exception of this trend is the study of Nam-Hai Chua and associates, from the Rockefeller University (NY) and the National University of Singapore (Bao *et al.*, 2001). These authors reduced the levels of α-tubulin by producing transgenic Arabidopsis with antisense sequences to α-tubulin. It appeared that, while there were no visible abnormalities in the aerial parts of the transgenic plants, the roots were strongly affected. One of the abnormalities concerned root hairs: in plants in which α-tubulin was reduced, there was an ectopic formation of root hairs and even root hair branching. Why is the presence of α-tubulin more important in roots than in shoots? The authors suggested that due to the much faster growth of roots than of shoots, the α-tubulin may be the bottleneck for normal growth and differentiation in roots.

J. Schiefelbein and associates of the University of Michigan (Ann Arbor, MI) pioneered in identifying genes involved in the determination of root epidermal cells to become either hairy or hairless (see review of Schiefelbein, 2003). It was found that the mutants of three genes, *TRANSPARENT TESTA GLABRA* (*TTG*), *GLABRA2* (*GL2*) and *WERWOLF* (*WER*), caused the formation of hairs on all epidermal cells. The W.T. genes thus appeared to suppress the hair formation in those epidermal cells that became hairless. The role of one of these genes as well as a gene (*CAPRICE*) that appeared to promote hair formation was studied further by a team of eight Japanese investigators and one US investigator (Wada *et al.*, 2002). These investigators found that the *CAPRICE* (*CPC*) gene is expressed in cells that will become hairless, but the CPC protein then moves from the hairless to the hair-forming epidermal cells. Once in the hair-forming cells, the CPC protein causes the repression of *GL2* and, by that, leads to the derepression of hairs by *GL2*. Wada *et al.* (2002) made also a postulation that the transcript (or the protein) of *TTG* moves from cortex cells to epidermal cells that are in contact with the former cells (they probably assumed that there is less transfer when the epidermal cell is located between two cortex cells). However, no experimental evidence was furnished to support this possibility.

The interaction of genes affecting root hair formation was intensively studied by the Schiefelbein team (e.g. Lee and Schiefelbein, 2002 and see also review of Schiefelbein, 2003). These investigators found that the WER protein causes the induction of expression of CPC and GL2 in cells located in specific locations (relative to cortex cells). Also, the CPC protein is involved in lateral inhibition. They suggested a model that claimed that in nonhair cells there are high levels of expression of the three genes, *WER*, *CPC*, and *GL2*, which together repress hair formation. Lower expression of these genes, in potential hair-producing cells, permits the differentiation of hairs.

In his review, Schiefelbein (2003) handled the formation of epidermal root hairs and epidermal trichomes in the leaves. There is a difference in the positioning of these hairs and trichomes. In the roots, the hairs are normally formed only in epidermal cells that are situated over the intercellular spaces, between two cortex cells. Those epidermal cells that face the cortex cell wall (but not the interspace between these cells) are hairless. In the leaf epidermis, there is no such control by the cells of another layer; rather, the location of future trichomes is probably controlled by the space between the trichome-producing epidermal cells. In spite of this difference, Schiefelbein finds similarities in patterning control between root hairs and trichomes. Here, I shall focus on the root hairs because trichome patterning shall be handled in a later chapter (Chapter 11). First, here are some characteristics of the main "players" in the differentiation (or suppression) of root hairs in Arabidopsis.

- The *GLABRA2* (*GL2*) gene encodes a homeodomain protein of the HD-ZIP type and is required for the specification of the nonhair epidermal cells. Accordingly, *GL2* is preferentially expressed in nonhair cells.
- The *WEREWOLF* (*WER*) gene encodes a transcription regulator protein of the R2R3 MYB class. *WER* is also required for the specification of nonhair epidermal cells. *WER* is actually required for the expression of the hair suppressor *GL2*.
- The *TRANSPARENT TESTA GLABRA* (*TTG*) gene encodes a protein that possesses WD40 repeats (which enable interaction with other proteins). The *TTG* is also required for the normal expression of *GL2* (there is a related gene *bHLH* in maize). Hence, *GL2*, *WER* and *TTG* are genes that interact to cause the specification of nonhair epidermal root cells.
- The *CAPRICE* (*CPC*) gene encodes a small protein that has a single MYB-like DNA-binding domain, but no transcriptional activation domain. There are indications that *CPC* is instrumental in specifying hair-forming epidermal cells because its encoded protein suppresses the expression of *GL2*. When *CPC* is overexpressed

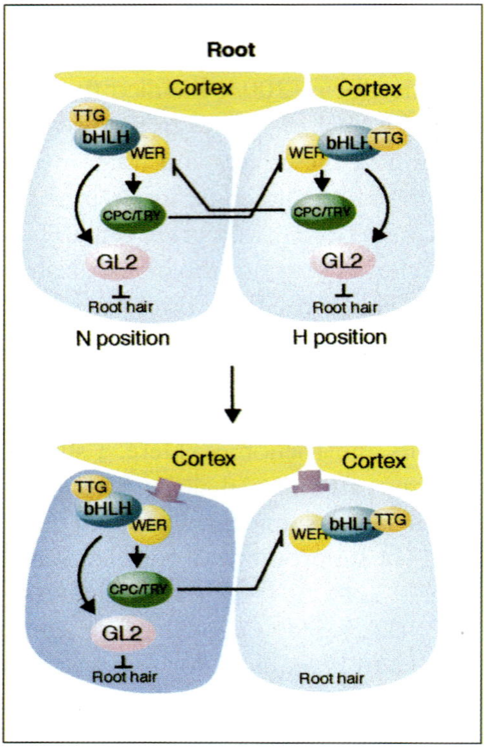

Fig. 36. Model for cell-type patterning in the root and shoot epidermis of Arabidopsis. The upper tier of illustrations shows cells with equivalent developmental potentials, whereas the lower tier shows cells that have been specified to adopt a particular fate. In the root, positional cues (indicated by the block symbols) bias the outcome of the patterning mechanism. A root epidermal cell in the "H position" (between two underlying cortical cells) invariably adopts the hair cell fate, whereas a cell in the "N position" (outside a single cortical cell) adopts the non-hair cell fate. (From Schiefelbein, 2003.)

by a transgene in which a strong promoter causes the transcription of *CPC*, there is an ectopic induction of root hairs.

Schiefelbein (2003) suggested a model (see Fig. 36) for cell type patterning in the root epidermis that intended to explain the various interactions between the proteins encoded in the genes participating in this patterning. One important assumption in this model is that *CPC* appears to be expressed nonautonomously, meaning that the gene product moves from cell to cell. This causes lateral inhibition between genes with a feedback (loop) system (e.g. *CPC* can inhibit *WER* and *WER* can induce *CPC*, as shown in Fig. 36). Clearly, positional clues must play a role because the final outcome of the cell-to-cell interactions is decided by the position of the epidermal cell relative to the underlying cortex cells. It is plausible that the positional clues start to play their role very early on, probably already in the embryo stage.

The situation became apparently more complicated by the results of a further study of the Schiefelbein team (Bernhardt *et al.*, 2005), where it was revealed that the genes *GL3* and *ENHANCER OF GLABRA3* (*EGL3*), which encode putative suppressors of hair-producing cells, are preferentially expressed in the developing hair cells. However, it was then found that the GL3 protein indeed moves from the hair cells to nonhair cells. What exactly causes this unidirectional movement has not yet been determined, but this team found that another gene, *SCRAMBLED* (*SCM*) is required in the epidermal root cells to correctly interpret their location. The *SCM* encodes a receptor-like kinase protein with a predicted extracellular domain of leucine-rich repeats and an intracellular serine-threonine kinase domain. It seems that *SCM* regulates the expression of *GL2*, *CPC*, *WER*, *EGL3* and *GL3* all genes that affect root hair patterning in Arabidopsis.

In a more recent publication (Ryu *et al.*, 2005), Schiefelbein and associates increased the resolution of the picture that was previously drawn for the patterning of hairs in the root epidermis of Arabidopsis. It was found that the WER protein, which serves as a kind of "master regulator" in this patterning, accumulates in the nuclei of the nonhair producing cells, supporting the role of the WER protein as transcription factor. Furthermore, by using the GR-inducible promoter (by DEX), this study showed that the transcription of *CPL* is directly regulated by WER. In an independent study of six Japanese investigators (Koshino-Kimura *et al.*, 2005), experimental evidence supported the claim that the WER protein directly regulates the transcription *CPC* (and *GL2*) by binding to the promoter regions of the latter genes.

A further aspect of the patterning of root hairs in young seedlings of Arabidopsis was the subject of the study by Costa and Shaw (2006) of the John Innes Centre in Norwich (UK). These investigators looked for the role of chromatin remodeling as the link between positional clues and differentiation of epidermal root cells into either cells devoid of hairs (atrichoblasts) or cells with hairs (trichoblasts). By a three-dimensional *in situ* hybridization of living cells (FISH), the investigators could detect in which nuclei the chromatin in the region of the *GL2* gene is "open" (accessible to a probe) and in which nuclei this chromatin is "closed." Indeed, in the nuclei of the cell files that are destined to become hairless, the chromatin around the *GL2* gene is open, permitting the transcription of this gene. In the nuclei of cells that will become hairy, this chromatin region is closed, reducing the transcription of *GL2*. Furthermore, when the position clue was experimentally changed (the change in chromatin around *GL2*), resetting of the chromatin conformation followed rather quickly. To manifest this change in chromatin state required only one round of mitosis (but not a phase of DNA replication). The change in chromatin state by changing the position clue was specific to the chromatin around *GL2*. It appears that *CPC* is required to cause the closed conformation of the chromatin. Also, the wild-type gene of *FAS2* is required for the correct chromatin conformation; and in *fas2* mutants the correct hair patterning is lost. The authors suggested that the capacity of the chromatin configuration to be reset by position clues may be, in general, the basis of the plant cell's ability to respond promptly to new stimuli, change the transcription of key genes and provide plasticity in cellular development.

Lateral and Adventitious Roots

Plant roots typically undergo a process of ramification that continues throughout the life of the plant and is instrumental in the absorption of water and nutrients from the soil. I shall focus here on the formation of lateral roots from the primary root of young seedlings; and, again, the roots of Arabidopsis will be handled in detail because most structural, genetic and molecular genetic studies were performed with the roots of young Arabidopsis seedlings.

Lateral roots in Arabidopsis: structure, genes and hormones

Lateral root formation in Arabidopsis seedlings, as known up to about 10 years ago, was nicely reviewed by Malamy and Benfey (1997b). I shall use this review as well as another publication of these authors (Malamy and Benfey, 1997a) to summarize the knowledge on lateral root formation in Arabidopsis that existed in 1997.

The primary root that is starting to grow after germination of Arabidopsis plants continues with its growth throughout the life of these plants. The formation of lateral roots from the primary root is a continuous process that happens at a certain distance from the tip of the root. Furthermore, after a lateral root starts to grow, it will produce its secondary lateral roots and the latter will again form lateral roots. The process of the formation of laterals from the primary root and the formation of secondary lateral roots is similar. I shall therefore deal only with the former lateral roots.

As already stated, the advantage of Arabidopsis roots as a model for lateral root differentiation is manifested by their simple structure. Each of the cell files — epidermis, cortex endodermis and pericycle — is composed of a single layer, and there are promoters that are specifically activated in each of these layers. Thus, by fusing the appropriate promoter with the code of a marker gene (e.g. GUS) and expressing this construct in transgenic plants, one can label each layer of the respective transgenic root. Malamy and Benfey (1997a,b) used such transgenic lines in combination with meticulous cytological procedures to follow the stages of initiation of lateral roots until after the protrusion of these roots from the primary root. These stages of lateral root development are detailed in the legend of Fig. 37. The figure itself is composed with Nomarski images of the pericycle and then the differentiating lateral root, until its emergence. The first indication that a lateral root is being initiated are anticlinal divisions of the pericycle. A schematic presentation of the lateral root primordium (LRP) that leads to the protruding of this root is provided in Fig. 38.

The LRPs are produced only in the pericycle cells. Moreover, the LRPs start to be visible (by anticlinal cell division) at a root region

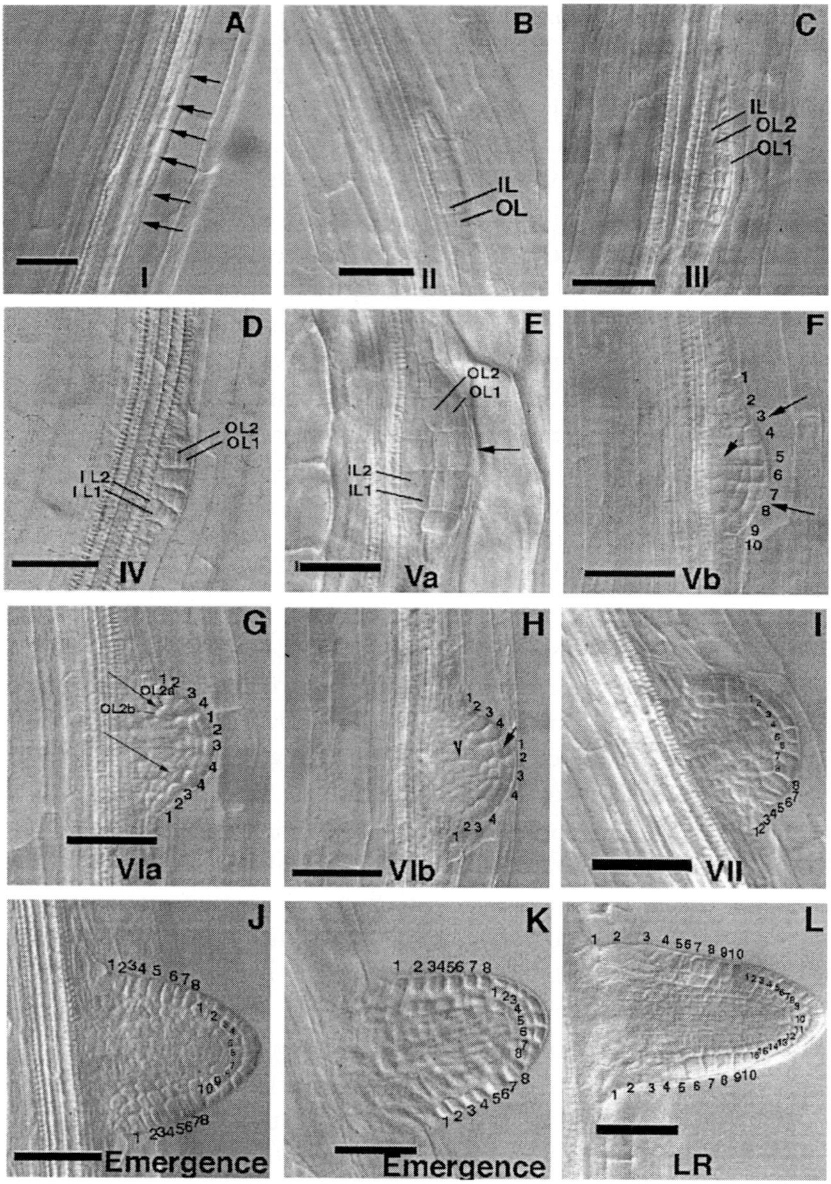

Fig. 37. Stages of lateral root primordium development. Shown are Nomarski images of cleared whole mounts of 2- to 6-week old roots. **(A)** Stage I. Arrows point to new cell walls indicating anticlinal division in the pericycle. **(B)** Stage II. A periclinal division

that is beyond (further away from the root tip of) the meristem region. This means that the pericycle first differentiates, stops further divisions (probably at the G2 phase of the cell cycle), and thereafter the cell division is resumed when LRP initiation takes place. Furthermore, not all the pericycle cells will produce such LRPs. Rather, LRPs are formed in those pericycle cells that are adjacent to the protoxylem poles. In spite of these limitations in the location of LRPs, the number of LRPs can be influenced by several factors, such as auxin, removal of the root tip, and various environmental effectors (e.g. nutrients in the soil). A certain gene that is involved in cell division (*cyc1At*) is probably essential for the division of root meristem cells, but the number of lateral roots was not specifically altered by *cyc1At* (Doerner *et al.*, 1996).

---

has divided the LRP into two layers, outer layer (OL) and inner layer (IL). **(C)** Stage III. A periclinal division in OL has created a total of three layers. **(D)** Stage IV. A periclinal division in IL creates a fourth layer. **(E)** Stage Va. Arrow indicates a single anticlinal division in the center cells of OL1 and OL2. **(F)** Stage Vb. Arrows indicate two additional anticlinal divisions in OL1 and OL2. The short arrow indicates the region in which cells of the IL2 undergo expansion and division, distorting the shape of IL1 and OL2. Cells in the outermost layer are numbered to indicate the constant organization at this stage. **(G)** Stage VIa. A periclinal division in all but the center cells of OL2 creates a new tier of cells (*arrows*). The two tiers are designated OL2a and OL2b. **(H)** Stage VIb. Four central cells in OL1 divided periclinally to create another new tier at the tip of the primordium (*arrow*). Numbering shows that there are now four cells in OL1 on either side of the four divided central cells. The OL1 is considered to be the inner of these two layers of divided central cells. The arrowhead points to the elongated cells reminiscent of vascular elements. The events shown in G and H may not necessarily occur sequentially; they are shown in two panels for clarity. **(I)** Stage VII. All the cells in the OL1 have undergone anticlinal divisions, as evidenced by the cell shape and the increased number of cells in this layer. This gives the characteristic 8-8-8 cell pattern, as indicated by cell numbering. In the central region of the tip of the primordium, numbers are on top of the relevant cells. **(J, K)** emerging LRP. Note that the cells in the OL1 are enlarged. The number of cells near the apex has increased slightly, potentially indicating the first divisions of meristematic initials. **(L)** Fully emerged LR. Note the gross increase in size that has occurred in the basal cells of the OL1. Cell numbering indicates that there are now more cells in this layer near the root tip, consistent with the presence of an active lateral root apical meristem. Bar = 50 μm. (From Malamy and Benfey, 1997a.)

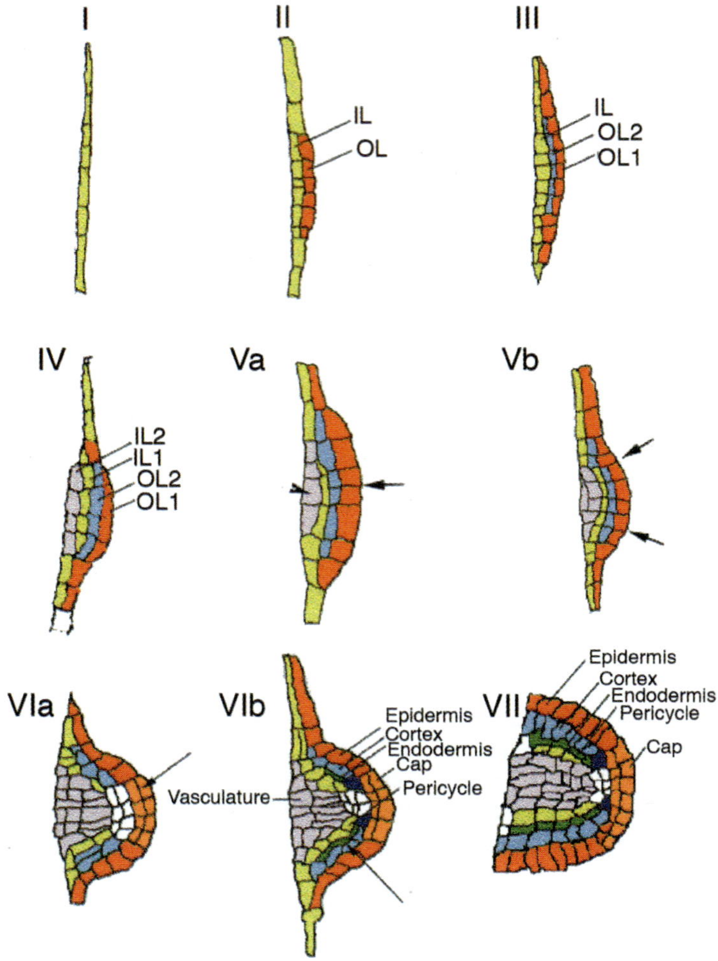

Fig. 38. Model of LRP development in Arabidopsis. Stages are indicated above each diagram. Images were created by tracing cell outlines from Fig. 37. Color coding shows the putative derivation of each tissue from Stage I through Stage VII, based on information from the histological studies and the marker lines. Colors correspond to the same cell-type as in Fig. 37. Note that by Stage VIb, all the radial pattern elements of the primary root are present in the LRP. The cluster of white cells near the LRP tip at Stages VI and VII cannot be clearly identified, but the position and lack of staining of these cells in differentiated cell-specific market lines suggest that they develop into initials and quiescent center. White cells at the base of the LRP could not be identified. (From Malamy and Benfey, 1997a.)

External application of auxin will increase the number of lateral roots: all the pericycle cells will render almost all these cells that are adjacent to protoxylem poles to LRPs. A similar increase in LRPs is triggered by two genes (*sur* and *rty*) of Arabidopsis that increase the level of IAA in the mutant plants. A further indication that auxin plays a role in the initiation of lateral roots comes from the evidence that auxin-resistant mutants (*axr1*, *axr4* and *aux1*) have a reduced number of lateral roots. The number of lateral roots can also be reduced by the application of chemicals that inhibit the polar transport of auxin. The reduction of lateral roots can be caused by mutants that are defective in auxin transport. However, why the LRPs are formed only from pericycle cells is not yet known (see Casimiro *et al.*, 2003; Marchant *et al.*, 2002).

Only a small number of Arabidopsis mutants were found that specifically lack lateral roots. One of these is *alf4* (*aberrant lateral root formation 4*). There is also a maize mutant, *rtcs* (*rootless for crown and seminal roots*) that produces a primary root, but no lateral roots.

Although auxin plays a major role in the initiation of the LRP, it appears that when the lateral root primordium reaches a certain stage, it becomes autonomous with respect to auxin.

A team of investigators from Gent, Belgium, that included M. Van Montagu and D. Inze (Himanen *et al.*, 2004) used novel procedures, such as transcript profiling, to trace changes in the expression of genes that precede the histological events (cell division in the pericycle cells) leading to LRPs. These events could be synchronized by first exposing the root to N-1-naphthylphthalamic acid (NPA), and then removing the cells from this inhibitor and providing 1-naphthalene acetic acid (NAA).

The team of N-H. Chua reported several years ago (Xie *et al.*, 2000) that auxin provides a signal for lateral root initiation. The evidence for that came from the *NAC1* gene. The expression of antisense *NAC1* cDNA blocks the TIR_1-induced lateral root formation, while ectopic expression of *NAC1* restores this formation. The TIR_1 is a gene required for auxin response. Since this publication, the involvement of auxin in lateral root initiation has been repeatedly reported (e.g. Casimiro *et al.*, 2001). A publication in this line came from China (Dai *et al.*, 2006). This study revealed a mutation in

Arabidopsis (*bud1*) that is involved in auxin action. In the presence of this mutation, far fewer lateral roots are formed.

Auxin seems not to be the only plant hormone involved in the formation of lateral roots. A team composed of investigators from the USA and Japan (Bao *et al.*, 2004b) found that brassinosteroids (BRs) interact with auxin in the promotion of lateral root development in Arabidopsis. These investigators revealed that BRs promote the acropetal (meaning from the base to the tip) auxin transport in the roots. This transcript can be inhibited by NPA (as indicated above). Hence, the auxin flow promoted by BR, but inhibited by NPA, is required for normal lateral root formation.

We should not be surprised to find that cytokins (CKs) also play a role in root architecture, possibly indirectly, by affecting auxin flow. Since the synthesis and flow of each plant hormone is affected by several genes, the total number of genes involved in the regulation of lateral root formation should be rather large. Indeed, in their review, Casimiro *et al.* (2003) listed about 30 genes in Arabidopsis that appeared to regulate lateral root development (see Table 1 in Casimiro *et al.*, 2003).

The impact of nutrients and environmental conditions on the formation of lateral roots and root hairs

Information on the impact of soil nutrients on root architecture, and more specifically on lateral root formation, exists mainly for nitrate, phosphate and sulfate. This information was reviewed by Lopez-Bucio *et al.* (2003). The absorption of nitrogen is problematic because of its low availability in the soil, and because nitrate can easily be lost from the soil in several ways. The problem with phosphate is that it is extremely insoluble in most soil types. Also, other elements may become limiting for plant growth due to soil compositions. It was found that the increase in nitrate availability causes reduction in lateral root *elongation*; whereas when plants are grown in low nitrogen but a section of the primary root is exposed to high nitrate, the number of lateral roots is increased. A gene, *NITRATE-REGULATED1* (*ANR1*), in Arabidopsis was found to be involved in the change of root architecture by different levels of nitrate: *ANR1* has a local stimulatory effect on the number of laterals. Auxin seems to be involved in the response to nitrate levels: *axr4* mutants do not respond to locally applied nitrate. Also, abscisic acid (ABA) plays a role in the response to nitrogen supply; ABA mediates the effect of reduced lateral root elongation by high NO_3^- levels in the soil.

Phosphate availability has a special effect on some plants (e.g. white lupin). In low P, the root system produces a special kind of root architecture: proteoid roots. These are clusters of very short lateral roots. It seems that even Arabidopsis roots react similarly to low availability of P. When the level of P was reduced to 50 μM in the medium, the root architecture changed drastically; and when P was reduced to 1 μM, the number of lateral roots increased fivefold over the number of laterals in optimal P concentration (1 mM). A French team from Montpellier (Al-Ghazi *et al.*, 2003) studied the effect of phosphate starvation on root architecture in Arabidopsis. These investigators reported that the impact of this starvation was different on the primary root than on the lateral roots. The starvation *decreased* the elongation of the primary root, while it *increased* the elongation of lateral roots. In addition, the phosphate starvation caused a decrease in the number of lateral roots that showed an arrest after initial growth. Also, there were indications that auxin plays a role in the manifestation of the responses to phosphate starvation.

An interesting interaction was reported between genes that code for phospholipase Ds (*PLDs*) *PLDζ1* and *PLDζ2*. *PLDζ1* was previously found to affect the initiation and maintenance of root hairs, and was suggested to be negatively regulated (in Arabidopsis) by the GL2 protein (Ohashi *et al.*, 2003). However, in a later investigation, Li *et al.* (2006) found that the knockout of both *PLDζ1* and *PLDζ2* genes did not affect root hairs. Further, the expression of both of these genes increased in response to phosphate starvation. The elongation of the primary root in the double-mutant knockout seedlings was slower than in wild-type Arabidopsis roots. The loss of *PLDζ2* expression, but not of *PLDζ1* expression, caused a decreased accumulation of phosphatidic acid in roots under phosphate-limited conditions. The full picture is awaiting further clarification, but it is already clear that the genes *PLDζ1* and *PLDζ2* have roles in regulating root development in response to P limitation.

A reduction of sulfate (SO_4^{2-}) supply also affects branching of Arabidopsis roots. Under SO_4^{2-} starvation, the initiation of lateral roots appears closer to the tip of the primary root than when the plants are furnished with normal levels of SO_4^{2-}. An interesting phenomenon regarding sulphur starvation was revealed by Kutz *et al.* (2002). Sulphur starvation led to the activation of a nitrilase gene (*NIT3*). This nitrilase converts indoleacetonitrile to IAA. This increase of nitrilase was focused at the primordia of lateral roots that were thus promoted.

A schematic presentation of the influence of levels of P, N and S on root architecture in Arabidopsis is given in Fig. 39.

Auxin seems to be involved in the responses of root development to nutrient levels (Lopez-Bucio *et al.*, 2003). The auxin transport inhibitors 2,3,5-triiodobenzoic acid (TIBA) and NPA inhibit the formation of the proteoid or lateral roots in response to low P levels in the medium. Also, applied CK suppressed lateral root initiation in low

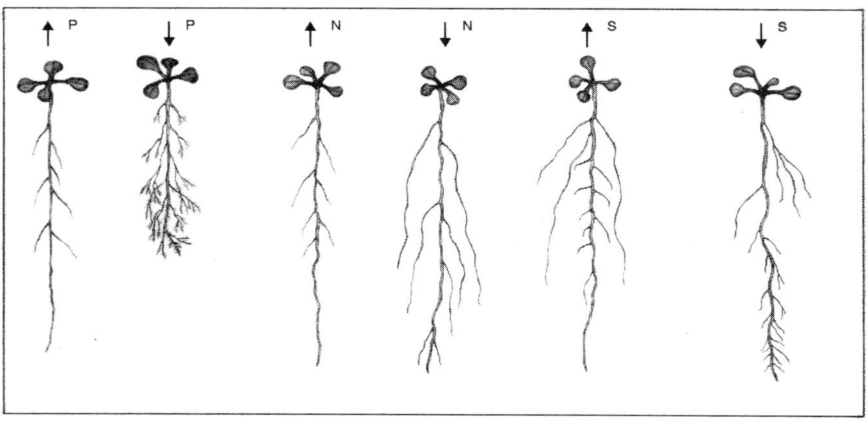

Fig. 39. Responses of Arabidopsis root systems to different nutrient supplies. The root systems were grown in nutrient-rich media with or without a high concentration of P, N and S. P and S deficiency can dramatically alter the primary and lateral root growth, modifying the overall root architecture. Contrasting N availability may alter lateral root elongation. (From López-Bucio *et al.*, 2003.)

P levels. In addition, exogeneous CK repressed the expression of low P-regulated genes, such as *Arabidopsis thaliana INDUCED BY PHOSPHATE STARVATION* (*AtIPS1*). There is an indication that the effect of nitrate on root architecture is not by its nutritional effect on the metabolism in the plant, but rather as a signal: in plants that are deficient in nitrate reductase and thus unable to metabolize nitrate, the impact of nitrate on root architecture still occurs. A similar phenomenon was reported by Little *et al.* (2005). In the latter study by J.E. Malamy and associates, the initiation of lateral roots in Arabidopsis was repressed by high external sucrose and low external nitrate; but in a *lin1* mutant, which is defective in a nitrate transporter, no such repression was revealed. The experimental results indicated that the repression of lateral root initiation is independent of actual nitrate uptake.

It was also revealed that the impact of low P is localized to the root region that is exposed to the low level of P. For example, local low P levels will cause local elongation of lateral roots.

A nitrogen compound that is located inside the plant, nitric oxide (NO), was investigated in tomato by a team from the University of Mar del Plata in Argentina (Correa-Aragunde *et al.*, 2004, 2006). These investigators found that application of sodium nitroprusside (SNP), which releases NO in the plant, will induce in tomato seedlings the emergence of lateral roots and increase the elongation of these laterals in a dose-dependent manner, while the elongation of the primary root is reduced. When the conversion of SNP to NO is blocked, the effect is eliminated. It seems that auxin signaling is involved in the NO effect. Further, it seems that NO exerts its effect directly on lateral root initials (but not directly on the emergence of the lateral roots from the primary root). The NO promotion of lateral roots was involved with cell division. For example, NO mediated the induction of the gene *CYCD3;1*, which is required for the completion of the cell division cycle. Also, the auxin-dependent cell cycle regulation was dependent on the presence of NO.

The effect of osmotic pressure on lateral root differentiation was studied by Deak and Malamy (2005) of the University of

Chicago. The increase in osmotic pressure was caused by the addition of mannitol, sucrose and/or salts (e.g. KCl). Arabidopsis plants were germinated on agar plates that were vertically incubated. Experimental evidence indicated that the reduction of water availability (i.e. by high osmotica) caused repression of lateral roots. The number of lateral root primordia was not affected by high osmotica, but the growth of the laterals from their primordia was strongly repressed. The osmotic-responsive regulatory mechanism required abscisic acid (ABA), and the wild-type gene *LATERAL ROOT DEVELOPMENT2* (*LRD2*). Hence, in mutants defective in ABA biosynthesis, high osmotica will not repress lateral roots. There also seems to be a requirement for ABA–auxin interaction in order to lead to the osmotic-potential-mediated lateral root repression.

Malamy (2005) reviewed the intrinsic and environmental response pathways that regulated root architecture. She emphasized the regulation of these effectors on the formation of lateral roots. Some of these effectors (as was indicated above) specifically affect one phase of lateral root formation, as the development of laterals from existing primordium. The number of lateral root primordia could also be affected by intrinsic or environmental effectors. Among the intrinsic effectors are specific genes. As noted above in Arabidopsis, only very few genes were revealed that specifically affect root architecture; while in maize, which produces several kinds of roots, mutants were found that lack only one kind of root.

Polar movement of auxin is not only a major player in lateral root formation, but there is also increasing evidence for the interaction of auxin with other hormones, such as ABA and BR, in this formation. Also, it seems that the mechanism of initiation and extension of lateral roots from the primary root of young seedlings is different from the mechanism that operates in older roots, although we do not yet fully understand either of these mechanisms. We do know one component of these mechanisms: in young Arabidopsis seedlings, a pulse of auxin that moves down from the shoot is involved in lateral root formation; while in older roots, the formation of lateral roots appears to be independent of the shoot-derived auxin.

The evidence on the impact of intrinsic and external factors on lateral root formation can now be divided according to three phases of this formation:

- Lateral root initiation;
- Lateral root primordia, emergence and meristem activation;
- Lateral root growth.

Adventitious roots in Arabidopsis

Under natural conditions, Arabidopsis plants do not produce adventitious roots. However, such roots are essential when investigators wish to propagate cuttings of these plants. Consequently, a group of investigators from Versailles and Gif-sur-Yvette in France and Umeå in Sweden (Sorin *et al.*, 2005) studied the formation of adventitious root eliminated in Arabidopsis, and identified the gene *AUXIN RESPONSE FACTOR17* (*ARF17*) as a possible major regulator of adventitious roots. They also found that the gene *ARGONAUT1* (*AGO1*) (a gene involved in RNA silencing) is required for adventitious root formation. In a subsequent study, these investigators (Sorin *et al.*, 2006) searched for proteins that are correlated with the development of adventitious roots in Arabidopsis. Using mutants and proteomic analyses, the investigators identified 11 proteins whose abundance was either positively or negatively correlated with endogenous auxin levels, the number of adventitious root primordia and/or the number of mature adventitious roots. One of these proteins was negatively correlated only with the number of adventitious root primordia, while two proteins were negatively correlated with the number of mature adventitious roots. Other proteins (among them putative chaperones) were positively correlated with primordia and mature adventitious roots. This study is the first step for the breeding of Arabidopsis lines with improved formation of adventitious roots.

Vascular differentiation in young Arabidopsis roots

The meristem of young Arabidopsis roots has a group of stem cells that are located above the QC of this meristem (meaning away from the root tip). These stem cells contribute cell files that differentiate into the cortex, the endodermis and the pericycle (single layers) as well as to the stele. The steles will then differentiate into the vascular cylinder. While the molecular genetic processes that operate in the shoot meristem, as well as those processes that lead to the differentiation of root hairs from the epidermis and lateral roots from the pericycle, were amply studied (as reported above), the processes that operate in the differentiation of vascular cells of the young root were relatively neglected. Mähönen *et al.* (2000), a team of five investigators from Finland and one investigator from the US, found that in Arabidopsis root meristems, xylem cell lineage is specified very early, whereas phloem and procambium are established later. One gene, *WOODENLEG* (*WOL*), has a putative role as a sensor for vascular morphogenesis.

This study was continued by 10 investigators (Mähönen *et al.*, 2006a, b) from Finland, the US, Japan and Sweden. They looked at the differentiation of protoxylem in the cell lineages that will form the transporting tissues (xylem and phloem). In young wild-type roots of Arabidopsis, the central axis of the protoxylem and metaxylem files in the center of the root stele are flanked by phloem files and intervening procambium cell files. The latter will constitute the cambium through periclinal divisions. Previous studies (e.g. Nishimura *et al.*, 2004) clearly indicated that for these cell divisions that will produce the cambium in the young root, functional receptors of cytokinin (CK) are required. When all three genes that encode histidine receptors for CK were mutated, the normal patterning of Arabidopsis seedling was disturbed. The periclinal division in the procambial files was markedly reduced in the triple mutants, and the number of vascular files was consequently strongly reduced. This phenotype could be copied by the depletion of CK. In other words, it was indicated that CK signaling through the CRE1 family of receptors is required for the normal proliferation of the procambium.

Mähönen *et al.* (2006a, b) asked what would happen in the stele when CK will be specifically eliminated. To target the depletion of CK, the investigators used the promoter of CRE1, but the activation of this promoter was under the control of estrogen induction; this promoter was fused with the coding sequence of a cytokinin oxidase. Hence, transformed Arabidopsis roots with this transgene could be treated with estrogen to reduce CK in a temporal- and spatial-specific manner. When this CK elimination procedure was performed at a proper time, all newly formed cell files were protoxylem. This indicated that CK signaling is required to promote (and maintain) all other cell identities in the center of the young root that are not protoxylem. Protoxylem differentiation thus seems to be the default identity, and CKs negatively regulate protoxylem identity.

The important role of CKs in the determination of cell identities in the stele of the young Arabidopsis roots was verified by several elegant experiments in which previously known mutants (such as *atrp6* that is defective in the CK-prompted transfer of phosphoryl groups and a mutant, *wol*, of the *WOODEN LEG* gene) were utilized. The picture that emerged from these experiments was the following: There is a regulatory circuit between cytokinin signaling and its newly identified inhibitor (*AHP6*) that specifies *meristematic* versus differentiated nature of procambial cell files. While this study brought the understanding of the differentiation of the vascular cells in the root one step further, major enigmas are retained. One of these is a common question in patterning: why is the AHP6 protein restricted to specific files in the developing root? The answers to this and similar questions may be provided following a better understanding of the positional information sensing by plant cells.

Trafficking of macromolecules between root cells

The knowledge that symplastic, intercellular movements of proteins and RNAs (as well as complexes of both, e.g. plant viruses) can take place in plants from one cell to its neighboring cell, in spite of the cell's walls, has existed for several decades (Tucker, 1982; Zambryski and Crawford, 2000). There are two main means for short-range communication in plant cells. One of these is the apoplastic pathway that is manifested by ligand receptors which pass an outside signal through the cell wall to the inside of the cell. Such ligand receptors were mentioned in previous chapters of this book because this is a common communication pathway of plant hormones. The other avenue of communication is the symplastic one, in which macromolecules are transferred through plasmodesmata.

During the first years after discovering plasmodesmata's (PD's) role in the transfer of macromolecules between plant cells, it was surmised that only proteins smaller than 1 kD could pass PD. Later, it became clear that PDs are dynamic and that their size exclusion limit (SEL) may reach 54 kD in some tissues (see review by Wu, Weigel and Wigge, 2002).

One of the early reports on the movement of a protein (a transcription factor) from cell to cell in the meristematic region of Arabidopsis roots was provided by the laboratory of P.N. Benfey of the New York University (Nakajima *et al.*, 2001). The product of the *SHORT ROOT* (*SHR*) gene was known to be required for the specification of the endodermis cell files. In the *shr* mutant, there are cortex cells, but no endodermis cells. These investigators found that the mRNA encoded by SHR is transcribed in certain cells of the stele of W.T. roots, but the translated protein (SHR) moved to a single layer of adjacent cells, where it enters the nucleus. Hence, the SHR protein clearly moves and serves as a signal for endodermis specification as well as for *SCARECROW* (*SCR*)-mediated cell division. In a further study by this team (Sena *et al.*, 2004), the movement of SHR was further verified and found to be rather specific from the stele to the endodermis (i.e. SHR will not move from the phloem to companion cells). There was also evidence that SCR may limit the movement of SHR. Interestingly, in neither of the mentioned publications (Nakajima *et al.*, 2001; Sena *et al.*, 2004) did the authors discuss PD as the gateway for the movement of SHR from cell to cell.

In a later publication, the Benfey team (Gallaghter *et al.*, 2004) provided additional information on the trafficking of the SHR protein via PD. The SHR protein is initially present in the stele cells in two locations, in the nucleus and in the cytoplasm; while in the endodermis of the young Arabidopsis root, the SHR is located only in the nuclei. Only the cytoplasmic-located SHR is able to move. Hence, SHR that reaches the endodermis does not move further to the cortex. Moreover, the W.T. SHR has a component that allows its passage through the PD. Therefore, the mere existence of SHR in the cytoplasm of cells is not sufficient for its movement across PD. There should be a controlling element for this movement.

A detailed discussion on the role of PD in the non-cell-autonomous pathways of proteins in plants was provided in a collaborative publication of researchers from Mexico and Vienna (Ruiz-Medrano *et al.*, 2004). Proteins that moved from cell to cell via PD were termed "non–cell-autonomous proteins" or NCAP. These authors suggested a detailed model for the NCAP movement across PD (see Fig. 40) and reviewed the proteins that were found to be required for the movement of NCAP via PD. These authors noted that the accumulated information indicated that PDs form a cytoplasmic network throughout the plant body and provide the symplasmic cell-to-cell transport. Proteins such as homeodomain transcription factors and small RNA molecules (as well as viral genome information) can move selectively to adjacent cells via PD. Obviously, PDs are also the gateway for the movement of small metabolites that we shall not consider in this section. The movement of transcription factors via PD is obviously not restricted to the apical meristem in roots. As mentioned above in the chapter on the shoot meristems, and that shall be mentioned again in the chapter on leaf patterning, it is probably a general strategy of plants to use PD for the movement of regulatory proteins. The movement of SHR mentioned above is a good example for the movement across the PD of a regulatory protein involved in root patterning. However, it is not the only example. A putatively similar movement is involved in the differentiation of hair-forming epidermal cells in the roots where the proteins CPL, GL3 and EGL3 are probably transferred through PD (Wada *et al.*, 2002 and see the section on root hairs above).

Up until now, the controlled trafficking across PD was experimentally analyzed for only two transcription factors in roots. We should not be surprised when the regulated crossing of PD of additional transcription factors, which control root patterning, will be revealed in future studies, because such crossings are putative components of positional sensing. Such sensings are required in the differentiation of specific cell types that are radiated from the stele cells (e.g. xylem, phloem and companion cells).

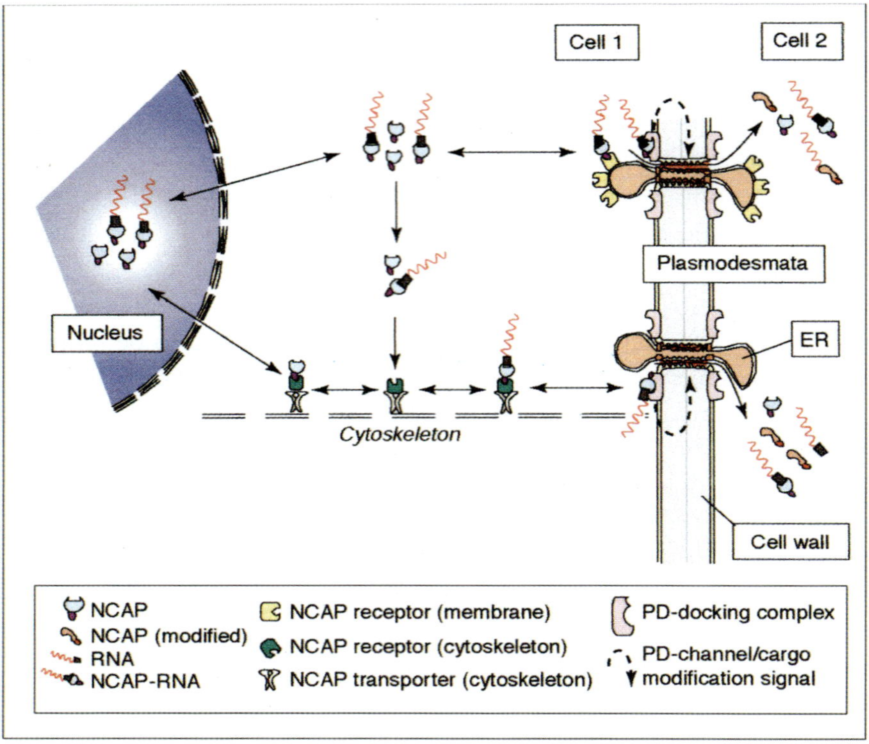

Fig. 40. A model outlining the two constitutive steps that are involved in non-cell-autonomous transport via plasmodesmata. A decision is made in the cytosol between nuclear import and the PD pathway. NCAP (non-cell-autonomous proteins) or NCAP-nuclei-acid cargo complexes assemble in the cytosol and expose targeting motifs, which are recognized by either the nuclear import system or the PD intercellular transport system. Depending on the prevalence of the NCAP transport signal motif, the NCAP or NCAP-nuclei-acid cargo is delivered into the nucleus or to the neighboring cell. An NCAP enters the PD-pathway either via an ER-associated initial PD pathway receptor (top middle) or via a cytoskeleton-associated initial PD pathway receptor (lower middle). Once at the proximity of the PD orifice, the complex is handed over to a PD-docking complex. This complex recognizes the presence of a cargo and mediates an increase in the PD size exclusion limit and the entry of the cargo into the PD micro channel. During the translocation process, the cargo is potentially modified, thereby altering its downstream transport capacity. (From Ruiz-Medrano *et al.*, 2004.)

An Update on Auxins and Root Patterning

The areal parts of plants (including Arabidopsis), especially the very young developing leaves, are considered to function as an important source of IAA for the rest of the plant. This young leaf-derived IAA is transported towards the root and moves further in the root towards the root tip (see Chapter 2 of this book). However, is it the only source of IAA in the roots? Several studies of Karin Ljung of the Swedish University of Agricultural Sciences in Umeå and associates found local sources of IAA in young roots of Arabidopsis. A comprehensive study of such local root sources of IAA was published recently by Ljung *et al.* (2005). This study was made possible due to innovations in the analytical methods for detecting minute amounts of IAA in small volumes of tissue. The analyses were performed on 0.2-to-0.5 mm slices of young primary roots. To detect IAA biosynthesis of submilligrams of plant tissue, the investigators used the incorporation of deuterium from deutrium oxide (^2H_2O) and a very selective and sensitive analytical technique with monitoring mass spectrometry.

The results of these analyses indicated that the entire young root has IAA synthesis capacity, but the IAA synthesis is especially high in the primary root tip and in the tips of developing lateral roots. Lower, but still significant IAA synthesis was also revealed in the root tissue away (upwards) from the tip. Hence, the root appears to have multiple auxin sources. The results of this study were schematically shown for roots of 4 days and 8 days after germination (see Fig. 41). The local IAA synthesis in Arabidopsis roots appears to be executed by the tryptophan-dependent pathway, because in double mutants, *cyp79B2* and *cyp79B3*, the synthesis of IAA in the roots was diminished. The enzymes encoded by *CYP79B2* and *CYP79B3* are required for the pathway that leads from tryptophan to indole-2-acetaldoxime. The latter compound is converted via indole-3-acetoitrile to IAA (see Fig. 10).

In a previous section of this chapter that deals with the formation of lateral roots, I mentioned the study of the team of N-H. Chua (Rockefeller University and the University of Singapore), in which the transcription regulator NAC1, the protein required for auxin

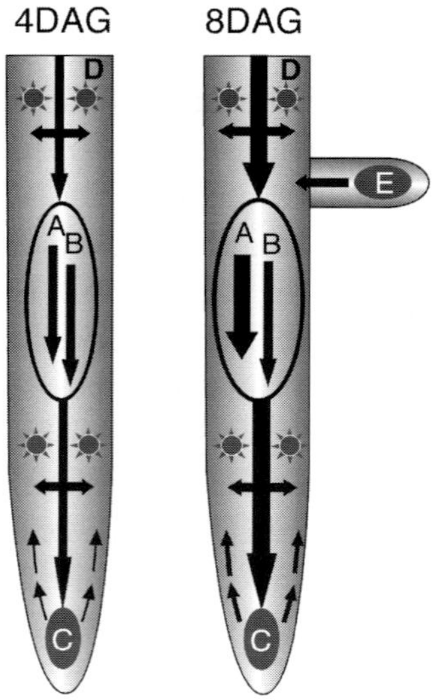

Fig. 41. Model showing how different sources supply the root with auxin during early seedling development. Sources and transport of *de novo* synthesized IAA in the root system of Arabidopsis. At 4 DAG, auxin produced in the aerial part of the seedling is transported to the root via two routes of equal importance: phloem-mediated transport **(A)** and polar transport **(B)**. At 8 DAG, the phloem-mediated transport is relatively more important than the polar transport. Already early in root development, two internal sources are established, one specifically localized to the outermost 0.5 mm of the root **(C)**, and one representing all tissues upward from the root tip with low but important synthesis capacity **(D)**. In the later stages of development, all the emerging lateral roots gain synthesis capacity **(E)** and will be important contributors to the auxin pool in the root system. (From Ljung *et al.*, 2005.)

response (TIR1), and auxin were found to interact in the formation of lateral roots (Xie *et al.*, 2000 and also Xie *et al.*, 2002). In a more recent publication of this team (now with the addition of two investigators from the Chinese Academy of Science in Beijing), this study went one step further (Guo *et al.*, 2005) and added the role of microRNAs (mir164a and mir164b) in the search for the process of lateral root development. The roles of microRNAs in plant patterning is an emerging subject. It was introduced in my book (Galun, 2005a) and in reviews by myself (Galun, 2005b) and by others (Zhang *et al.*, 2006). At the time of this writing, 872 microRNAs were identified in plants, but many more are expected to be identified in the future. These are about 20–24 nt small RNAs that are not coding for proteins, but have extensive sequence complementarity to their target mRNAs. Consequently, the targets are either cleaved or their translation is repressed. Many of the target mRNAs of microRNAs are transcription factors; hence, microRNAs are expected to be involved in several aspects of plant patterning. Mutants of genes coding for transcription factors can be rescued from degradation when the mutation changes the sequence that is recognized by the respective microRNAs. Guo *et al.* (2005) found that this happens with NAC1, the transcription factor involved in lateral root formation (by transducing auxin signals for lateral root emergence).

Cleavage of NAC1 was blocked in NAC1 mutants with a change in the sequence that is normally the target of microRNA164. These authors also reported that auxin itself can induce microRNA164 (mir164), and that a temporary increase of mir164 by a transgene which can be activated at will led to a decrease of NAC1 mRNA level. It is clear that mir164 negatively regulates lateral root development and that this regulation is associated with a reduced level of NAC1. However, the entire picture of this, as well as of other patterning in plants in which auxin is involved, may be further clarified by future studies.

An additional report on the interaction of a specific microRNA and auxin in root patterning came from Shanghai, China (Wang *et al.*, 2005c). These investigators focused on the root cap. The root cap mediates the direction of growth of the root tip and also protects the internal root cells. Root cap cells are continuously projected from the distal stem cells of the root, which are located below the QC. Two auxin response factors of Arabidopsis were identified: *ARF10* and *ARF16*. The (controlled) expression of these ARFs is essential for the development of normal root tips. Both of these ARFs are targets for microRNA160 (mir160). When mir160 is overexpresed (in transgenic Arabidopsis plants), the expressions of *ARF10* and *ARF16* are repressed and the root tip has an unorganized phenotype. This phenotype is similar to that of a *arf10/arf16* double mutant, showing an uncontrolled cell division and blocked cell differentiation in the distal region of the root with a tumor-like root apex, and the root loses gravity response. In wild-type roots, *ARF10* and *ARF16* play a role in restricting the stem cell niche and promote columella cell differentiation. The expression of *ARF10* and *ARF16* is regulated independently by mir160 and auxin; e.g. the expression of *ARF16* was found to be auxin-inducible. It appears that mir160 helps to keep the expression of ARF genes at a required level. It also appears that mir160 has additional targets outside of the root tip differentiation.

2005 was a good year for studies on the involvement of auxin in root patterning. Another pioneering team of root research, composed of investigators from Utrecht in the Netherlands and from Freiburg and Tübingen in Germany (Blilou *et al.*, 2005), joined the effort of elucidating the regulation of root patterning by auxin. Blilou *et al.* (2005) studied five *PIN* genes that together control auxin distribution in roots, and found that these genes regulate cell division and cell expansion in the primary root of Arabidopsis. It was also found that the joint action of two or more of these genes is involved in setting the auxin maximum at a specific location, and that these genes restrict the expression domain of another important family of genes: *PLETHORA* (*PLT*) genes. The latter genes encode transcription factors that contain an auxin-inducible domain. The *PLT* genes are considered to be major determinants of stem cell specification in the root. On the other hand, *PLT* genes were found to be required for the transcription of *PIN* genes (such as *PIN3* and *PIN7*). These results suggested an interaction network of auxin transport facilitators (i.e. by cell efflux movement, controlled by *PIN* genes) and root cell fate determinants that control patterning and growth of the root primordium. Interestingly, the combinations of several specific *PIN* mutants had different consequences in root patterning. For example, the triple mutation *pin3pin4pin7* caused only a reduction of the elongation zone of the young root, while the mutant combination of *pin1pin3* caused a different phenotype.

It should be noted that the auxin/PIN/PLT interactions are already established during embryogenesis, but they continue to operate in the primary root of Arabidopsis seedlings. Blilou *et al.* (2005)

also noted that not only auxin flow from outside the root's meristem may have a role in patterning. Local sources of auxin that are synthesized in the root tip may participate in root patterning as well.

The study of Blilou *et al.* (2005) was supplemented by a study of another team of experts on plant patterning (Vieten *et al.*, 2005), composed of scientists from Gent, Belgium, Turo, Poland, Vienna, Austria, as well as J. Friml from Tübingen, who also participated in the study of Blilou *et al.* (2005). Vieten *et al.* also investigated the interactions of PIN proteins with auxins, but put emphasis on interactions between individual PIN proteins and the possible synergisms between these proteins.

Blilou *et al.* (2005) accepted the model that proposes that increased auxin levels in different cells activate a signaling cascade which, via a network of AUX/IAA (a protein involved in auxin influx into cells) transcriptional repressors and ARF factors, leads to the expression of a specific set of genes and to activation-specific developmental programs. The study involved mainly the *PIN* genes: *PIN1*, *PIN2*, *PIN3*, *PIN4* and *PIN7*. The respective proteins have typically different locations. PIN1 is typically located in the lower (towards the tip) side of the stele and the endodermis of roots. PIN2 is commonly found in the lateral root caps and in older epidermis and cortex cells. PIN4 is found in the central root meristem with a subcellular localization pointing towards the columella initials. PIN3 and PIN7 are localized in largely overlapping patterns in the columella and the stele of the elongation region of the root. In spite of these different distributions of the PIN proteins, there is a clear redundance between these proteins (with the exception of PIN2: in mutants of *PIN2*, there is agravitropic root growth that cannot be compensated by another PIN protein). Furthermore, it was found that when one *PIN* gene is mutated, another *PIN* gene may become ectopically expressed. The replacement of one PIN protein by another is not only with respect to the region, but also in the subcellular polar location. For example, PIN2 and PIN4 were basically localized when upregulated in the endodermis or stele of the *pin1* mutant. It thus seems that functional redundance of PIN proteins involves cross-regulation of *PIN* gene expression and polar targeting in a cell-specific manner.

Interesting results with respect to *PIN* gene expression were obtained by interference with auxin flow. This interference was achieved by the auxin transport inhibitor NPA. For that, the promoters of the different *PIN* genes were fused to mark the gene GFP. Several of the *PIN* genes were found to be upregulated by the inhibition of auxin flow, clearly indicating that the flow of auxin has an impact on the expression of *PIN* genes. In other words, auxin transport can modulate *PIN* gene expression in a way similar to that observed in *pin* mutants. As for a direct effect of supplemented auxin, the situation seems to be somewhat complex. The auxin was applied as IAA, NAA or 2,4-D with similar results. Such additions of auxins increased the transcription (manifested in the increase of GUS fused to the respective *PIN* promoters) of *PIN2* and *PIN4*. However, the levels of auxins required to cause maximal impact on the increase in transcription was different for different *PIN* genes. The speed of the reaction to the supply of outside auxin was different (e.g. *PIN1*, *PIN3* and *PIN7* increased already 15 minutes after auxin treatment, which in *PIN2* and *PIN4* took 1 hour), and the rate of increase in transcription was also different. For example, *PIN7* expression could be increased 22-fold by auxin treatment, while *PIN2* was increased less than twice. While auxin provided at a certain range of levels could increase PIN proteins, the investigators also revealed that there is an auxin-dependent posttranscriptional *downregulation* of PIN protein, especially when high auxin levels were applied. However, this was not happening with all PIN proteins; there was no visible decrease in the amount of PIN4:GFP following auxin treatment.

Interestingly, auxin-induced increase of PINs does not require *de novo* synthesis of any additional factor — it can take place in the presence of cycloheximide. Actually, cycloheximide by itself caused an increase in the expressions of *PIN1*, *PIN2*, *PIN3* and *PIN4*, suggesting that these expressions are normally repressed by a yet unknown cellular factor. The auxin-dependent degradation of the AUX/IAA transcriptional repressors may well be involved in this process; possibly auxin utilizes AUX/IAA-dependent signaling to regulate the expression of *PIN* genes.

In summary, this study indicated that auxin, together with cell type-specific factors, can positively control *PIN* transcriptions, which involve the activity of AUX/IAA transcriptional repressors. The results also hint at the possibility that *PINOID*, a primary auxin-response gene, plays a role in this system: auxin may mediate changes in cellular PIN polarity via the control of *PINOID* expression. There are still unanswered questions in the detailed system of root patterning due to auxin flow, specific PIN protein and internal as well as external clues.

Phenomena such as phototropism and gravitropism are outside the scope of this book; but a study on gravitropism (Swarup *et al.*, 2005) has to do with PIN and AUX and auxin flow, and shall thus be mentioned briefly. In this study, the rapid asymmetric release of auxin from gravity-sensing columella cells at the root tip that results from reorienting Arabidopsis seedlings was analyzed. Targeted expression of the auxin influx facilitator, AUX1, demonstrated that the root gravitropism requires auxin to be transported via the lateral root cap to all elongating epidermal cells. Evidence was furnished suggesting that AUX1 causes the majority of auxin to accumulate in the epidermis. Selective disruption of the auxin responsiveness of the expanding epidermal cells, by expressing a mutant form of the AUX/IAA17 protein, abolished root gravitropism. Hence, the authors concluded that gravitropic curvature in Arabidopsis roots is primarily driven by the differential expansion of epidermal cells in response to an influx-carrier-dependent auxin gradient, as schematically shown in Fig. 42.

As described above in the section on cereal roots, the primary root of the very young seedling of cereals, such as maize and rice, is replaced by other roots for the establishment of the main root system. In rice, the crown roots initiated from an underground (or underwater) node constitute such a replacement. A team of investigators from Nagoya and Tokyo, Japan (Inukai *et al.*, 2005), investigated the genes involved in the formation of crown roots of rice. These investigators first characterized a mutant that was defective in crown root initiation. The respective gene was termed *Crown rootless1* (*Crt1*). The *crt1* mutant had additional abnormalities that appeared to be auxin-related, such as a lower number of lateral roots. They also found by sequencing that the coding region has similarities to the protein family of *ASYMMETRIC LEAVES2/LATERAL ORGAN BOUNDARIES*. Also, the expression

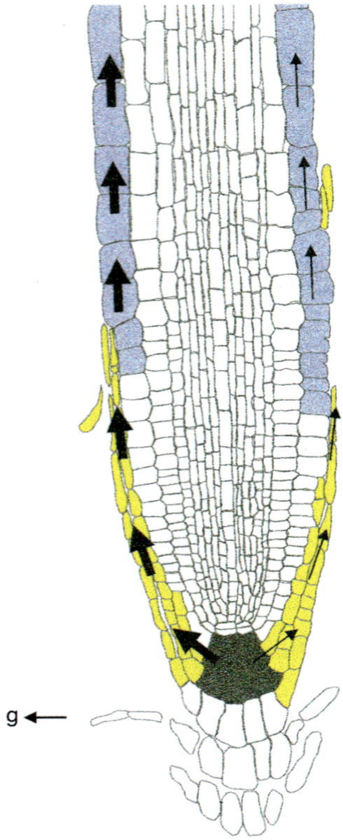

Fig. 42. A schematic model for auxin-regulated root gravitropism. A gravity stimulus (denoted by *g*) causes auxin to be asymmetrically redistributed (denoted by arrow width) at the root apex due to the gravity-induced retargeting of PIN efflux facilitators, such as PIN3, to the lower side of the gravity-sensing columella cells (colored in gray). The resulting lateral auxin gradient is mobilized through the combined action of auxin influx and efflux facilitators AUX1 and PIN2 via lateral root cap cells (colored in yellow) to expanding epidermal cells in the elongation zone (colored in lilac). The lateral auxin gradient inhibits the expansion of epidermal cells on the lower side of the root relative to the upper side, causing a differential growth response that ultimately results in root curvature downwards. (From Swarup *et al.*, 2005.)

of *Crt1* in wild-type rice was localized to sites where crown roots and lateral roots are initiated. The expression of *Crt1* was found to be increased by auxin, and such an increase required the degradation of AUX/IAA proteins. The analysis of the promoter of *Crt1* showed that it contained two putative auxin response motifs. One of these motifs could specifically interact with a rice AUXIN RESPONSE FACTOR (ARF) serving as possible *cis*-motif enhancers of *Crt1* expression. In summary, a system emerged that has similarity to root patterning in Arabidopsis: the existence of a gene (*Crt1*) that encodes a positive regulator of crown and lateral root formation, the expression of which is directly regulated by an ARF in the auxin-signaling pathway.

The Identification of Genes Involved in Root Patterning

The claim that progress in understanding biological systems is strongly dependent on technological innovations is also verified in the subject of root patterning. Clearly, a prerequisite for understanding root patterning is the identification and characterization of genes that affect this patterning. Two extremely different approaches may be followed for collecting such genes. One approach will start with a mutant that has a clear phenotype. It will proceed with the molecular characterization of the gene(s) involved in this mutation, and is followed by the search for other genes that interact with the characterized gene. The study may then go on and search for additional genes that are affected by the interacting gene, and so on, until a whole group of genes that are involved in a specific patterning (e.g. lateral root formation from the pericycle cells of a primary root) is identified. A broad picture of molecular control of patterning will thus emerge.

The other, extremely different, approach is to screen for all or almost all genes that are expressed in an organ and to identify the proteins that are encoded in these genes. Then, one can look at changes in gene expression levels occurring during patterning and, finally, design experimental interference with the patterning process by environmental, biochemical or genetic means. Examples of the former approach were provided in previous sections of this chapter, where different root patterning processes were handled. The latter approach is in the realm of systems biology.

A step in this latter approach was undertaken by Benfey and associates, a team that included mathematicians and biologists from several US universities (Birnbaum *et al.*, 2003). These investigators came up with a "Global Map of Gene Expression" in the Arabidopsis root. They thus localized the expression of over 20 000 genes in the various cell types of the Arabidopsis root. To do so, Birnbaum *et al.* (2003) went through the following steps. They chose a few genes that were known to be expressed in specific tissues of the root (e.g. *WOL* for endodermis, *SCR* for endodermis and cortex, *JO571* for epiderm, and *J3411* for epidermal atrichoblasts). Then, they used the respective promoters of these genes and fused them with the coding sequence of a clearly visible marker: the green fluorescent protein (GFP). Thereafter, transgenic Arabidopsis plants were obtained, each including a transgene that had a different promoter. The 1-cm tips of roots from 5-day-old seedlings were macerated to obtain protoplasts. Protoplasts with the fluorescence (that were isolated by a cell sorter) should also contain the specific promoters that will cause expressions only in cells of the specific tissue. Moreover, the mRNAs isolated from such protoplasts should represent mRNAs that are typically expressed in cells of the given tissue. These mRNAs were then analyzed with the use of microarrays. This led to the Gene Expression Map for the Arabidopsis root. Clearly, this basic procedure can be supplemented in several ways, such as by following changes in gene expression after hormone treatments. Once genomic sequences are identified as being transcribed in specific root cells and under various conditions and developmental stages, the way is open for reverse genetics. In the latter procedure, siRNAs and synthetic microRNAs can be used to repress, specifically, these genomic sequences. The phenotypes that will result from such specific repression will reveal additional genes (in the fully sequenced Arabidopsis genome), and thus increase the arsenal of genes involved in patterning.

An example of a study aimed at a better understanding of the involvement of auxin and genes in plant development is that of the team of J. Friml (of Tübingen), B. Scheres (of Utrecht) and their associates (Xu *et al.*, 2006a). The approach of these authors was to investigate the events that take place during a local regeneration response in Arabidopsis roots, which were obtained by a laser-induced wounding in the root tip. They found that the activity of specific genes is required to cause the changes in cell fate during this response.

These genes were *PLETHORA* (*PLT*), *SHORTROOT* (*SHR*) and *SCARECROW* (*SCR*). These genes that encode transcription factors regulate (as noted above in this chapter) the expressed and polar positions of PIN auxin efflux-facilitating membrane proteins to reconstitute the auxin flow in the healing root tip.

A recent publication by Hardtke (2006) reviewed the isolation and characterization of genes involved in root patterning, focusing on the role of auxin and endogenous microRNAs.

I expect that the number of characterized genes that are involved in the patterning of Arabidopsis roots and the roots of some other model plants will be vastly increased in the coming years. I shall therefore not tabulate the presently characterized genes that affect this patterning. However, I shall note two examples of lists of genes involved in Arabidopsis root patterning. Nawy *et al.* (2005), a team in which J. Malamy, K. Birnbaum and P.N. Benfey also participated, presented a list of about 290 genes, whose expression is enriched in the quiescent center (QC) of the primary root. This was achieved by a procedure similar of the one used by Birnbaum *et al.* (2003), in which labeled mRNA from QC cells was hybridized to microarrays. Another team (Bassani *et al.*, 2004), from the laboratory of S. Gepstein of the Israel Institute of Technology in Haifa, searched for genes involved in elongation of root cells (root growth-related genes) in maize. About 150 such genes were identified; of these, 41 were further analyzed and tabulated.

This chapter on root patterning shall be concluded with a prophecy from Ueda *et al.* (2005): "The successful marriage of smart ideas and novel technologies will clarify the molecular mechanisms of root patterning in the near future." I shall not provide my remarks (which I have) on this prophecy because it is politically correct to refrain from arguments against prophets.

Chapter 10

THE VEGETATIVE SHOOT AND TRANSITION TO FLOWERING

In this section, I update the main knowledge that has accumulated, especially during the last 10 years, with respect to structural and molecular genetic issues related to the patterning of the shoot. The definition of "shoot", for this chapter, will be the angiosperm organ that is derived from the apical meristem of the embryo. Because the available information is mainly on the shoots of young plants, this chapter will discuss developmental issues of such shoots. I shall also handle the lateral outgrowths of young shoots that include the formation of leaves and axillary meristems as well as the formation of branches from the latter meristems. This chapter shall also deal with later development, such as the transition from juvenile shoots to more mature vegetative shoots and the subsequent transition from the vegetative phase to the formation of the inflorescence. Detailed studies on the patterning of shoots were performed with only a minute fraction of the angiosperm species. These are the model plants, such as Arabidopsis, tomato, petunia and pea of the dicots, and maize and rice in the monocots. In recent years, studies on the mechanisms of differentiation of the shoots and their lateral outgrowths focused on Arabidopsis and rice, since the full genomic sequences of these plants are now freely available. Subsequent chapters deal specifically with the patterning of leaves and angiosperm flowers. Consequently, I avoid the discussion of these patternings in the present chapter.

There are considerable differences and similarities between the patterning of dicot shoots (e.g. Arabidopsis, tomato and pea) and the patterning of monocot shoots (e.g. maize and rice). These differences and similarities are noted in this chapter.

Structural Aspects of the Angiosperm Shoot

While the final form of the above-ground plant, which develops from the shoot apical meristem (SAM) of the embryo, can reach a great complexity of primary and higher-order axillary branches, the very early phase of the growth of the shoot is rather simple. It can be described as consisting of phytomers. Each phytomer is a cylinder with a node at its base. The node bears one or more leaves or leaf-like structures, and an axillary meristem may reside in the axil of the leaf. Above the node is the internode, which may attain very different lengths. These lengths are different not only in different plant species, but also in the various phases of shoot development. In Arabidopsis, the lower phytomers have very short internodes at the region of the rosette leaves; but with vegetative maturation, the internodes elongate considerably. Structural and genetic aspects of the shoot of two dicots, Arabidopsis and tomato, were provided by Schmitz and Theres (1999), who updated and reviewed the information on this subject up to the end of 1998. The reader should be aware of the fact that nomenclature used by different authors can be misleading. What is commonly termed *phytomers* (e.g. by McSteen and Leyser, 2005) is termed by the former authors as *metamers*.

Shoots of dicot plants

Some plants, such as Arabidopsis, have an indeterminate shoot apex meristem. This SAM maintains its activity throughout the life span of the plant, while the lateral branches project from leaf axils that are at some distance from the SAM. The latter branches may then behave as the primary shoot and form secondary branches from their leaf axils. The ramification can go on and attain an elaborate branching system. Such plants are said to have *monopodial* growth. A schematic presentation of the monopodial growth habit of Arabidopsis is shown in Fig. 43.

In Fig. 43 the monopodial growth of the Arabidopsis shoot is compared with the sympodial growth of tomato. In tomato, the young shoot will develop several nodes with leaves, and vegetative

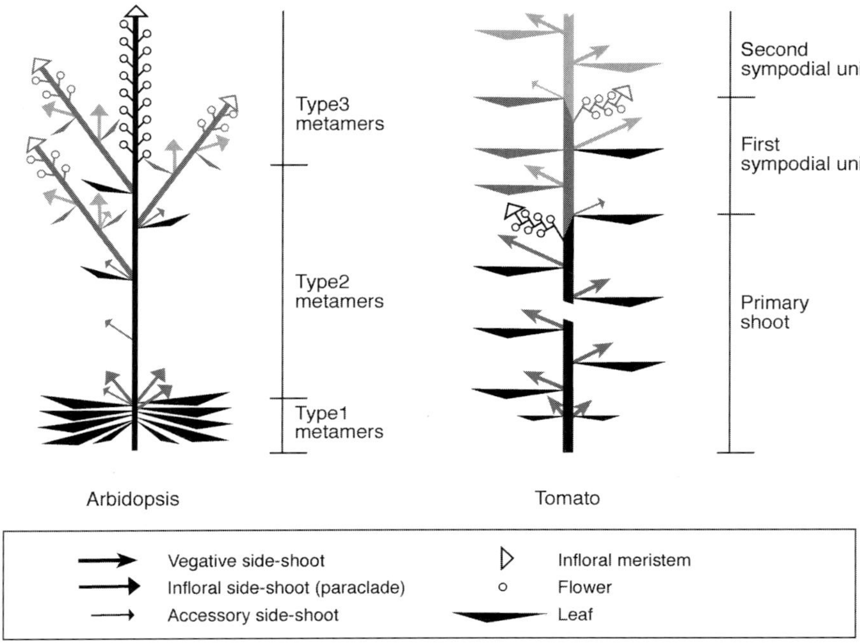

Fig. 43. Schematic representation comparing shoot development in Arabidopsis and tomato. The primary shoot is illustrated in black lateral axes of successive order in grey. In Arabidopsis, the primary meristem forms the main plant axis, consisting of a rosette, a bolting stem and an indeterminate inflorescence. Side-shoots originating from type 1 and type 2 metamers (paraclades) are infloral and repeat the development of the bolting stem. Accessory side-shoots are often formed between the primary side-shoot and its subtending leaf. In tomato, the primary shoot meristem forms the first segment of the sympodial stem terminating in the first inflorescence. In all the leaf axils, vegetative side-shoots are formed. The side-shoot in the axil of the youngest leaf primordium grows vigorously, pushing the inflorescence to a lateral position, and transferring its subtending leaf to an elevated position above the inflorescence. After the formation of three nodes, the sympodial shoot also terminates in an inflorescence. In wild-type tomato plants sympodial shoots of progressively higher order continue the elaboration of the main axis indefinitely. Inflorescences consist of a series of terminal flowers originating from sympodial inflorescence axes of progressively higher order. Accessory side-shoots can be found regularly between the sympodial side-shoot and its subtending leaf, but only rarely in other leaf-axils. (From Schmitz and Theres, 1999.)

side shoots (branches) will develop from the axils of these leaves. Then, the shoot apex will form a terminal inflorescence, the inflorescence will be pushed sidewise, and a new shoot will carry on the upward growth further, developing again a few nodes with vegetative side shoots. Thereafter, again, a terminal inflorescence will develop and a further sympodial unit will emerge. The same process that is going on in the main sympodial shoot will then go on in the side branches. This simplified description of the two main types of shoot development in dicots, which is exemplified by the monopodial Arabidopsis shoot and the sympodial tomato shoot, ignores several details. Some of these details are mentioned below when the molecular genetics of branching is discussed. However, here one example from tomato, concerning the development of inflorescences and flowers, shall be noted.

Generally, botanists gave the term *cymose* pattern to the tomato inflorescence. The primary inflorescence meristem is transformed into a floral meristem, and a new lateral meristem will originate from the pedicle of the first flower. This meristem will continue the inflorescence development. It was observed that already at the primary inflorescence meristem, there appears a "division" that yields two entities: one is a flower, while the other is the meristem that continues the inflorescence development.

Clearly, the actual ramification of dicot plants results in very diverse plant morphologies, even among annual dicots, and this diversification is even greater in perennial dicots such as trees. Models of shoot branching that result in the branching of trees were provided by Sussex and Kerk (2001), and are shown in Fig. 44. Another way to look at the axillary branching was adopted in the detailed review of McSteen and Leyser (2005). These authors presented schemes of axillary branching, not only in Arabidopsis and tomato, but also in two monocots: maize and rice (the two latter plants belong to the Poaceae that was previously termed Gramineae). Figure 45 shows these schemes as well as the respective units of inflorescence in Arabidopsis, tomato, maize and rice.

Conceptually, one may regard the phytomer as the basic unit of shoot branching architecture. As noted above, the ideal phytomer

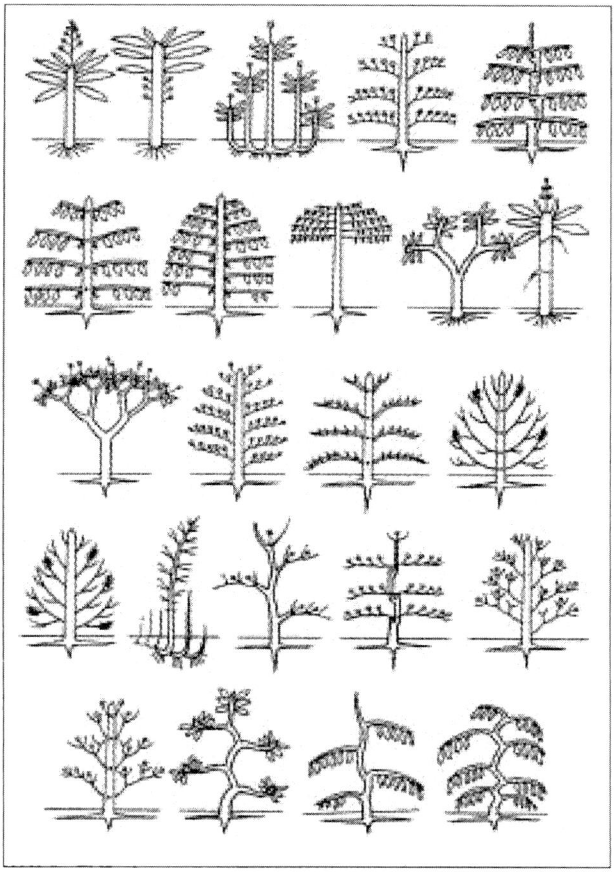

Fig. 44. Models that describe shoot architecture in the vascular plants. (From Sussex and Kerk, 2001.)

consists of a basal node from which a leaf may develop, an internode and an axillary meristem (AM) that may develop into a vegetative shoot and/or an inflorescence. Modifications of the phytomer components can thus lead to very different architectures. For example, as in the rosette region of the young Arabidopsis shoot, the phytomers have very short internodes; the leaves grow to an extensive size, but the axillary meristems do not develop. On the other hand, in shoots of Arabidopsis that passed the transition to flowering, the internodes are

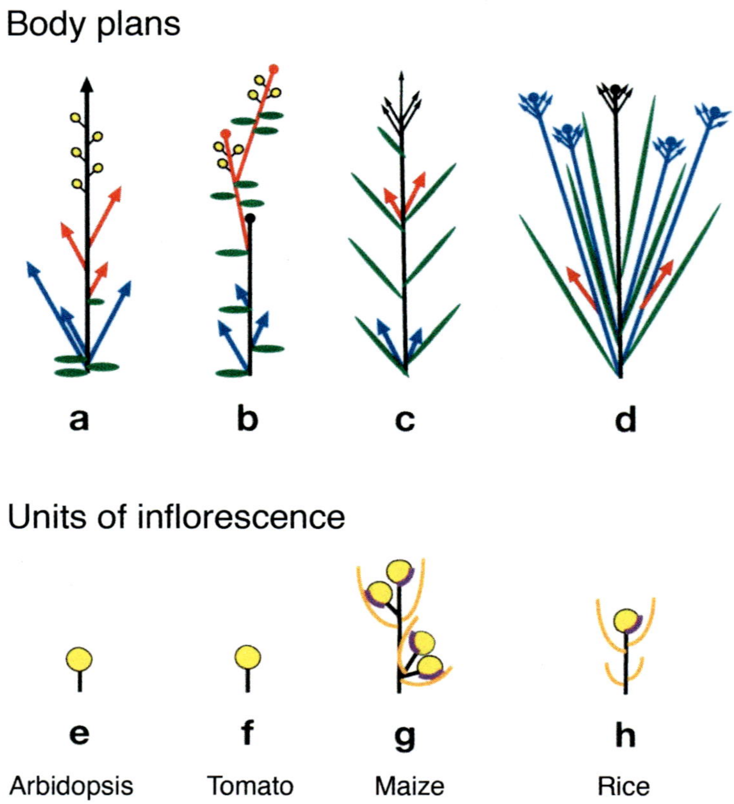

Fig. 45. Axillary branching in dicots and monocots. The schematic indicates the activity of axillary meristems during development. Arrows represent indeterminate meristems and circles represent determinate meristems. The figures show body plans of **(a)** *Arabidopsis*, **(b)** tomato, **(c)** maize, and **(d)** rice. The black lines indicate the main axis. The green lines indicate leaves. The blue arrows indicate the most basal-arising branches. The red arrows indicate axillary meristems arising later during development. The yellow circles represent flowers. Maize and rice also produce lateral branches in the inflorescence as indicated by the black or blue lines with smaller arrowheads. The units of the inflorescences are shown in **(e)** Arabidopsis, **(f)** tomato, **(g)** maize, and **(h)** rice. The yellow circles represent flowers. Maize and rice flowers are enclosed by small leaf-like organs called glumes (orange lines). The purple lines indicate the position of the lemma, which is the bract subtending the flower. The rice spikelet consists of one flower and four glumes. Maize produces spikelets as pairs, each consisting of two flowers and two glumes. (From McSteen and Leyser, 2005.)

long, while the leaves are minute; however, in their axils, their meristem develops into a side branch that terminates with an inflorescence.

It is common to mention an AM in each leaf axil. However, there is good evidence for the potential existence of more than one AM in each leaf axil (Grbic and Bleecker, 2000). This tendency to form more than one AM per leaf axil is stronger in the reproductive phase of the main shoot of Arabidopsis than in the axils of rosette leaves. There are other differences in the formation of axillary meristems in the axil of leaves in juvenile versus mature shoots. For example, while no such axillary meristem can be discerned in the axil of the rosette leaves of Arabidopsis, such meristems develop in the axils of all the later leaves. These axillary meristems, in the vegetative shoot of tomato, will develop into rather extensive side branches, resulting in a bushy growth habit of normal tomato cultivars. It appears that quantitative changes in the components of the phytomer can lead to vastly different final growth habits. We can now add a process that shall be dealt with in detail below: the transition from a vegetative meristem into a floral meristem. Altogether, the latter transition and the quantitative changes in the phytomer's components can then lead to the full range of growth habits of angiosperms.

Clearly, these changes and transitions are controlled by a combination of external and internal signals that we shall discuss in subsequent sections. The major external signals are prompted by environmental conditions, such as photoperiods, light intensity, and thermal regimes. Internal signals are prompted by the interactions of numerous genes that encode transcription factors and other proteins as well as by several plant hormones. These signals and effectors are spatially and temporarily regulated, meaning, for example, that the location on the shoot and the shoot's juvenility or maturation play an important role in these regulations. Axillary meristems may have very different fates. In the axil of Arabidopsis rosette leaves, such meristem may be initiated, but further development is arrested soon after initiation. When the environmental and/or internal conditions are changed, these meristems may develop into branches. This can also happen to other axillary meristems: they may either stay dormant or start to grow out under changing conditions. Since prehistorical times, horticulturists knew that pruning of deciduous trees and vine shoots can cause dormant buds to "wake up", grow and form branches. All this sounds like a rather complex situation; and although we are only beginning to understand these regulations, what we already know is indeed complex. Before we start to discuss these regulations, let us take a closer look at the shoots of certain monocots.

Shoot of monocot plants

Our information on the architecture of monocot shoots is based primarily on plants belonging to the grass family (Poaceae, previously termed Gramineae), and even in this family most studies were performed with two species: *Zea mays* (maize) and *Oryza sativa* (rice). Clearly, there are many families of monocots, but our knowledge on the regulation of their shoot patterning is meagre or nonexistent. Moreover, there is a vast number of genera in the Poaceae with a total of about 10 000 species (see Fig. 46). Therefore, handling the branching systems in these species is far beyond the scope of this book. On the other hand, by focusing on only two species, we are actually "looking for the coin under the lamp post."

If we accept the phytomers as the building stones of shoots, and allow a large range of variation in the development of their components (e.g. length of leaf, length of internode), then there is a similarity between the shoot of maize and the shoot of Arabidopsis. Moreover, the shoot of maize undergoes similar developmental stages as the shoot of Arabidopsis. These stages, for the maize shoot, were schematically summarized by Irish and Nelson (1991), as shown in Fig. 47. The shoot of maize that emerges from the germinating embryo passes through a vegetative phase, during which leaves are initiated as lateral outgrowth from the indeterminate apical meristem of the vegetative shoot. After a certain number of nodes with leaves, the apical meristem is rendered determinate, forming a terminal inflorescence that develops

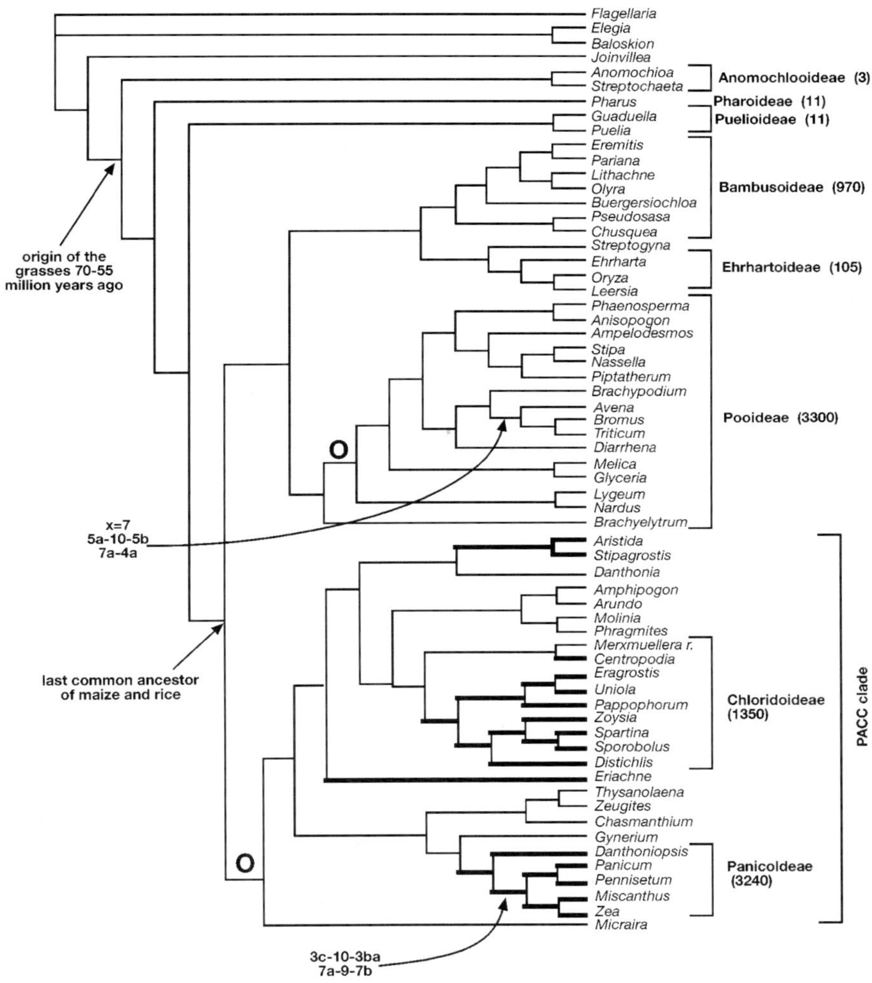

Fig. 46. Phylogeny of the grass family based on combined data from chloroplast restriction sites, *rbcL*, *ndhF*, *rpoC2*, phytochrome B, ITS, GBSSI, and morphology. Heavy lines indicate C_4 photosynthesis; numbers with hyphens indicate chromosomes formed from the joining of the appropriate rice chromosomes; O indicates shift to open habitats. Numbers in parentheses indicate approximate numbers of species. (From Kellogg, 2001.)

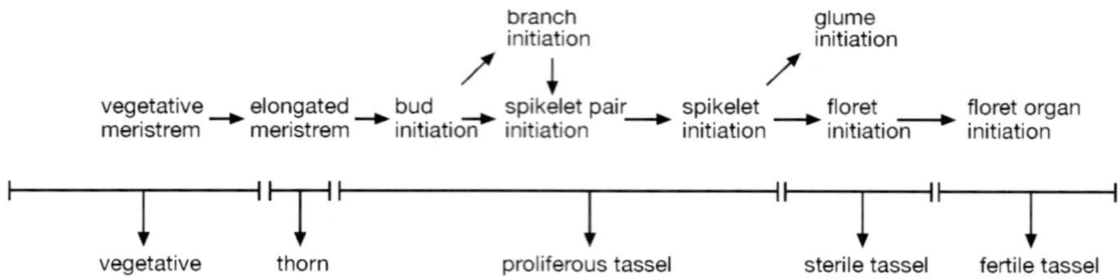

Fig. 47. A schematic summary of the stable developmental phases of the transition from indeterminate, vegetative development to determinate, floral development. The upper portion represents the morphologically identifiable stages in tassel development, in which the vegetative meristem becomes elongated and initiates buds. The buds may develop into branches bearing spikelet pairs or directly to spikelet pairs. Spikelet pair primordia give rise to spikelets, which initiate glumes and florets. Each floret initiates a lemma, palea, and floral organs. The lower portion represents the type of development observed when meristems of those stages are cultured. (From Irish and Nelson, 1991.)

into the tassel, which is the staminate inflorescence of maize. The number of leaves that precede the conversion of the indeterminate apical meristem into a terminal one is typical for each maize cultivar, and thus controlled by maize genes.

As shall be detailed below, many genes were identified that control this standard patterning of the maize shoot. Mutation of certain genes will, for example, change the single-shoot growth habit of maize cultivars to a bushy plant, similar to the growth habit of teosinte. There are several clear differences between the indeterminate shoot of maize (and other species of the Poaceae family) and the indeterminate vegetative shoot of Arabidopsis. The leaves of the former shoot differ considerably from the leaves of latter shoots: maize leaves contain a blade and a sheet. No such sheets exist in dicot leaves. Also, the maize embryo, before germination, already has several leaf initials, much more than the number of leaf initials in the dicot embryos. Interestingly, the number of vegetative nodes in the shoots of a given maize cultivar is rather fixed and is not changed by *in vitro* culture of the shoot.

Shoots cut and cultured, after developing between 11 to 16 nodes, all develop further vegetative nodes until they form the standard number of vegetative nodes (leaves) that is typical to the given cultivar. In experiments conducted by Irish and Nelson (1991), the cultured shoots terminated their vegetative growth phase after 16 nodes were formed. Unlike Arabidopsis, the shoot of maize that is in the juvenile phase may form adventitious roots in the lower four to six nodes. Higher nodes will not produce such roots. Growth in short-day photoperiods will reduce, at least in some cultivars, the number of vegetative nodes. Hence, the maize plant "knows to count" (nodes) and to react to environmental cues. The molecular genetic basis of this "counting capability" is still obscure, but clearly other monocot plants such as the banana also have such counting capability, forming their terminal inflorescence (in banana, the inflorescence is bisexual) after a certain number of leaves are formed.

In the Poaceae, there are two types of branching from the main shoot. During the vegetative growth of the shoot, the first type of branches are termed *tillers*. These may develop from the basal nodes and grow out in acropetal sequence. The basal phytomers that develop these tillers have characteristically short internodes. The tillers, which usually either do not exist or are very rare in maize, are frequent in the maize ancestor, teosinte, as well as in, other genera of the Poaceae (such as *Oryza* and *Triticum*), where the many tillers cause a bushy habit. The other (rare) type of branching in maize is termed *axillary* or *secondary branches*. These branches grow out much later in the development and are located further up on the main stem or as branches from the tillers. The phytomers from which these branches develop have longer internodes and bear leaves with adult characteristics. These branches have a delayed apical meristem activity and a basipetal sequence of branch outgrowth. The ear of maize (the pistillate inflorescence) is formed from such axillary branches that are located a few nodes below the tassel.

The Molecular Genetics of Branching

As reported in the detailed review of McSteen and Leyser (2005) on shoot branching, there exist two opposing explanations for the origin of axillary meristems (AMs). One explanation is based on the "classical" publications of M. and R. Snow, claiming that AMs are initiated *de novo* in leaf axils. The other explanation claims that the AMs are derived from (the flanks of) the SAM at the time of leaf initiation, and never lose their meristematic identity (even though they do not show the morphology of a meristem). There are good arguments for both explanations. For example, in the relatively large SAM of Arabidopsis shoots, there are locations on the flanks of the SAMs subtending to the emerging leaf, from where the new AM can be initiated. On the other hand, there is no morphologically distinguishable meristem in the axils of some young leaves. In more apical nodes of the Arabidopsis shoot, AM can be initiated further up the petiole, meaning it is clearly initiated from the leaf itself. Rather than resolving the question which of the explanations on the origin of AM is the correct one, we shall briefly review the knowledge on the expression of genes involved in AM initiation and outgrowth, and on the interactions between the products of these genes. Conceptually, one can divide this issue into three phases. One is the incipial initiation of the AM; the other is the further growth of AM (e.g. into a lateral branch); and the third is the possible *arrest* of the initiated AM.

Information from dicots

As for the initiation of AMs in dicots, the reader is referred to Chapter 7 and Chapter 8 of this book that discuss apical meristems. In the SAM of Arabidopsis embryos and young shoots, the gene *SHOOT-MERISTEMLESS* (*STM*) is expressed and its product is required for SAM activity. However, *STM* is downregulated where leaves are initiated. Methods that enabled the identification and cellular localization of *STM* expression indicated that there is an intensification of *STM* activity immediately preceding the morphological appearance of an AM. The gene *REVOLUTA* (*REV*) was identified by Talbert *et al.* (1995), who found that this gene is necessary for apical meristem development and for limiting cell divisions in the leaves and stems of Arabidopsis. The schematic presentation of branching in wild-type and *rev* mutants of Arabidopsis is provided in Fig. 48. Later studies showed that *REV* is involved in several controls of plant patterning; but with respect to AM initiation, it should be noted that *REV* is also upregulated in the leaf axil, and this upregulation precedes the upregulation

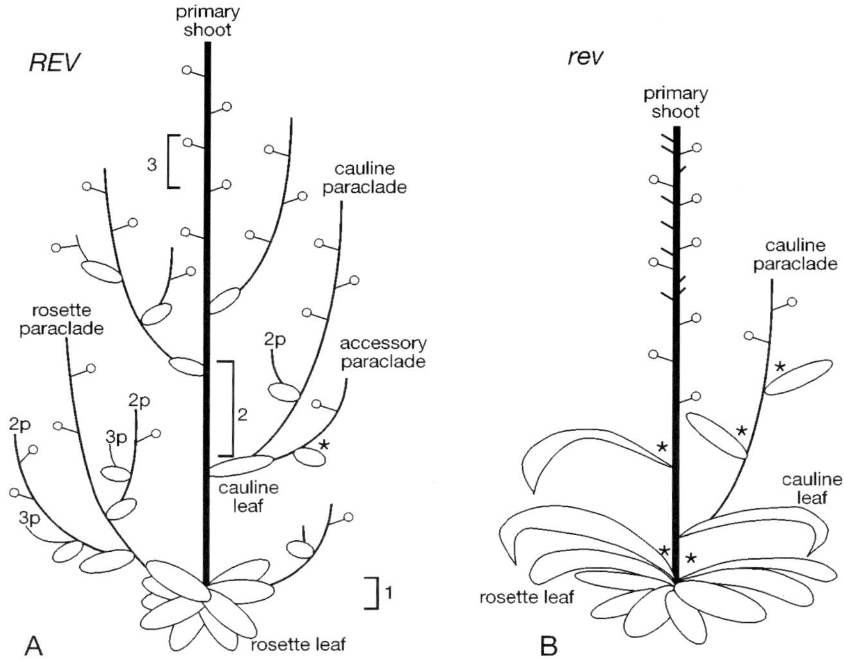

Fig. 48. Schematic representation of wild-type *REV* and mutant *rev* morphologies. **(A)** Wild-type, showing phytomer types (1, 2 and 3) and branching pattern. **(B)** *rev*, showing reduced growth of primary shoot, paraclades and flowers, and increased growth of leaves and stems. o, flowers; 2p, second-order paraclade; 3p, third-order paraclade; *, axil lacking a paraclade. (From Talbert *et al.*, 1995.)

of *STM*. In *rev* mutants, the AMs are not initiated. Notably, *REV* expression is also required during the initiation of the determinate floral AMs, which emerge in the axils of cryptic leaves that develop on the primary shoot after floral transition. This supports the notion that there is a similarity between vegetative and floral AMs.

Two other genes come into the play: the *LATERAL SUPPRESSOR* (*LS*) of tomato and its ortholog, *LAS*, in Arabidopsis. These genes encode a VHIID transcription factor of the GRAS family. Tomato plants with a loss-of-function mutation in *LS* have reduced branching due to failure to develop AMs in the respective nodes. Similarly, the Arabidopsis *las* mutants have reduced branching. The phytomers of the rosette phase have a poor capability to develop AMs. In wild-type plants, *LAS* transcripts are localized to the axils of *all* leaf primordia that are derived from the primary SAM. The transcripts begin to accumulate rather early during the initiation of leaves from the SAM, at the stage when the primordium begins to form a bulge. The beginning of this upregulation of *LAS* expression precedes the focusing of *STM* expression and continues until the upregulation of *STM* takes place. However, in *las* mutants, neither *REV* nor *STM* is upregulated in the axils of rosette leaves. The floral meristems that arise immediately after a lateral organ is initiated on the primary shoot are not affected in these mutants, but the AMs that are to be initiated later are lost. This fits the finding of Greb *et al.* (2003) which indicated that *LAS* affects axillary meristem in a different manner in the growth phases of the shoot: during vegetative development the *las* mutant fails to initiate AMs, while after the floral transition the AM initiation in these mutants is not abolished. It appears that during the reproductive phase the function of the *LAS* product is not required, while it is required for AM initiation during the vegetative phase. It also appears that the *LAS* product acts upstream of the product of *REV*.

The control of lateral branching, i.e. the elongation of AM, is still somewhat mysterious. Stirnberg *et al.* (2002) found that two Arabidopsis genes, *MAX1* and *MAX2*, retard the branching from the vegetative nodes of the main shoot. The respective mutants of these genes started to branch at these nodes, resulting in a bushy phenotype. The mutants had additional phenotypes, such as fasciated lateral shoots and changes in leaf shape. The authors suggested that the wild-type genes, *MAX1* and *MAX2*, repress axillary shoots and are involved in ubiquitin-mediated degradation of as yet unidentified proteins that activate axillary growth. But, no evidence for this suggestion was furnished by this study.

Additional information on *MAX* genes emerged from the work of Schwartz *et al.* (2004). These investigators studied genes that encode carotenoid dioxygenase (e.g. *MAX3* and *MAX4*). Such enzymes are involved in the biosynthesis of certain compounds such as abscisic acid (and vitamin A in animals). The Arabidopsis mutants, *max3* and *max4*, are defective in carotenoid dioxygenase. Interestingly, the mutants that were found to have this defect displayed a dramatic increase in lateral branching. The authors concluded that the mutants are defective with respect to an as yet unidentified factor which suppresses AM development in wild-type plants. The process of further development of the already initiated AM is still enigmatic. It is possible that the *MAX3* and *MAX4* genes of Arabidopsis will provide a lead to understanding this process.

A relatively early study by Long and Barton (2000) of the University of Wisconsin at Madison analyzed the formation of AM in Arabidopsis from another angle. They reviewed the literature on the role of subtending leaves in the initiation of AM, and pointed out that the AM develops in close association with the *adaxial* base of the subtending leaf. When the abaxial leaf fates are transformed into adaxial fate by the *PHABULOSA* mutant (*phb*), ectopic meristems arise on the abaxial side of the leaf. Moreover, it was reported long ago that when the subtending leaf was surgically excised, no AM appeared.

There is evidence for the notion that the SAM and the adaxial half of a young leaf constitute the same positional cue: when the gene *PINHEAD* (*PNH*) is expressed in the SAM, it is also expressed in the adaxial half of the young leaf. Long and Barton (2000) provided evidence suggesting that the *PNH* gene causes meristem-forming competence to the adaxial side of the young leaf. However, there is a dilemma: some AMs (in Arabidopsis) develop in the absence of a subtending leaf. This is true for AMs that form floral meristems, not only in Arabidopsis, but also in other members of the Brassicaceae (previously termed Cruciferae). In the latter cases, a floral meristem is apparently initiated without a subtending leaf, as though there is a switch changing a leaf into a floral meristem.

Here, we should recall what was said in previous chapters of this book: the typical SAM-localized *STM* transcript is "extinguished" when cells at the flank of the SAM are specified as leaf primordia. The *AINTEGUENTA* (*ANT*) gene causes almost exactly the opposite change in expression: it is expressed in the young primordia (including leaf primordia). By following the changes in expression of *STM* and *ANT* in axillary meristems, it was suggested that while there is no clear manifestation of subtending leaves adjacent to floral meristems of Arabidopsis, such leaves do exist but in a highly reduced form. Did Long and Barton (2000) mean that floral AM develop from the adaxial side of virtual leaves? It was already previously reported (Janssen *et al.*, 1998) that overexpression of the tomato *LcT6*, a class I *knox* homeobox gene, causes adventitious meristems on the adaxial leaf surface.

In a more recent publication, Schmitz and Theres (2005) reviewed the roles of several genes in the different stages of AM initiation up to the outgrowth (elongation) of the lateral bud. In the early stages of the initiation, several genes such as *LAS* (which was mentioned above) are required for the establishment of the new AM between the shoot and the adaxial side of the leaf primordium. The expression pattern of the *LAS* gene at this stage is similar to the expression of several other genes, such as *CUP-SHAPED COTYLEDON* (*CUC*) and *LATERAL ORGAN BOUNDARIES* (*LOB*). These localizations indicate that the site of the future AM indeed provides a specific positional clue for the activation of several genes. There is an interesting issue that concerns the localization (boundary domain) of gene expression. There are three partially redundant *CUC* genes: *CUC1*, *CUC2* and *CUC3*. Laufs *et al.* (2004) from INRA in Versailles, France, found that when the gene encoding mir164 is overexpressed, the translations of *CUC1* and *CUC2* are downregulated in Arabidopsis. On the other hand, changes in the code for *CUC2*, so that it is no longer affected by mir164 or reducing the level of mir164, cause the enlargement of the boundary domain. It thus appears that mir164 can degrade *CUC1* and *CUC2* mRNAs.

Antirrhinum has a gene, *CUPULIFORMIS* (*CUP*), which is homologous to the *CUC1* and *CUC2* genes of Arabidopsis. A mutation in *CUP* (i.e. *cup*) leads to misformed leaf axils that lack lateral buds (Weir *et al.*, 2004). This indicates that basic mechanisms of AM initiation are conserved in dicots.

Later in the AM evolution, the *LAS* gene starts to play a role in the outgrowth of the AM, and subsequently the genes *REV* and *PHB* mentioned above also seem to be involved. The gene *PHAVOLUTA* (*PHV*) is also involved in this outgrowth. The abovementioned *MAX* genes seem to be involved in the further elongation of the lateral bud. A similar situation emerges in the development of AM in tomato. In the latter plant, there is a gene encoding a *MYB* transcription factor, termed *BLIND* (*BL*), which seems to be essential for the development of the AM. The *bl* mutant is compromised with respect to the formation of lateral shoots and inflorescence meristems.

Further evidence for the conservation of branching strategy in angiosperms was furnished by the recognition that the pea gene *RMS1* and the petunia gene *DECREASED APICAL DOMINANCE1* (*DAD1*) are functional homologs of *MAX4*, and the rice *D3* is a homolog of *MAX2*.

It is reasonable to assume that further outgrowth of an AM into an inflorescence or a lateral branch requires the vascular differentiation that will connect the outgrowth with the vasculature of the main shoot. The *INTER FASCICULAR FIBERLESS1* (*TFL1*) gene is homologous to *REV*, which was independently isolated. The *TFL1* gene was found to be required for the correct patterning of the vascular system in Arabidopsis.

Information from monocots

While this section carries the heading "monocots", the information on the molecular genetics of branching is restricted to the Poaceae, and among these almost exclusively to maize and rice. The branching of laterals from the main shoots of maize and rice is schematically presented in Fig. 45 (in a previous section of this chapter). There are clear similarities in the mechanisms of branching between dicots and monocots; however, the names of homologous and orthologous genes are different, and so are some other terms. For example, the term *lateral branches* is given to the outgrowths from the lower part of the main shoot of Arabidopsis. In maize and rice, these lower outgrowths are termed *tillers*. Monocots of the Poaceae also produce lateral branches higher up on the main stem. These are then termed axillary or secondary branches. The phytomers of the latter branches have relatively long internodes.

Other differences, real or apparent, between the dicot main shoot and the main shoot of Poaceae plants were summarized in the previous section on the structural aspects of the angiosperm shoot. I already hinted above that there are similarities between the molecular genetics of branching between dicots and monocots. One such example came to light in the study of rice tillering by Li *et al.* (2003). These authors isolated the *MONOCULM1* (*MOC1*) gene, which encodes a transcription factor of the GRAS family and is orthologous to the *LAS* of Arabidopsis and *LS* of tomato. The *moc1* mutants have no tillers: they fail to initiate AM (as the *las* and *ls* mutants). Increase in *MOC1* expression increases the number of tillers in rice. It was presumed that *MOC1* is required for both the initiation and the outgrowth of rice tillers. However, there is a difference: *LAS* and *LS* affect the vegetative outgrowth with little or only slight effect on the branching of the inflorescence; the *MOC1* also seems to have an important role in inflorescence branching of rice.

In barley there is also a gene (*UNICULM2*) that, when mutated (*cul2*), will have defects in branching; the AMs are initiated, but further development of the AMs is retarded. Also, in this case the mutant (*cul2*) affects not only the vegetative branching, but also the spikelet differentiation of the mutated barley plants (Babb and Muehlbauer, 2003).

During many years in the 20th century, there were bitter debates with respect to the origin of maize and the role of teosinte in the evolution of cultivated maize. The prime rivals were P.C. Mangelsdorf and George Beadle. The claim of the latter was finally accepted. It was agreed that teosinte is the wild ancestor of maize (this debate is an interesting story that was detailed in the outstanding biography of Beadle, titled *George Beadle, An Uncommon Farmer* by Paul Berg and Maxine Singer, 2003). We know now that there are only a few genes that cause the difference between teosinte and maize, and the two should be considered as variants of the same species. One of these genes is *TEOSINTE BRANCHED1* (*TB1*). Maize plants that are mutated in this gene (*tb1*) have many axillary branches (as well as secondary and tertiary branches and elongated internodes in the branches). The phenotype of this maize mutant thus resembles the bushy architecture of teosinte (Hubbard *et al.*, 2002).

In wild-type maize, *TB1* is expressed in AMs as soon as they become visible; while in teosinte, *TB1* is not expressed in the AMs.

On the other hand, *TB1* is expressed in the ear predicellate spikelets of teosinte, causing their suppression. An ortholog of *TB1* was also reported in rice (Takeda *et al.*, 2003), where it was found to negatively regulate lateral branching, and the gene was termed *OsTB1*. Foxtail millet (*Setaria italica*) is another Poaceae species. This cultivated species has relatively few tillers. It was proposed to be derived from the wild *S. viridis*, which has many tillers. The suppression of tillers was suggested to be at least in part from the effect of a *TB1*-like gene, although additional quantitative trait loci (QTLs) are also involved in the suppression of tillers (Doust *et al.*, 2004).

The gene *LAX1* of rice was one of the first genes controlling AM initiation in the inflorescence from a Poaceae (Komatsu *et al.*, 2003). In *lax1* mutants, there is a reduced number of long branches in the inflorescence of rice and a reduced number of spikelets. Moreover, those spikelets that do appear are defective. There is also another mutant of the *SMALL PANICLE1* (*SPA1*) that causes the reduction of long inflorescence branches as well as a reduced number of spikelets. The double mutants of rice, *lax1/spa1*, show a stranger phenotype: in addition to the lack of branches and spikelets, they also lack tillers. It was thus suggested that *LAX1* and *SPA1* are at least in part redundant and have roles in AM initiation.

The highly branched inflorescence of grasses is the result of several types of AMs. This was reviewed in some detail by McSteen *et al.* (2000) for maize, as schematically shown in Fig. 49. Numerous genes were found to be involved in the patterning of the male (terminal) inflorescence of maize (see Table 7). Of these there are two types of branch meristems: one type that is produced at the base of the inflorescence leads to the long branches of the tassel, and another where meristems that arise on the main spike and on the lateral branches of the meristem (spikelet-pair meristems) give rise to two spikelet meristems. Spikelet meristems then lead to floral meristems.

In total, there are therefore four types of AMs: branch, spikelet-pair, spikelet and floral. Although the phytomer concept can also be applied in the tassel, one should note that all the internodes in these phytomers are very short, and the subtending leaves of the AM are suppressed or actually absent. Moreover, the branch meristem is an indeterminate

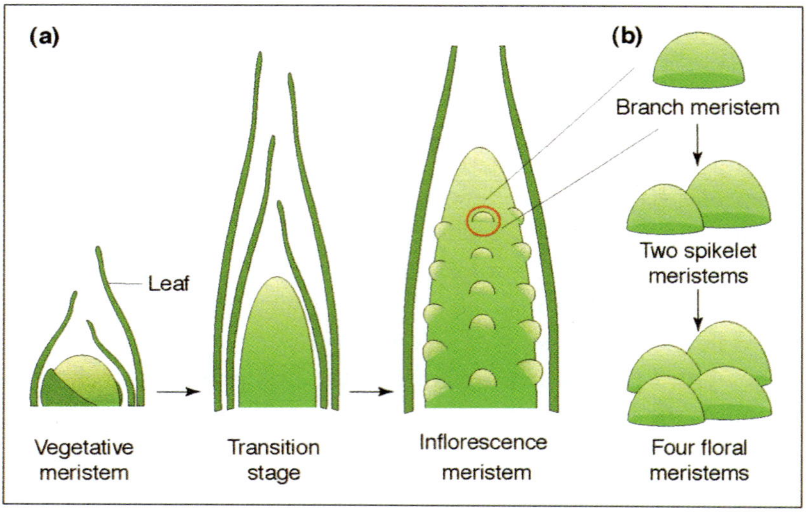

Fig. 49. Meristem transitions during maize development. (**a**) Transition from a vegetative shoot apical meristem (SAM) to an inflorescence meristem. During vegetative growth, the SAM initiates leaves that have an axillary meristem in the axil of each leaf (not shown). At the transition to flowering, cessation of leaf initiation is accompanied by an increase in the rate of cell division followed by a morphological restructuring of the SAM (transition stage). The inflorescence meristem then initiates files of axillary meristems called branch meristem (smaller domes). (**b**) The maize inflorescence produces three types of axillary meristem. Each branch meristem produces two spikelet meristems. The two spikelet meristems branch to form two floral meristems. Thus, four floral meristems arise from a single axillary branch meristem. (From McSteen *et al.*, 2000.)

AM; whereas the spikelet-pair, the spikelet and the floral meristems are determinate AM. Although the flower of maize appears quite distinct from the Arabidopsis flowers, there are parallel floral members in these two flower types. The names of these parallel members are different. The bracks, sepals and petals of Arabidopsis are termed lemmas, pleae and lodicules, respectively, in the maize flower. More details on floral members of the Poaceae shall be provided in Chapter 12.

About 100 years ago, a peculiar maize plant was found in a commercial maize field (Gernet, 1912). It was so peculiar and bred true, that it was assumed to represent a new species: *Zea ramosa*, or at least

Table 7 Meristem Mutants Define Stages in Maize Development

Meristem	Mutant	Phenotype
Vegetative ↓	*id1*, *dlf1*, *Lfy1*	Extra leaves, delayed flowering
Inflorescence	*Fas1*, *fae1*, *td1*	Fasciated, more spikelet pairs
	kn1	Fewer branches and spikelet pairs
↓	*tls1*	Short tassel, fewer spikelet pairs
Branch	*ra1*, *ra3*	Spikelet pair to branch
	ts4	Branch meristem reiterates
	ifa1	Extra spikelets
↓	*ub1*, *lg2*	Fewer branches
	ba1, *Bif1*, *bif2*, *Sos1*	Fewer branches and spikelets
Spikelet	*bd1*	Spikelet meristem reiterates
	ids1, *rgo1*, *ifa1*, *Ts6*	Extra florets
↓	*ra2*	Spikelet to branch
	bif2	Fewer florets
Floral	*ifa1*	Proliferative meristem
	zag1	Extra carpels
	bif2	Fewer stamens
	si1	Lodicules to palea, stamen to carpel

During development, meristems of different identity are produced. Upon transition to flowering, the vegetative meristem is converted to an inflorescence meristem. The inflorescence meristem then gives rise to axillary meristems, called branch meristems, which produce spikelet meristems and finally floral meristems. Mutants with defects at each stage define genes required for meristem identity, determinacy and structure. Abbreviations: *ba1*, *barren stalk1*; *Bif1,2*, *barren inflorescence1,2*; *bd1*, *branched silkless1*; *dlf1*, *delayed flowering1*; *fae1*, *fasciated ear1*; *Fas1*, *Fascicled1*; *ifa1*, *indeterminate floral apex1*; *ids1*, *indeterminate spikelet1*; *id1*, *indeterminate1*; *kn1*, *knotted1*; *lg2*, *liguleless2*; *Lfy1*, *Leafy1*; *ra1,2,3*, *ramosa1,2,3*; *rgo1*, *reversed germ orientation1*; *tls1*, *tasselless1*; *ts4,6*, *tasselseed4,6*; *ub1*, *unbranched1*; *si1*, *silky1*; *Sos1*, *Suppressor of sessile spikelet1*. (From McSteen *et al.*, 2000, where references were provided).

a new subspecies of *Zea mays*. It was later established that this peculiar maize, which has ramifications of the basal lateral branches in the tassel and branchings in the ear, is caused by a mutation in the gene *RAMOSA1* (*RA1*) (see Vollbrecht *et al.*, 2005 and McSteen, 2006 for details). The different grass genera differ from each other in their arrangement of florescence branches, spikelets and florets. The majority of grasses bear spikelets singly on the main axis and the branches of the inflorescence. This is the case in rice, where the inflorescence has many long branches that bear single spikelets.

The production of short branches that bear two spikelets is typical in the grasses belonging to the tribe *Andropogoneae* (e.g. maize, sorghum, sugar cane). In maize, there are long branches and a main axis that constitute the tassel; while a somewhat different branching occurs in sorghum, where there is a multiple order of branching in the inflorescence. Studies by Vollbrecht *et al.* (2005) and Bortiri *et al.* (2006), who isolated the *ra1* and *ra2* genes, respectively, suggested to McSteen (2006) that the *ramosa* pathway plays a critical role in imposing determinacy on the spikelet pair in the tribe *Andropoganeae*. It should be noted that the studies of Vollbrecht *et al.* (2005) and Bortiri *et al.* (2006) were not independent endeavors; R. Martienssen and E. Vollbrecht participated in both studies.

In the tassels of maize with the *ra1* mutation, there is a transformation of the spikelet pairs (that represent short branches) into long branches that bear single or paired spikelets. The *ra1* mutation also affects the ear. Instead of being unbranched in the wild-type, it becomes highly branched in the *ra1* mutant. In the study of Vollbrecht *et al.* (2005), the *RA1* gene was identified and cloned by transposon tagging. For that, the transposable element *Spm* (see Galun, 2003) was introduced into wild-type plants. When the *Spm* is inserted in a given gene, it causes a mutation in this gene. When this mutation has a typical and unique phenotype, it is possible to look for a DNA sequence that contains this *Spm* element and then sequence the flanks of the *Spm*. The effect of the "jumping" of an *Spm* element in cells that will form the tassel or the ear is rather clear. For example, when one finds an ear that has one "patch" of a branched ear, this patch is a possible area in which *Spm* "invaded" the *RA1* gene, causing it to behave as *ra1*. DNA samples from the same ear outside and inside the patch will verify whether indeed the *Spm* invaded the *RA1*.

Interestingly, *RA1* is not expressed in all AMs of the maize inflorescence. It is not expressed in the branch meristem (mentioned above), but begins to be expressed as the spikelet-pair meristem is initiated. The *RA1* probably imposes its effect by causing determinancy on the spikelet-pair meristems. Hence, in the *ra1* mutants, the spikelet-pair meristem causes indeterminancy and produces additional spikelets. There is a complication: there is evidence that *RA1/ra1* functions by controlling a mobile signal. Is it because the *RA1* protein itself can move from cell to cell, or is it possible that hormones are involved? There is no clear answer to this question; however, clearly the application of GA can affect tassel branching in maize, and auxin also affects the number of tassel branches.

The *ra2* was also revealed many years ago (1935), and it causes a similar phenotype as *ra1*. However, there are several clear differences. First, the ears of *ra2* are less affected than *ra1* ears, and the branches in the *ra2* tassel are borne at a more upright angle than in *ra1* tassels; furthermore, the *ra2* spikelets are borne on more elongated pedicles in the tassel than in *ra1*. The cloning of *ra2* by Bortiri *et al.* (2006) was not performed by transposon tagging, but rather by map-based cloning, in which the colinearity of the known sequence of the rice genome was utilized. Ortologs of *ra2* were found in sorghum, rice and barley (meaning outside of the tribe *Andropogoneae*). *RA2* is expressed earlier in the maize inflorescence than *RA1* and acts upstream of the latter gene. From double mutants (*ra2/ra1*), it appears that both genes operate in the same pathways; also, in *ra2* the expression of *RA1* is reduced, hence *RA2* may be involved in the transcriptional activation of *RA1*. However, there seem to be additional factors that activate the expression of *RA1*.

The closest homolog of *RA2* in Arabidopsis is *ASYMMETRIC LEAVES-LIKE2*. Notably, the C-terminus is very different in the two genes. It seems that the C-terminus of RA2 exists only in the Poaceae. The *RA2* is already expressed during the early initiation of the branch meristem of the inflorescence, but its expression disappears as the meristem grows. McSteen (2006) discussed the role of the expression of the *RA* genes in patterning the inflorescence of the various *Andropogoneae* genera. For example, the *Sorghum* inflorescence is more branched than the maize tassel. The inflorescence of the former genus produces spikelet pairs only after several rounds of branching, and this delay is correlated with the delay in the peak of *RA1* expression in

Sorghum. Is this change in the timing of expression regulated by the *cis*-acting elements that are upstream of the coding sequence or is it due to *trans*-acting elements? Clearly, the answer could be that both *cis*- and *trans*-acting elements are involved, but there is no answer yet to this question. As for *RA2*, it is clear that its expression pattern is conserved not only in maize and *Sorghum*, but also in rice and barley. It was therefore suggested by Bortiri *et al.* (2006) that *RA2* is essential for the initial steps in the architecture of the inflorescence in the Poaceae (grasses) family.

A relatively recent and rather detailed review on rice plant development was provided by Itoh *et al.* (2005). These authors insisted that in order to understand this development, a detailed categorization of the development into phases and stages is essential. They indeed went on to divide the development of rice into numerous phases and stages, although they noted that some developmental transitions are gradual. The embryogenesis was divided into 10 stages, the development of the inflorescence into 9 stages, the development of the leaves was divided into 6 stages, the development of the crown roots into 7 stages, the development of the spikelets into 8 stages, the development of the ovule into 10 stages, and the development of the anther into 8 stages. Itoh *et al.* (2005) also tabulated the known genes that are expressed in the various stages. Clearly mentioning all these stages and genes is far beyond the scope of this book; however, the information is obviously useful for experts, especially for those who enjoy an elaborated "chest of drawers."

A team of seven investigators from Japan and Palmerton North of New Zealand, headed by Yasuo Nagato of the University of Tokyo [some of them coauthored the publication of Itoh *et al.* (2005)] investigated the role of the gene *PLASTOCHRON2* (*PLA2*) in the architecture of the inflorescence of rice (Kawakatsu *et al.*, 2006). The mutants of *PLA2* (*pla2*) exhibit shortened plastochrons (i.e. shorter time intervals between the initiation of new leaves on the shoot). The *PLA2* was also identified as an ortholog of the maize gene *TERMINAL EAR1*, and was found to cause precocious leaves during the vegetative growth as well as ectopic shoot formation during the reproductive phase. The *PLA2* product is a

MEI2-like RNA-binding protein, and the *PLA2* expression is predominantly located in young leaves' primordia. The *PLA2* protein seems to retard the rate of leaf maturation. This later effect is independent of the effect of another protein (that has a similar name), *PLA1*. The latter protein is a member of the P450 family (hence, a very different protein). The phenotypes of shoots in rice plants with the mutation *pla1* and *pla2* are similar. The authors proposed a model in which the plastochron is determined by signals from immature leaves (e.g. *PLA2*) that act non–cell-autonomously in the shoot apical meristem to inhibit the initiation of new leaves. Note that when the term "non–cell-autonomously" is used, the authors mean that a factor is flowing from cell to cell.

Now, we return to the theme of this section: the branching of the rice inflorescence. Here, the mutants of *PLA2* (e.g. *pla2-1*) have an outstanding phenotype (that, again, have a similar phenotype to the inflorescence of *pla1* mutants): vegetative shoots emerge from the primary rachis instead of the secondary rachis branches and spikelets. Furthermore, the bracts that are adjacent to the vegetative shoots are enlarged. In brief, the experimental results of Kawakatsu *et al.* (2006) lead to two suggestions: the (wild-type) *PLA1* product functions by regulating the rate of leaf maturation (affecting the plastochron), but it also has an important function in the suppression of the vegetative program in the reproductive phase (i.e. in inflorescence).

The Transition from Vegetative to Reproductive Growth

The transition from vegetative shoots to the reproductive phase varies considerably among angiosperms. In some plants such as cucumber (*Cucumis sativus*), there is no abrupt change from one phase into another. Flowers may develop in axils of normal-size leaves. A similar mixture of normal foliage and flowers is common in many plants; while in other plants such as the monocots wheat, rice and banana, there is a sudden change from the lower vegetative phase to a terminal inflorescence. Still, in some other plants there is a gradual transition. It is beyond the scope of this book to describe the morphology of the transition to flowering in angiosperm. Moreover, it is intended in this section to highlight the present knowledge on the molecular genetics of this transition and the internal and external factors that are active in controlling it. We shall therefore handle mainly model plants in which experimental approaches yielded the relevant information. True, even before novel genetic and molecular approaches were utilized, certain physiological and genetic knowledge on the initiation of flowering was obtained. For example, in 1920 it was already established that short days facilitate flowering in certain plants, while long days promote flowering in other plants. Also, the requirement for a cold period was found to be necessary for the bolting of some plants. An example that is derived from my own work is the flowering of cucumbers (see Frankel and Galun, 1977, where my studies with cucumber during the 1950s and 1960s are summarized).

In brief, the flowers of cucumber are borne on very short inflorescences in the axils of leaves, starting from a few nodes above the base of the shoot. Under certain conditions (short days and relatively low temperatures), the flowers start to appear at lower nodes than under other environmental conditions (such as long days and higher temperatures). Furthermore, the sex expression is changed by environmental conditions. Most cucumber cultivars are monoecious; they bear mainly staminate flower on lower nodes, and then the frequency of nodes with pistillate flowers increases until, at the upper end of mature plants, most flowers are pistillate. Environmental conditions, such as short days and low temperatures, will increase the ratio of pistillate to staminate flowers. There are also genes that will (almost) eliminate staminate flowers, while other genes will eliminate pistillate flowers. Moreover, hormonal treatments such as ethylene and gibberellins (GA) will also strongly affect the sex expression in cucumber. It appears that the change of sex expression in the cucumber, as induced by GA treatment, does not cause a developing flower to *change* its sex expression. The evidence indicated that the short inflorescence produces pistillate flowers at the base of the inflorescence, while distally there are potential staminate flowers. When the hormonal treatment (e.g. GA) suppresses the lower (pistillate) flower initials, the upper (staminate) flowers (that are otherwise suppressed by the pistillate flower) will develop into mature flowers. However, treatment with antiethylene compounds of nodes that normally develop pistillate flowers will have a different effect. In this case, the anther initials of the embryonal pistillate flowers will grow, and the flowers will become either bisexual or staminate. Unfortunately, cucumber is not a good model plant for molecular genetic studies, and thus the promotion of flowering and the sex determination of cucumber flowers (that are bisexual at the very early stage of development) is still awaiting clarification. The terms assigned to the different types of inflorescence may confuse some of the readers; therefore, several types are schematically presented in Fig. 50.

In the following paragraphs of this section, I handle mainly model plants such as snapdragon and Arabidopsis, as well as some grasses,

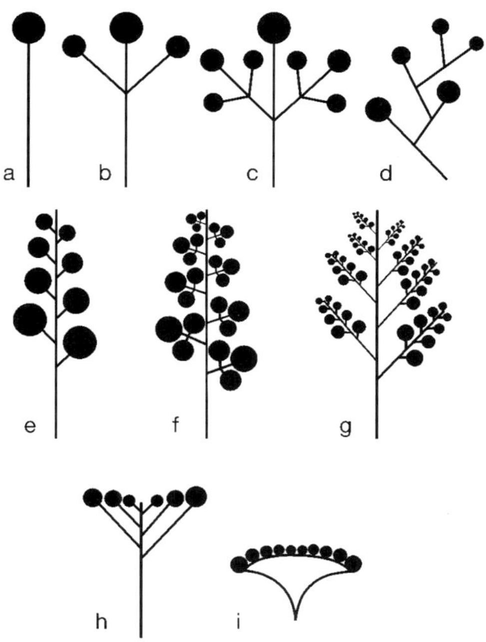

Fig. 50. Diagrammatic illustration of various inflorescence types. **(a)** Solitary terminal flower; **(b)** simple dichasium; **(c)** compound dichasium; **(d)** cyme; **(e)** raceme; **(f)** thyrse; **(g)** compound raceme; **(h)** corymb; **(i)** capitulum. Flowers are indicated by solid circles. Bracts are not shown. (From Coen and Nugent, 1994.)

where the transition to flowering is rather clear and molecular genetic methodologies are easily applicable. Once the understanding of the transition from vegetative growth to the reproductive phase becomes clearer in the model plants, it may be possible to gain useful information from other plants.

As in the previous section on shoot branching, the model dicot plants are handled in a separate paragraph from the model monocot plants. I also discuss the interactions between genes, hormones and environmental conditions in the process of transition from vegetative growth to the reproductive phase.

Transition to flowering in dicots

Two very useful early reviews on the evolution of flowers and inflorescences and the transition to flowering in dicots came from the John Innes Centre in Norwich, UK: Coen and Nugent (1994) and Levy and Dean (1998). The names of the two first authors of these reviews cannot pass without notice by an author who himself is of the Levy tribe: the Coens and the Levys were the stewards and the chorus members, respectively, in the Temple of Jerusalem until 70 AD.

The first of the above-mentioned reviews (Coen and Nugent, 1994) could already mention several important genes that affect inflorescence development. One of these, studied in *Antirrhinum* by E.S. Coen, R. Carpenter and associates, was *FLORICAULA* (*FLO*). The wild-type *Antirrhinum majus* (snapdragon) first produces a vegetative main shoot, where two opposite leaves develop from each node. After the switch to the reproductive phase, single flowers develop from the axils of the bracts and a raceme is produced. In the *flo* mutant, the flowers are replaced by shoots. The latter may further ramify, but no flowers are produced. It appeared that instead of becoming a floral meristem, the meristems of the reproductive phase of the shoot in the mutant produce vegetative meristems. This can be viewed as a change from terminal meristems in the wild-type to indeterminate meristems in *flo* plants.

The transcripts of *FLO/flo* are indeed found first in their subtending bract primordia. At a later phase (in wild-type), they are expressed transiently in all floral member primordia, except stamens. Clearly, a change in the timing of the expression of *FLO* would result in abnormal phenotypes; for example, a slight delay in expression could result in clusters of flowers rather than a single flower subtending a bract, whereas a further delay in *FLO* expression could result in a highly branched inflorescence. The indeterminate, wild-type inflorescence could be converted into a determinate one if the upper meristem is converted into a floral meristem. This happens in snapdragons with a mutation (revealed many years ago by H. Stubbe, in Germany) in the gene *CENTRORADIALIS* (*CEN*). Indeed, *cen* mutants have a short raceme, terminated with a single flower. It is possible that *CEN* prevents the apical meristem from developing into a flower; however, other modes of action could not be excluded. A mutant of the gene *TERMINAL FLOWER* (*TFL*) in Arabidopsis (i.e. *tfl*) has similarities to *cen*, but *tfl1* flowers early. The homolog of *flo* in Arabidopsis (*fly*) was studied intensively. In the *fly* mutants of Arabidopsis, the flowers are replaced by secondary shoots subtended by caulin leaves, although in older mutant plants, the normal phenotype may be restored. Hence, in spite of the great phylogenetic distance between snapdragon and Arabidopsis, the two plants share controlling features of their inflorescences.

Another "old" gene was revealed by H. Stubbe, but this time in tomato that has a sympodial growth: *FALSIFLORA* (*FA*); in *fa* mutants of tomato — there are no flowering inflorescences, but the mutants still maintain the sympodial growth. This indicates that the sympodial growth is independent of the flower production. While in wild-type Arabidopsis (as in other species of the Brassicaceae family) the flowers are not subtended by bracts (as they are in snapdragon), in the *fly* mutants of Arabidopsis the floral primordia are subtended by bracts.

During the years 1994 to 1998, there was a blooming of identification of genes affecting flowering in Arabidopsis. The update of Levy and Dean (1998) listed over 40 genes and mutations that affect the flowering time in Arabidopsis, and their phenotypes were described. However, it was already then estimated that many more genes are involved in the transition to flowering. These 40 genes were found to

Table 8 Cloned Arabidopsis Genes that Affect Flowering Time (a)

Gene	Sequence Similarity and Probable Function
Promoters of flowering	
ADG-1	ADP glucose pyrophosphorylase, involved in starch metabolism
CO	Putative transcription factor with two zinc fingers
DET2	Steroid 5α-reductase, an enzyme involved in brassinolide biosynthesis
FCA	RNA binding protein with a protein-protein interaction domain
FHA	Cryptochrome 2, a flavin-containing blue light photoreceptor
FPF-1	Novel protein that may be involved in signaling or response to GAs
FT	*TFL1* homolog(b)
GA1	*ent*-kaurene synthetase A, an enzyme involved in GA biosynthesis
GAI	Member of a novel family of putative transcription factors
GI	Novel protein with putative membrane-spanning regions(c)
LD	Glutamine-rich homeobox transcription factor
PGM	Phosphoglucomutase, involved in starch metabolism
PHYA	Light-labile R-FR light photoreceptor
Repressors of flowering	
CCA1	MYB-related transcription factor: *LHY* homolog
CLF	Homology to *Enhancer of Zeste*, a Drosophila polycomb-group gene
ELF3	Novel protein(d)
ESD4	Novel protein
LHY	MYB-related transcription factor; *CCA1* homolog
PHYB	Light-stable R-FR light photoreceptor
SPY	*O*-linked *N*-acetylglucosamine transferase, involved in modification of proteins
TFL1	Similarity with phosphatidylethanolamine binding proteins
WLC	Novel protein(e)

(a) For up-to-date information on the cloning of genes involved in flowering time, refer to "The Flowering Web" (http://www.salk.edu/LABS/pblo-w/flower_web.html). (b) T. Araki and D. Weigel, personal communication. (c) K. Lee, G. Coupland, S. Fowler, and J. Putterill, personal communication. (d) D.R. Meeks-Wagner, personal communication. (e) C. Hutchison and C. Dean, unpublished data. (From Levy and Dean, 1998.)

be distributed in all the five chromosomes of Arabidopsis. Twenty-two of these genes were cloned and their sequences were analyzed (see Table 8). Moreover, the authors could already propose a scheme of the genetic pathways that control the flowering time in Arabidopsis

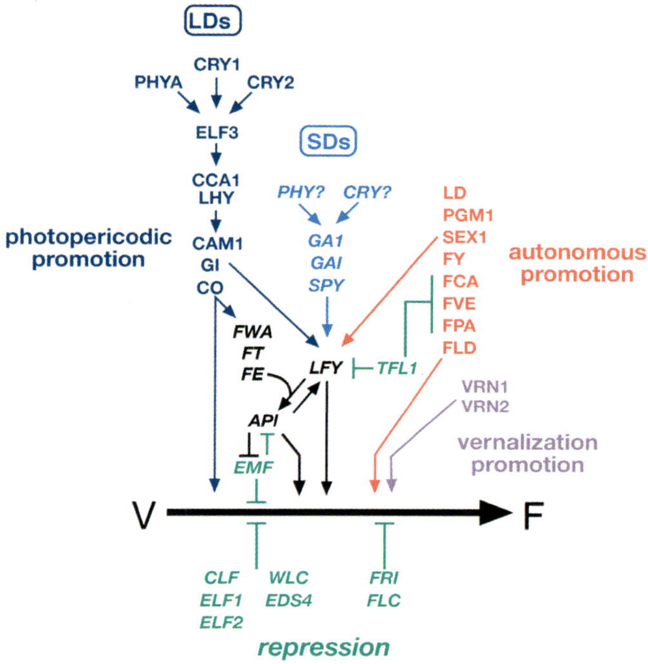

Fig. 51. Genetic pathways that control flowering time in Arabidopsis and proposed interactions among some of the genes involved. The horizontal line symbolizes the vegetative (V) to floral (F) transition, with the promotive and repressive pathways exerting their influence on this switch. Four pathways are shown: repression (green), autonomous promotion (red), photoperiodic promotion under long days (LDs; dark blue) and short days (SDs; light blue), and vernalization promotion (pink). Genes that influence both floral meristem identity and flowering time are shown in black. Promotive (*arrows*) and repressive (T-bars) interactions are based on genetic epistasis experiments and analysis of gene expression in mutant and overexpressing lines. Not all interactions have been tested directly, and little is known about how the floral repressors interact with the various promotive pathways: thus, most of the repressors have simply been represented below the horizontal line. (From Levy and Dean, 1998.)

and suggest interactions among genes involved in the autonomous (internal control), photoperiodic (external control) and vernalization (external control) promotions of flowering, as well as the repression of flowering (see Fig. 51). It should be noted that this scheme is based mainly on genetic information; the molecular biology of these interactions had to await further studies. Also, this scheme does not show where exactly these genes act and interact (i.e. in which cells of the meristems, leaf primordia and leaf axils activations and the interactions take place). Nor does this scheme inform us during which developmental phases these activities take place. In other words, the spatial and temporal aspects of the activities of the genes were not included in this scheme. Even more important, the file V → F in this figure conceals a "black box" that appears in several schemes on the regulation of the induction to flowering, such as in the following scheme of Grennan (2006) on the induction of flowering in Arabidopsis.

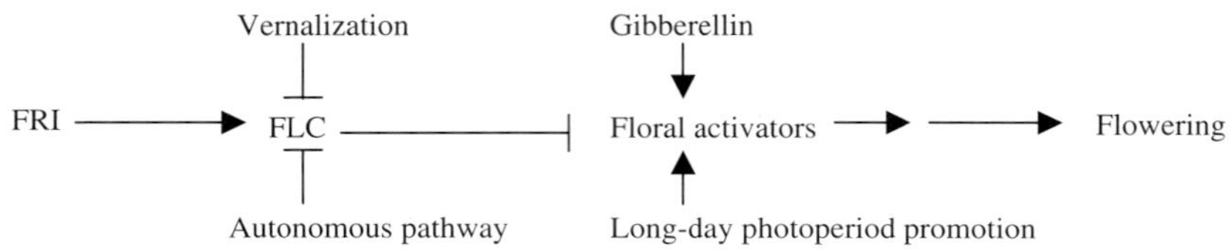

(In this scheme, arrows mean promotion and bars mean inhibition.)

In this scheme, the floral activators and the two arrows between floral activators and flowering are indeed a "black box". To better explain what I mean by this black box term, let us consider the following: the incipial axial meristem has a choice between two "blueprints". One blueprint gives exact instructions on how to build a vegetative meristem, and the other blueprint gives exact instructions on how to build a floral bud (or an inflorescence). We know that there is such a choice, and are beginning to understand which genes as well as external and hormonal factors affect the choice; however, the blueprints themselves are unknown to us. They are still unknown now as this book goes to print. This is obviously a general problem in patterning, and only in very few cases have the blueprints been at least partially resolved. We shall see one such case in the patterning of floral members in snapdragon and Arabidopsis, due to the pioneering studies of Z. Schwarz-Sommer, E. Coen, E.M. Meyerowitz and their associates. This is detailed in Chapter 12 that deals with patterning of floral organs.

Let us turn back to studies on the molecular genetics of the transition from vegetative growth to flowering. From 1998 to 2003, there was an intensive activity by molecular geneticists to reveal more and more genes that are involved in this transition. These activities were updated by several good reviews, such as Smith (1999), Pidkowich *et al.* (1999), Samach and Coupland (2000), Araki (2001), Simpson and Dean (2002), and Mouradov *et al.* (2002). These reviews dealt with a plethora of genes, but a few of these genes attracted the main attention of the investigators. Among the latter were *LEAFY* (*LFY*), *APATELA1* (*AP1*), *CAULIFLOWER* (*CAL*), *TERMINAL FLOWER1* (*TFL1*), *FRIGIDA* (*FRI*), and *FLOWERING LOCUS C* (*FLC*). A significant step to provide a clear picture of the molecular basis of flowering time was provided by the important study of Gazzani *et al.* (2003) from the laboratory of C. Dean at the John Innes Centre that focused on two genes: *FRI* and *FLC*.

Before describing the essence of the study of Gazzani *et al.* (2003), it is helpful to recall the growth and flowering habits of Arabidopsis ecotypes. The ecotypes that are commonly used for molecular genetic studies are the so-called "rapid cycling" ecotypes, such as Columbia (Col) and Landsberg *erecta* (*Ler*); physiologists call these "summer annuals." They do not require a period of cold (vernalization) (the term jarravization was coined by Trofim Lysenko, who for many years headed the USSR Academy of Agricultural Sciences and sent many Soviet geneticists to the colds of Siberia; this term was then translated to vernalization) in order to attain the flowering phase. Other ecotypes are slow- or late-flowering and respond to vernalization to facilitate flowering. By 2003, it was already established that *FLC* and *FRI* interact synergetically to confer the winter annual growth habit. In fact, some summer annual ecotypes seemed to contain a defective *FRI*. Since a functional *FRI* was known to be required for an active *FLC*, this deficiency in *FRI* could explain the rapid cycling of the respective Arabidopsis ecotypes. Moreover, it was suggested that rapid cycling ecotypes evolved from winter annuals that lost an active *FRI*. In March 2003, Michaels *et al.* (2003) of the laboratory of R.M. Amasino in the University of Wisconsin submitted a paper that showed that some annual ecotypes have a functional *FRI*, but they have an allelic variation in *FLC*. A very weak *FLC* could thus be the reason for their rapid cycling. Such ecotypes do not have null mutations in *FLC*, but rather have a low steady state of transcript (mRNA) of *FLC*.

At about the same time the Michaels *et al.* (2003) paper was submitted, the team of Caroline Dean from the John Innes Centre (Gazzani *et al.*, 2003) submitted their analysis of the molecular basis of flowering time variations in Arabidopsis ecotypes and also focused on *FRI* and *FLC*. There is, as expected, a similarity between the results of the two independent studies. Gazzani *et al.* (2003) acknowledged that many early-flowering ecotypes carry loss-of-function *fri* alleles that explain their rapid cycling habit. However, the analyses of five rapid cycling ecotypes (Cvi, Shakhadara, Wil-2, Kondara and K3-9) had only changes in this gene, but the gene was not eliminated. For example, the *FRI* allele of Cvi has a number of nucleotide differences (relative to wild-type), one of which causes an in-frame stop codon in the first exon. The other four ecotypes had *FRI* genes with nucleotide changes that caused amino acid substitutions in the encoded protein, thus reducing the functionality of this gene. Notably, there was no clear correlation between the *FRI* transcript levels and flowering time in the five analyzed ecotypes. Such a correlation did exist with respect to *FLC* transcripts.

The Dean team also analyzed the allelic variations at the *FLC* gene in several ecotypes of Arabidopsis. They revealed in the Landsberg *erecta* ecotype (a rapid cycling ecotype) a weak allele of *FLC*, probably resulting from a novel nonautonomous *Mutator*-like transposon. This transposon was located in the first intron of the gene. The intron region is required for normal *FLC* regulation. This transposon does not exist in other rapid cycling ecotypes.

It should be noted, and it was indicated in the publication of Gazzani *et al.* (2003), that several studies conducted before 2003 in several laboratories revealed that ecotypes differed in the alleles of *FRI*, and it was concluded that this gene was under selective pressure in various locations in the world to adapt the local ecotypes to the respective environmental (climatic) conditions. The difference in time to flowering among ecotypes is rather great: the common laboratory ecotype Landsberg *erecta* flowers after about 7 leaves, while the Kondara ecotype from Tajikistan flowers after about 58 leaves (when exposed to 16/8 hours of light/dark).

The picture that emerged after the studies of Michaels *et al.* (2003) and Gazzani *et al.* (2003) is the following: when a sufficient level of a functional FLC protein is produced, the transition to flowering in Arabidopsis is inhibited. This inhibition is repressed when the level of this protein is low or when it is not functional. The activation of the *FLC* gene is under control of *FRI*. When the latter's transcript is low or its product is not functional, *FLC* will not be activated to a level that would suppress the transition to flowering.

In the previous discussion of the transition to flowering in dicots, I neglected important components of this subject. Three of these components are (1) the cold signal that in Arabidopsis winter annuals is required for this transition; (2) the photoperiod regime that in Arabidopsis can facilitate flowering (i.e. long days); (3) that when the FLC-induced suppression of the transition is eliminated, several factors have to enter into play in order to realize the development of inflorescence.

We should now look at these neglected components. Let us start with the cold requirement. First, looking at the scheme shown above, there is no cold requirement when *FLC* is not fully functional or its encoded product is not produced at a sufficiently high level. However, when *FLC* is potentially active and at a sufficient level, there should be a signal resulting from a cold exposure that will suppress *FLC*. Sung and Amasino (2005), who provided an excellent review on the cold requirement for the transition to flowering, used a motto from the movie version of *The Wizard of Oz*, in which the scarecrow declares that he has no brain. They then asked how a plant can remember a past exposure to a prolonged winter if it has no brain. The question seems justified at first sight; but one soon recognizes that many organisms have neither a brain nor a nerve system (e.g. prokaryotes), yet they cannot only remember but even respond correctly to environmental cues. Their "wisdom" is stored in their respective genomes during a billion years of evolution. This is also true for plants, where this wisdom potentially exists in every one of their living cells.

One may claim that for responses to expected situations, there is no need for a brain; the information stored in the genome can cope with such situation. The brain is useful to cope with unexpected situations, especially when an organism has to fight against another organism that also has a brain. The short answer to the question of how plants remember the cold is that they do it by chromatin remodeling [for a background on this modeling, see the respective Appendix in my book, Galun (2003)]. For a more detailed answer to this question, we shall look closer at the vernalization phenomenon, and how the memory of it is stored and thereafter involved in the transition to flowering.

We may start with the classic experiments of Anton Lang and Georg Melchers with henbane (*Hyoscyamus niger*). The biennial henbane has two requirements in order to reach the flowering phase. It has a requirement for a cold period (vernalization) and then a requirement for an exposure to long days (inductive photoperiod). Unless exposed to the low-temperature regime, the henbane will never flower. Moreover, even after such an exposure, when the plants are then maintained in short days, flowering will be suppressed. However, vernalized henbane plants that are exposed to an inductive photoperiod will flower readily. The vernalized henbane plants will remember the cold treatment for a certain period (during which they are kept in a noninductive photoperiod). When then shifted to long days, they will readily flower. The length of this memory differs among winter annuals. It can be only a few days or last for many weeks. Many physiological experiments indicated that the shoot apex is the site of cold perception (and memory) during vernalization. Grafting experiments did not furnish an unequivocal answer as to whether or not the signal resulting from vernalization is transmissible.

What was repeatedly revealed, however, was that only dividing cells are capable of perceiving (and remembering) the exposure to cold. Once perceived, the memory is kept in the cell progeny of those cells that were the recipients of the cold exposure. This transmission of memory, through several cycles of mitosis or mitotically stable cellular memory, is typical of epigenetic transmission of traits. One more note: in addition to vernalization, plants may also have the ability of cold acclimation. The latter phenomenon is rather different from vernalization; it requires a very short exposure to cold (less than a day), involves many genes, and is instrumental for the resistance of the acclimatized plant to withstand severe cold. There are also cold periods that are required by perennials (e.g. deciduous fruit trees, such as peach and cherry) to break the bud dormancy. Both cold acclimation and the breaking of bud dormancy will not be considered in this book. Also, the handling of vernalization in grasses (*Poaceae* species, such as wheats) will be delayed to a subsequent section.

The two genes that were found to be involved with vernalization in Arabidopsis (*FRI* and *FLC*) were noted above; similar genes were identified and cloned in other Brassicaceae (*B. rapa* and *B. napus*). The cloning of *FLC* (Michaels and Amasino, 1999; Sheldon *et al.*, 1999) provided important information on this gene. The gene encodes a MADS-box transcription regulator. *FLC* was found to be predominantly expressed in mitotically active cells, such as shoot apices, which are the location of cold perception. It became clear that vernalization repressed *FLC*, and by that derepressed the transition to flowering. However, there could also be *FLC*-independent components that derepress flowering. It should also be noted that in winter annual Arabidopsis ecotypes, the cold requirement is facultative. These ecotypes may ultimately flower even without vernalization, although vernalization will accelerate flowering considerably. This ultimate flowering can be helpful in self-pollination of certain mutants that are winter annuals.

Now, here comes a warning: "mine danger." There are *VRN* genes in cereals, but there are also *VRN* genes in Arabidopsis. These identical designations can be confusing. We shall now deal with the *VRN* genes of Arabidopsis. Genetic studies with Arabidopsis to reveal mutants that impair the vernalization response led to the identification of three genes: (1) *VERNALIZATION2* (*VRN2*); (2) *VERNALIZATION1* (*VRN1*); (3) *VERNALIZATION INSENSITIVE3* (*VIN3*). *VRN2* encodes a Polycomb group (PcG) transcriptional repressor, i.e. a chromatin modifier. *VRN1* encodes a plant-specific DNA-binding protein, and *VIN3* encodes a PHD finger-containing protein (see Sung and Amasino, 2005 for details and literature). PHD finger proteins are often found in protein complexes involved in chromatin remodeling. The involvement of *VRN2* and *VIN3* in chromatin remodeling clearly indicated that chromatin remodeling is required for the process of winter memory (in the brainless Arabidopsis).

A Polycomb group (PcG) and the trithorax group (trxG) proteins were found to play a major role during the differentiation of animals such as the *Drosophila* fly in repressing and derepressing, respectively, the gene activity by chromatin remodeling. However, the PcG and trxG proteins are operative in animals by *internal* controls of the animals' development. The chromatin remodeling imposed through *VRN2* is activated by *external* stimulation (cold). But, the process in animals and in plants also has similarities: (1) both are stable through mitotic cell divisions; (2) both are relieved during meiosis so that the repression of the genes in the chromatin must be reset in each new sexual generation; (3) the repression concerns genes in the euchromatin; and (4) the animal PcG and the PcG-like proteins involved in the repression of *FLC* expression involve the Zeste12 (Su(z)12) complex.

The Enhancer of zeste (E(z)) of the PcG complex is a histone methyltransferase, an enzyme that can modify the amino-terminal tails of histones. There are several types of these modifications; but one of them, dimethylation of lysine9 (k9) in histone H3, is a common remodeling that represses gene activity. Therefore, the finding that — in the mutants *vrn1*, *vrn2* and *vin3*, which impair the vernalization process — there is no dimethylation of the k9 of H3 is a clear indication that these genes are involved in the remodeling of chromatin, and that the realization of the vernalization effect requires chromating remodeling. In other words, the vernalization causes specific regions of the chromatin (i.e. where the *FLC* resides) to change the histone/DNA complex from active euchromatin to inactive heterochromatin.

The full picture of the relation of *FLC* expression and chromatin remodeling is more elaborate than summarized above, but a detailed account is beyond the scope of this book. Nevertheless, one more note: while the dimethylation of k9 in H3 represses the expression of a gene that resides in the respective region of the chromatin, the trimethylation of k4 in H3 can facilitate the transcription from the respective chromatin region and thus can enhance the expression of *FLC*. Hence, different chromatin remodeling could either repress or enhance the transcription of *FLC*. A scheme for the maintenance of active transcription of *FLC* (in the absence of vernalization) and the repression of *FLC* expression by vernalization, as affected by chromatin remodeling, is provided in Fig. 52.

As indicated above, in rapid-cycling Arabidopsis ecotypes, the *FLC* is either not functional or is not transcribed in sufficiently large amounts to repress the transition to flowering. A similar situation

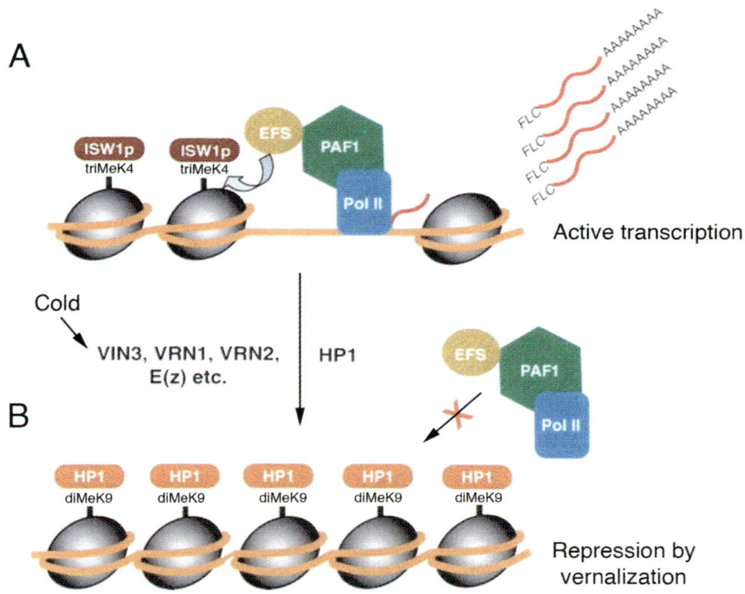

Fig. 52. Maintenance of active and repressed states of *FLC* transcription by chromatin modification. **(A)** Activation of *FLC* transcription by either the presence of *FR1* or loss of autonomous pathway genes requires a suite of proteins that are involved in chromatin remodeling. **(B)** Vernalization results in heterochromatin formation at the *FLC* locus. Heterochromatin formation requires the induction of VIN3 by a prolonged period of cold exposure, and involves a series of histone modification, such as deacetylation and methylation. VRN2 and VRN1 are also involved in the methylation of K9 of histone H3. Enhancer of Zeste (E(z)) is histone methyltransferase that is responsible for methylation on histone H3 in PcG-mediated repression; there are three homologs of E(z) in *Arabidopsis* that might play a role in this process. HP1 may recognize and bind to dimethyl K9 on histone H3, to participate in maintaining the heterochromatin-like state of FLC through mitotic cell divisions. (From Sung and Amasino, 2005.)

probably takes place in other annual dicots that do not require a cold period for the transition to flowering. A more recent scheme for the pathway to flowering in Arabidopsis was proposed by R. Amasino (2004) (see Fig. 53).

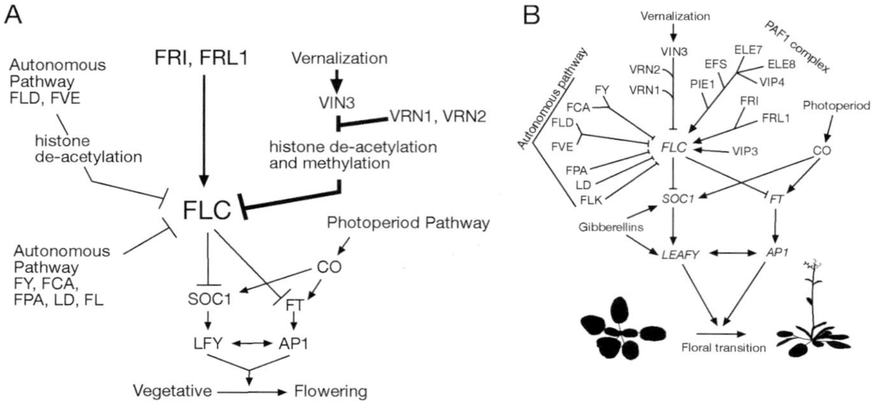

Fig. 53. **(A)** A scheme of flowering induction in Arabidopsis. This is a simplified model that does not contain all of the genes involved in flowering time control in Arabidopsis. The thickness of the lines indicates the hierarchy of *FLC* regulation: *FRI* overrides the repressive effect of the autonomous pathway and vernalization overrides the effect of *FRI.* (From Amasino, 2004.) **(B)** A more elaborated scheme of the induction to flowering. The autonomous pathway regulators and vernalization repress *FLC* expression. FRI and FRL1 upregulate *FLC* expression, and the H3-K4 trimethylation mediated by the PAF1 complex (ELF7, ELF8 and VIP4) activates *FLC* expression. FLC represses expression of the "flowering-time integrators" *SOC1* and *FT*, whereas the photoperiod pathway promotes expression of these integrators. *SOC1* and *FT* expression leads to the induction of floral-meristem-identity genes such as *LEAFY* and *AP1*, and thus of flowering. Lines with arrows indicate upregulation (activation) of gene expression and lines with bars for gene repression. Abbreviations: *AP1, APETALA 1; CO, CONSTANS; EFS, EARLY FLOWERING IN SHORT DAYS; ELF7, EARLY FLOWERING 7; ELF8, EARLY FLOWERING 8; FLC, FLOWERING LOCUS C; FLD, FLOWERING LOCUS D; FLK, FLOWERING LOCUS K; LD, LUMINIDEPENDENS; PIE1, PHOTOPERIOD INDEPENDENT EARLY FLOWERING 1; SOC1, SUPPRESSOR OF OVEREEXPRESSION OF CONSTANS 1; VIN3, VERNALIZATION INSENSITIVE 3; VIP3, VERNALIZATION INDEPENDENCE 3; VIP4, VERNALIZATION INDEPENDENCE 4; VRN1, VERNALIZATION 1; VRN2, VERNALIZATION 2.* (From He and Amasino, 2005.)

In this pathway, the genes *LFY* and *AP1* are required for the transition to the flowering phase. *FLC* was given a central role as an inhibitor of this transition. The expression of *FLC* can be repressed primarily by the vernalization process mentioned above. Also, the autonomous pathways that include genes such as *FLD*, *FY*, *FL* and *FCA* can repress *FLC*. Here, I shall divert again to indicate that FCA is not an acronym that stands for a long name of a gene (such as *FLC*), but rather is an arbitrary name given to a gene in Arabidopsis that, after its mutation, was defective in flowering (Koornneef *et al.*, 1983).

Photoperiodism and the transition to flowering

In the scheme of Fig. 53, the photoperiod pathway is indicated. The subject of photoperiodism became a major issue in plant physiology after 1920, when it was unequivocally found that some plants require short days in order to flower, while other plants require long days (still other plants are not affected by day length with respect to flowering). Again, a comprehensive discussion of photoperiodism is far beyond the scope of this book. I shall therefore handle only certain aspects of photoperiodism that are essential to the patterning process that causes the transition from vegetative growth to flowering. Because most experimental work on the molecular genetics of the impact of photoperiods on the transition to flowering in recent years was performed with Arabidopsis, it should be noted that Arabidopsis is considered to be a long-day-requiring plant. Indeed, most of the Arabidopsis ecotypes flower sooner when exposed to long days, such as daily cycles of 16 hours of light and 8 hours of darkness. There are several major questions regarding the impact of photoperiods on the transition to flowering:

- Which organs of the plant sense the photoperiod?
- How does a plant measure the light and dark periods, and thus how does the plant "know" that it is maintained in either long days or short days?

- Once perceived, how is the signal resulting from the photoperiod transferred to the respective meristems that manifest the transition to flowering?
- How are hormones involved in the transition to flowering, and do they interact with the photoperiod?

A recent short review by Blazquez (2005) provides a summary of the known answers to the above questions, with references to recent studies on the subject. For readers who are interested to look into the sources of publications that dealt with this subject, this review is indeed recommended. On the other hand, one can skip the arguments on why it is of benefit for the reproduction of a plant to be able to sense its environment (e.g. the thermal conditions) because these are rather obvious.

The answer to the first question, which organ perceives the photoperiod, came from numerous physiological studies. The photoperiod is clearly perceived by leaves (only or mainly). A study by my colleagues and myself (Galun *et al.*, 1964) performed more than 40 years ago with *Pharbitis nil* gave a clear answer. This plant can be exposed to a single inductive day a few days after germination, and the induced floral bud will be visible within a week. When the leaves are removed soon after induction, there will be no floral bud. Also, a treatment that will prevent mRNA synthesis abolishes floral induction. Due controls negated the "grasshopper experiment" possibility (meaning that the removal of leaves could have other effects, not just the elimination of the photoperiod signal). The grasshopper experiment is a joke about young experimentalists who trained a grasshopper to jump by order. When they removed one leg and gave the order, the grasshopper still jumped. This was repeated after more legs were removed; the grasshopper still jumped, but when all six legs were removed and the order was given, the grasshopper did not jump. The experimentalists recorded their results: "When all six legs were removed from the grasshopper it did hear the order." Amazingly, the experimentalists were right because the hearing organ of grasshoppers is on their legs! Regardless, it is presently agreed by most scholars of this phenomenon that after perception of the photoperiod by the leaf, a signal is transferred to the meristems, causing the change to flowering there.

The measurement of day length by plants is associated with the phenomenon of circadian rhythms that are found not only in angiosperms, but probably in all (or at least most) organisms: bacteria, fungi, flies, mammals and angiosperms. All these have circadian clocks ("circadian" means approximately 1 day long), where the running of the clock is based on feedback loops in which cycling gene products negatively control their own levels of expression (see Dunlap, 1999, for a general review on circadian clocks; and Millar, 1998, as well Yanovsky and Kay, 2001, 2003, for reviews on plants' clocks). I cannot refrain from noting the association with *Alice in Wonderland* by Lewis Carroll (Rev. C.L. Dodgson), where the rabbit is running, but his clock stays put — in the circadian rhythms of plants, the plants do not move, but their clocks are running.

The circadian clocks of organisms are of benefit to the regulation of many activities by synchronizing them with the day/night cycles in which they live. In plants, the circadian clocks are required not only to regulate the transition from vegetative growth to flowering, but also for leaf movements, opening and closure of stomata and the expression of genes involved in photosynthesis. The circadian clocks are adjusted by the signals of the solar cycle. Typically, the circadian rhythm is largely unaffected by temperature. Already in the 1930s, Erwin Bünning provided a hypothesis that time measurement in seasonal responses relies on a circadian oscillator, that a circadian clock derives a rhythm in a light-sensitive process and that photoperiodic responses are promoted (i.e. in long-day plants) or inhibited (in short-day plants) when the illuminated part of the day overlaps with the most sensitive phase of this endogenous rhythm.

This basic idea was amended during the last 70 years, especially after numerous genes were identified through mutants that are defective in their photoperiodic response, in Arabidopsis and other plants. The respective genes were cloned and characterized as summarized by Millar (1998) and by Yanovsky and Kay (2002, 2003). I shall only mention some of these: *CHLOROPHYLL A/B-BINDING PROTEIN2* (*CAB-2*), *CIRCADIAN CLOCK ASSOCIATED1* (*CCA1*),

CONSTANS (*CO*), *EARLY FLOWERING3* (*ELF3*), *LATE ELONGATED HYPOCOTYL* (*LHY*), *ZEITLUPE* (*ZLT*), *GIGANTEA* (*GI*), *TIMING OF CAB EXPRESSION1* (*TOC1*), *FLOWERING LOCUS T* (*FT*) and recently *FD* was added. The latter was already listed (as a mutant coined *fd* by Koornneef *et al.*, 1983). However, these are just the "tip of the iceberg", because recent studies indicated that in Arabidopsis about 2000 genes are either activated or repressed after a single long day! (Wigge *et al.*, 2005). Luckily for the readers, only a handful of these genes were found to play major roles in the circadian clock and the photoperiod response.

A proposal (see Yanovsky and Kay, 2003) for the roles of some genes of Arabidopsis says the following: There is an interaction between *TOC1*, *LHY* and *CCA1*. The *CCA1* and *LHY* products are MYB-related transcription factors. The mRNA and protein levels of the two respective genes peak at dawn. The overexpression of *CCA1* or *LHY* (used as transgenes with a strong promoter) leads to the downregulation of each other's expression, implying that *CCA1* and *LHY* are partially redundant components of a negative feedback loop that is essential for generating and sustaining the circadian rhythm. These authors further suggested that during the late evening, *TOC1* and *ELF4* activate the expression of *CCA1* and *LHY*. The proteins encoded by these two latter genes peak at dawn (when phytochromes that are activated by light move to the nucleus, where the transcription factors are upregulated). However, high levels of *CCA1* and *LHY* repress the transcription of *TOC1*. During the day, the levels of *TOC1* and presumably also *CCA1* and *LHY* levels are gradually reduced; thus, the repressive effect on *TOC1* transcription is released and the level of *TOC1* peaks at dusk. The *TOC1* accumulates, and by that again promotes expression from *CCA1* and *LHY*. This is clearly a simplification of the situation, but it features the feedback mechanism in the circadian system.

We should take into consideration another component of the circadian system: the perception of light. We may claim that plants can see, but their vision is with zero resolution. The intensity and quality of light can thus be perceived. For quality, plants have four types of chromophores (phytochromes for red/far-red, and three chromophores for blue/UV-A). Clearly, the quality of light changes during the day and plants can probably sense these changes (especially during dawn and dusk). It seems that plants are able to use the different phytochromes to perceive different qualities and intensities of light. For example, phyA may serve for clock adjustment under very low-intensity red light and high-intensity far-red; whereas phyB, phyD and phyE mediate clock entrainments in response to high-intensity red light. Cryptochrome1 may be important for clock resetting by high-intensity blue light. It is not yet fully understood how the light can set the clock, but a short light pulse could induce the expression of *CCA1* and *LHY* by the interaction of a chromophore with another protein. Clearly, when one component of the clock is not functional (e.g. a mutation in *TOC1*), the whole feedback will malfunction.

Another important gene in this system is *CONSTANS* (*CO*). Mutations in *CO* delay the flowering of Arabidopsis in long days, but do not affect the flowering time of plants maintained in short days. The *CO* encodes a transcriptional regulator that accelerates flowering time on long days through the upregulation of *FLOWERING LOCUS T* (*FT*). The upregulation of *FT* then promotes the expression of (flowering) meristem identity genes. Here comes another complication: the *CO* expression is regulated by the circadian clock mostly during the dark period of plants maintained in short days, while high levels of *CO* mRNA overlap with the light part of the day at dawn and dusk under long-day regimens. We can now look at what may happen during short days and during long days in the two model plants, Arabidopsis and rice (which are long-day and short-day plants, respectively). In Arabidopsis, the light is perceived by phyA and cry2 and activates *FT* through the *CO*, and this promotes flowering in long days; during short days, there is not enough accumulation of *CO* and consequently the *FT* remains low. In rice, light is perceived by a stable phytochrome and represses the expression of *FT*-like genes through *CO*, whereas *CO* activates the *FT*-like genes in the dark. However, the full molecular genetic difference that causes the different responses of long-day plants and short-day plants to light/dark regimens is not yet available.

Similar phenomena as revealed in Arabidopsis probably occur in other dicots, but the molecular biology of the circadian clock and its relation to floral induction in the latter plants was much less explored. A relatively recent study with Petunia (Angenent *et al.*, 2005) revealed similarities in the genetics of meristem differentiation and transition to flowering between Petunia and Arabidopsis, although these two plants belong to phylogenetically distanced angiosperm families. Hence, many of the molecular and genetic control mechanisms are probably conserved between distinct angiosperms. In-depth studies with additional plant species may provide important information, possibly even surprises.

Photoperiodism and florigen — lore and reality

As indicated above, during the vegetative phase in the shoot of Arabidopsis, every new meristematic primordium is initiated at the flank of an existing apical meristem. During the vegetative phase, this primordium will form a leaf. However, after the transition to the flowering phase, the newly emerging primordium differentiates into a flower. It was also indicated above that while the plant feels the cold directly at its apical dome, it sees the daily light/dark regime by its leaves; but the consequence of this seeing takes place far away, in the apical dome of the shoot (in the emerging primordium). How is this seeing and the consequence bridged? Miguel Blazquez, in a perspective feature (Blazquez, 2005), reviewed three studies in the journal *Science* that presented experimental work on this subject. A fourth publication, by E. Lifschitz, Y. Eshed and their collaborators from the Israel Institute of Technology in Haifa and the Weizmann Institute of Science in Rehovot, respectively, made a more recent contribution to this question (Lifschitz *et al.*, 2006). Very briefly, Abe *et al.* (2005) and Wigge *et al.* (2005) reported on the interaction between FT, a protein that is encoded by a gene expressed in the leaves, and FD, a bZIP transcription factor that is present only in the shoot apex, where it triggers the expression of floral identity genes such as API1.

The third publication, in a subsequent issue of *Science* (Huang *et al.*, 2005), reported on how the FT transcript travels from the leaf to the shoot apex. The Wigge *et al.* (2005) team consisted of seven investigators from La Jolla, CA, and Tübingen, Germany. They showed that the transcription factor FD, which is formed at the shoot apex, waits for an interaction with the FT that has to arrive to the apical meristem from its source in the leaves. The interaction between FT and FD (these proteins also interact *in vitro*) will then trigger floral identity genes at the flank of the SAM. As discussed above, the CO protein accumulates under favorable light conditions in the leaves. The CO protein can directly activate other genes, such as *FT* and *SUPPRESSOR OF OVEREXPRESSION OF CONSTANS* (*SOC1*). However, there are also additional targets of CO, such as a structural gene for ethylene biosynthesis. The Wigge team found that of the many genes that change expression as the result of a single long day, there is only one gene, *FT*, that will not respond at all to day length when the *CO* is mutated to *co*. Hence, *FT* is a major target of CO. The FT and the FD were suggested to form a combined transcription factor that activates floral identity genes. However, the authors assumed that, in addition to the *CO-FT-FD* interaction, there are other possible interactions that may promote flowering, such as those where *SOC1* plays a role.

Similar results and conclusions were provided by Abe *et al.* (2005), a team of 10 investigators from various locations in Japan. The latter team stressed that *FT* is expressed in the phloem tissues of cotyledons and leaves, and was therefore considered to be a long-distance signal in flowering — a characteristic bestowed many years ago to florigen, although at that time florigen was believed to be another plant hormone, such as GA, auxin, etc. From neither Wigge *et al.* (2005) nor Abe *et al.* (2005) is it clear whether the protein FT itself moves from the leaves to the shoot apex or the mRNA of *FT* travels and is translated at the shoot apex. The third *Science* paper mentioned above (Huang *et al.*, 2005) claimed to have a clear answer: a single leaf of Arabidopsis exposed to an inductive photoperiod produces the mRNA of *FT* in the leaf and sends this transcript to the apex of the shoot. These investigators also suggested that "the FT mRNA is an important component of the elusive 'florigen' signal that moves from leaf to shoot apex."

The authors of the last of the four abovementioned publications (Lifschitz *et al.*, 2006) intended to obtain a broader picture of the systemic signal for flowering. Therefore, they added the tomato (*Solanum lycopersicum* L.), which is a photoperiod-insensitive plant, to their experimental system, in addition to Arabidopsis, in which flowering is facilitated by long days, and the tobacco cultivar Maryland Mammoth (MM), a short-day cultivar. They first focused on the tomato gene *SFT* that was found to be an ortholog of the Arabidopsis *FT*. Tomato plants that were mutants of this gene (*sft*) had lesions in the coding region of *SFT* and were late-flowering plants (the late-flowering in tomato is expressed by a high number of leaves produced on the main shoot before the first inflorescence, and thereafter in more nonflowering nodes between nodes that produce inflorescence). For example, tomato plants with the *sft-k* allele produced 15–20 leaves on the main stem before flowering, compared to 8–12 leaves in the wild-type tomato plants. Transgenic tomato plants that overexpressed *SFT* flowered earlier (and had some additional peculiar phenotypes). Moreover, using such overexpressing plants in grafting experiments showed that the graft component with the overexpression of *SFT* complements the graft component that was mutated (*sft*). When the grafted SFT-expressing branch was later eliminated, the effect of SFT complementation discontinued.

When the chimeric construct *35S:SFT* was transferred to the short-day tobacco cultivar, the transgenic tobacco flowered irrespectively of exposure to day length. Such an induction to flowering were also achieved when transgenic donor (*35S:SFT*) shoots were grafted to leaf petals of MM plants. Enhancing the SFT in Arabidopsis leaves was made possible by using a chimeric gene, in which *SFT* was activated by a leaf-specific promoter. This caused earlier flowering in the respective transgenic Arabidopsis plants. The investigators analyzed the existence of the *SFT* transcript in various parts of tomato plants and found expression of *SFT* in leaves, stems, shoot apices and flowers, but not in roots. The promoter region of *SFT* was used to activate the GUS reporter gene to follow the developmental regulation of *SFT* expression. Interestingly, detailed localization indicated that no expression of transcript was detected in the shoot apical meristem itself, where floral transition should take place.

However, when a rather sensitive examination was performed to answer whether or not the *SFT* transcript could move from a *35S:SFT* scion to a *sft* stock, the answer was that this transcript *did not* move into the stock tomato! In plain words, the authors furnished evidence indicating that *SFT* transcripts did not cross graft unions at detectable levels. On the other hand, the SFT protein of tomato seems to interact with SPGB (a homolog of the Arabidopsis FD protein) already in the leaf!

Is the systemic (florigen) signal in tomato the result of a complex that is produced in the leaves (with an FD-like protein), and does this complex cause the systemic floral induction? In summary, the florigen story is not yet finished. I cannot resist the temptation of mentioning a meeting with the late Folke Skoog in front of the main lecture hall of an Annual Meeting of the American Society of Plant Physiologists, about 20 years ago. Skoog had to tell me a joke (probably his own invention) and it went like this: Soviet and American plant physiologists were arguing about who is the greatest plant physiologist. The Soviets claimed that they have the greatest plant physiologist (Chailachian) because he identified a flowering hormone (florigen) that does not even exist. The Americans answered that the greatest plant physiologist is their's (A. Lang) because he found an antidote for the nonexistent flowering hormone. Obviously, the relations between F. Skoog and A. Lang were never amicable.

The transition to flowering in Poaceae species and in other monocots

The monocots comprise an extensive taxonomic group of plants, among them major crops such as rice, wheat and maize that belong to the Poaceae. This group also includes numerous other plant families, some of which have value as food (Musaceae, the banana family) and the family of Orchidaceae that has a great value for the aesthetic value of the flowers in many genera of this family and is the source of vanillin. The genetics and molecular genetics of the transition to flowering has been investigated in monocots much less than in dicots, but this trend has been changing in recent years. Rice was used as a model plant due to its known genomic sequence, and the great amount of genetic information that accumulated in maize also drew the attention of plant biologists to this crop plant. While the gross morphology of the inflorescences and the floral structure of monocots differs considerably from the inflorescences of dicots, a detailed comparison also reveals similarity. The general structure of the main shoots of maize and rice as well as the units of inflorescences of maize and rice are schematically shown in Fig. 45, and the scheme for the development from the vegetative phase to spikelet and floral organs in maize is presented in Fig. 47. More details on the florets of maize and their comparison with a typical dicot flower are shown in Fig. 54. Some floral organs of maize have different names than the floral organs in dicots (e.g. lemma, palea and ledicule of maize can be considered bract, sepal and petal of the dicot flower);

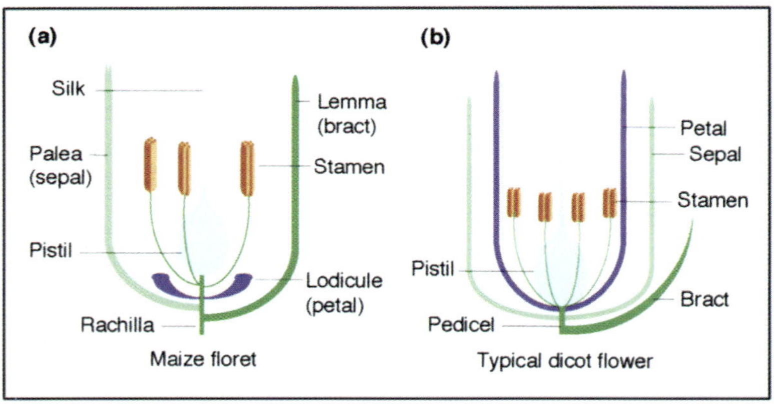

Fig. 54. A maize floret is similar to a typical dicot flower. **(a)** Stylized view comparing the floral organs of a maize floret with a typical dicot flower. The maize floret is composed of the lemma, palea, lodicules, and reproductive organs, stamens and pistil. During sexual differentiation, the stamens are aborted in the florets of the ear whereas the pistil is aborted in the florets of the tassel. **(b)** A typical dicot flower is composed of four whorls of organs: sepals, petals, stamens and pistil. The organs of each flower are color-coded to indicate analogous structures. The lemma, palea and lodicule of a maize floret might be equivalent to the bract, sepal and petal of a typical dicot flower. (From McSteen *et al.*, 2000.)

however, these floral members are considered to be equivalent, having a common phylogenetic origin. Also, as in dicots, the main stem of Poaceae species can be regarded as built of phytomers. Many monocot genera have a terminal inflorescence, such as spikes (in *Triticum*) or panicles (in *Pannisetum* and *Setaria*). However, maize (*Zea mays*) is a man-made species that has two kinds of inflorescences: the terminal tassel that is built of a main spike and lateral branches, and the lateral ears.

The basic similarity between dicot and monocot phytomers was revealed, at least in certain cases, by genetic information. For example, the cryptic leaf of the flower-bearing node in Arabidopsis is invisible morphologically; however, the gene *AINTEGUMENTA* (*ANT*) can render this leaf detectable, as it is in Poaceae species. In this example, the expression of a single gene can close the morphological gap between dicots and monocots with respect to phytomer structure. But, assigning equivalences for meristem types of Poaceae to those of Arabidopsis is problematic. Poaceae seem to have more types of meristems related to flowering than Arabidopsis. In the grass inflorescences, there are branch meristems, spikelet-pair meristems, spikelet meristems and floret meristems; while only branch meristems and flower meristems were revealed in Arabidopsis. McSteen and Leyser (2005) discussed this problem in some detail. Furthermore, there are mutual inhibitory interactions between the inflorescence meristem identity gene *TFL1* and the floral meristem identity genes *LFY* and *AP1* in Arabidopsis.

Mutations in the latter two genes cause increased branching and the replacement of flowers by an indeterminate axillary meristem. There are (partial) homologous genes in Poaceae that cause similar changes in architecture in the inflorescence. The *ZEITLUPE* (*ZFL1*) is such a partial homolog (in Poaceae) to the dicot *LFY*. In maize, there was apparently a gene duplication of the *LFY*-like gene, so that there are *ZFL1* and *ZFL2* on separate maize chromosomes. Both of the latter genes are strongly expressed in spikelet-pair meristems, floral meristems and floral organs. In Arabidopsis, *FLY* is most strongly expressed in floral meristems and floral organs. Does this mean that the spikelet-pair meristems and the spikelet meristems of maize have floral identity? Again, some observations indicate that the homology is not a simple one: the double mutants *zfl1:zfl2* have peculiar phenotypes that are not truly equivalent to phenotypes of *LFY* mutants. There is also a rice gene (*RFL*) that is an ortholog of the Arabidopsis *LFY*. When a chimeric gene having an *LFY* promoter and an *RFL* coding sequence was introduced into transgenic *lfy* Arabidopsis, the *RFL* partially compensated the *fly* mutant. However, overexpression of the Arabidopsis *FLY* gene in rice enhanced flowering, while overexpression of the rice *RFL* in Arabidopsis caused defects in leaf and general plant development.

Several genes were characterized in maize that, as implied by their mutant phenotypes, affect meristem identity and/or determinancy of AM types in the maize inflorescence. These were discussed by McSteen and Leyser (2005), and only a few will be noted below. The gene *INDETERMINATE SPIKELET1* (*IDS1*) encodes an *AP2*-like transcription factor. Loss-of-function mutants of *IDS1* produce 3 to 10 florets per spikelet instead of the normal 2 (Chuck *et al.*, 1998). Similarly, a mutant of the maize gene *REVERSED GERM ORIENTATION1* (*RGO1*) has three florets instead of two. There is a synergism between *ids1* and *rgo1*, and in the double mutant the spikelet meristem reverts to a spikelet-pair meristem after a number of glumes are initiated. The phenotype of the double mutant led to a possible model in which combinational interactions specify meristem identity.

Another notable maize gene is *INDETERMINATE FLORAL APEX1* (*IFA1*). In the double mutant *ids1:ifa1*, spikelet meristems in the ear are converted to branch meristem identity. In the tassel, spikelet meristems are converted to spikelet-pair meristem identity. This again points towards combinational interactions for meristem identity. A more recently cloned gene from maize is *BRANCHED SILKLESS1* (*BD1*). The functional gene was found to be required for the normal transition from spikelet to floral meristem identity. In the ears of the mutant *bd1*, the spikelet meristem produce branched structures that never progress to develop flowers on the ear (silkless). It is plausible that *BD1* encodes a transcriptional repressor of branch meristem identity, but the possibility that the product of *BD1* acts positively to activate floral meristem identity cannot be ruled out. An ortholog of *BD1* was identified in rice and termed *FRIZZYPANICLE* (*FZP*).

A team of five Japanese investigators (Murai *et al.*, 2003) studied the wheat homolog (*WAP1*) of *AP1*. Before summarizing the results of these studies, here are a few general remarks. First, we are dealing with the hexaploid bread wheat (*Triticum aestivum*), which is derived from three diploid genomes. Thus, genes can be triplicated. Then, as noted above, it is remarkable that the cultivation of the three most important crops started almost simultaneously, but independently, in three different regions of the globe. Bread wheat, rice and maize started to be cultivated about 6000 to 8000 years ago in the Middle East, China and Mexico, respectively. When bread wheat cultivation moved to Europe, the climate conditions required certain adaptations. Heading until after the cold winter avoided losses of yield. Indeed, genes involved in response to vernalization were identified in all three bread wheat genomes: *VER-A1*, *VER-B1* and *VER-C1* were located on chromosomes 5A, 5B and 5D, respectively. Also, genes for acceleration of heading under long days were revealed (*PPD1*, *PPD2* and *PPD3* on chromosomes 2D, 2B and 2A, respectively).

In addition to genes involved in heading time acceleration by vernalization and by long days, there is also a narrow-sense earliness (earliness *per se*) that is probably regulated by several genes. In Arabidopsis, the products of *FLY* and other genes activate the floral meristem identity gene *AP1* (as noted above). The latter is a MADS-box gene that promotes the formation of a floral meristem. Now, back to Murai *et al.* (2003) who found that the *AP1* homolog of wheat (*WAP1*) is involved in narrow-sense earliness. When *WAP1* was silenced by cosuppression in transgenic wheat, the narrow-sense earliness was negated, causing a delay in shooting. This suppression of *WAP1* did not affect the requirement for day length and vernalization. The authors thus concluded that WAP1 is a key component in the vegetative-to-reproductive phase transition, and acts downstream of the effects of day length and vernalization. As expected from a hexaploid, three homologous *WAP1* genes were revealed in bread wheat, and genes with very similar sequences were found in other Poaceae plants (e.g. *BM5* in barley, *LtMADS1* in *Lulium temulentum* and *RAP1B* in rice).

Further molecular genetic studies on the regulation of flowering in wheat by a gene (*VRN2*) involved with vernalization were performed by a team composed of nine investigators from several laboratories in the US (Yan *et al.*, 2004). *VRN2* is a vernalization gene that inhibits flowering, but it is downregulated by cold treatment. Loss-of-function mutants of *VRN2* resulted in wheat that had the characteristics of spring wheats, which do no require vernalization for the transition to flowering. These authors mapped, genetically and physiologically, the *VRN2* on chromosome 5A of *T. monococcum*, and revealed sequence similarity with equivalent genes in *Triticeae* species of temperate climates, but not in rice. They also furnished evidence showing that a previously isolated gene *ZCCCT1* (for zinc-finger domain in the first exon, and a CCT domain in the second exon) is identical to *VRN2*. Because in (most) wild *Triticeae* species vernalization accelerates flowering, the authors suggested that spring-type bread wheat evolved from cold-requiring plants through the loss-of-function mutation at vernalization loci. There is a problem with this suggestion. There is good evidence that wheat cultivation moved from the relatively warm climate of the Middle East to temperate Europe, and not from Europe to the Middle East. Hence, the suggestion of Yan *et al.* (2004) at least requires an amendment, such as that when tetraploid wheats moved southwards they lost the vernalization genes' function, or that further hexaploidization happened several times and in several different locations — but at this point, we dive into speculations that are outside the scope of this book.

Chang-Hsien Yang and his associates in Taiwan (Hsu and Yang, 2002; Hsu *et al.*, 2003) studied the molecular genetics of flowering time in Orchidaceae, a monocot family that is phylogenetically far removed from the Poaceae. They isolated from an orchid (*Oncidium* Gowen Ramsey) a gene that had similarities to the *AP1/AGL9* group of MADS-box genes. The gene, *OMADS1*, had extensive homology to the Arabidopsis *AGAMOUS-like6* (*AGL6*) gene. The expression of *OMADS1* was detected in the apical meristem of the orchid and in floral members. By *in vitro* analyses, the investigators found that the protein encoded by *OMADS1* interacts with the protein encoded by another orchid gene involved in floral initiation and flower formation in orchids. The orchid *OMADS1* was then expressed in various Arabidopsis lines, causing the development of transgenic plants. It was found that this ectopic expression of *OMADS1* promoted flowering in the transgenic Arabidopsis, and positively activated in Arabidopsis the flowering-time genes *FT* and *SOC1* as well as the flowering-initiation genes *LFY* and *AP1*. The results point towards the conservation of the mechanism of flower initiation in angiosperm since before the bifurcation into monocots and dicots. Another orchid MADS-box gene (*OMADS3*) that strongly affected floral member differentiation (Hsu and Yang, 2002) had sequence homology to some other B-group MADS-box genes in monocots and dicots, but its expression in orchid was low. This study also pointed to an ancestral gene involved in flower formation that existed before the bifurcation to monocots and dicots.

Plant hormones and the transition to the flowering phase

Reports on physiological experiments in which GA application facilitated flowering (e.g. in Brassicaceae, where GA caused the shooting of rosette plants) appeared half a century ago, but detailed information on the roles of plant hormones in the transition from vegetative growth to flowering is meagre. The rather detailed review of Zeevaart (1976) on the physiology of flower formation mentions only briefly the impact of application of the hormones ethylene, cytokinins, gibberellins and abcisic acid on flowering, and the extractions of these hormones that could provide information on the roles of plant hormones on flowering time. In Chapter 2, I devoted a paragraph to absisic acid (ABA) and noted there that details on the perception and signal transduction of ABA are still enigmatic. This is beginning to change, and recent publications from Canada (Razem *et al.*, 2006a, b) herald this change. In their publications, the investigators reviewed what is known about the physiology and molecular biology of ABA in plants, and pointed out that an important link in the signal transduction of ABA was missing: the identity of ABA receptors. They thus went to find this missing link. Razem *et al.* (2006b) actually found a receptor for ABA, and the finding was an unexpected one. For many years, information has accumulated for the involvement of ABA in several physiological phenomena, such as response to environmental stress, function of stomata and inhibition of germination. The Canadian investigators found that one receptor, the product of the *FCA* gene, binds ABA specifically. The FCA protein (as noted above) is a nuclear RNA-binding protein that is involved in *flowering time*.

The investigators screened a barley complementary DNA-expression library and identified a protein that they termed ABAP1, which bound ABA *in vitro*. They then revealed that ABAP1 was actually similar to the protein encoded by *FCA*. They undertook a series of experiments to assure that this binding is specific and that it affects flowering, as follows:

- The binding took place only when the active (+)-ABA was used; the analogs *trans*- and (−)-ABA showed no *in vitro* binding.
- In the process of flowering control, the FCA protein interacts with the 3′ RNA-processing factor FY; the investigators found that the interaction of FY with FCA was markedly inhibited by (+)-ABA, but not by (−)-ABA. It seemed that ABA either disrupts the FCA–FY complex or, in the course of association/dissociation, prevents the reassociation.
- ABA treatment phenocopied the disruption of FCA *in vitro*; for example, treatment of wild-type Arabidopsis with ABA led to the disruption of the processing of *FCA* pre-mRNA (which is required to maintain the effect of FCA on flowering).
- The application of ABA delayed the timing of the switch to flowering in wild-type Arabidopsis: there was an increase of days until bolting and an increase in the number of rosette leaves; again, this effect took place only after (+)-ABA application, but not after the application of (−)-ABA or *trans*-(+)-ABA.
- The ABA effect could disrupt the function of *FCA* not only in wild-type Arabidopsis, but also in ABA mutants (*abi-1*, *abi-2*); but ABA had no effect in plants that were mutations of *FCA* (e.g. *fca*-1).

All the above supports the claim that FCA is a receptor of ABA, but clearly there could be other plant proteins that will be revealed as ABA receptors.

Chapter 11

THE ANGIOSPERM LEAF

This chapter deals with the incipial initiation of primordia of leaves from the periphery of the shoot apical meristem (SAM); the growth of leaves from these primordia; and the further development of leaves and their components such as veins, trichomes and stomata. This book is confined to angiosperms; among these, the focus shall be on dicot leaves because there is more molecular genetic information on leaf patterning in dicots than on monocot leaves. Nevertheless, there is also ample information on the patterning of monocot leaves; hence, a special section shall be devoted to the latter leaves.

Initiation of leaves was already narrated briefly in previous chapters of this book (e.g. Chapters 7 and 8); however, this process is dealt with in a more comprehensive manner in the present chapter. For example, the involvement of hormones and microRNAs in the differentiation of leaves is discussed, and the control of leaf asymmetry is handled in some detail.

The epidermis of angiosperm leaves harbors two entities: trichomes and stomata. Two sections shall be devoted, respectively, to these entities. Plasmodesmata are gates for intercellular communication through which macromolecules such as proteins, nucleic acids and nucleoproteins (and plant viruses) can be trafficked between neighboring cells. Plasmodesmata are not confined to leaves, but they have important roles in the differentiation of leaves. The information on the molecular genetics of plasmodesmata patterning is meagre; therefore, it is not discussed in this book.

When Andrew Fleming (2005) of the University of Sheffield (UK) summarized the formation of leaf primordia and phyllotaxy, he started his review with a citation from the Austrian poet and dramatist Hugo von Hofmannsthal (1874–1929):

> *"Wüsst' ich genau wie dies' Blatt aus seinem Zweig hervorkam, schweig ich auf ewige Zeit still, denn ich wüsste genug."*
>
> *("If I knew precisely how this leaf grew out of its branch, I would remain silent forever, because I knew enough.")*

The poet of the Psalms (chapter 104) — which were rehearsed in the Temple of Jerusalem by the Levites about 3000 years ago — did not ask, "How?" — this poet only expressed bewilderment: "O Lord, how manifold are thy works! In wisdom hast thou made them all, the earth is full of thy riches." Well, von Hofmannsthal became "silent" before he or anyone else knew precisely how a leaf grows from a branch, but this chapter is intended to provide information on the present knowledge on structural and molecular genetic aspects of leaf initiation and patterning.

Innovations in methodological procedures are keys for the acquisition of knowledge. Such innovations as the use of lenses caused a leap in astronomy in the hands of Galileo Galilei (1564–1642), and other lenses were applied in pioneering structural studies with plants and animals by the members of the Academy of the Linceans (the first post medieval academic society established by Prince Federico Cesi in 1603). A recent leap in understanding patterning, including the control of the initiation of new lateral members (e.g. leaf initials), can also be traced to an optical innovation, but in this case it was combined with molecular genetic innovations. As recently reviewed in a perspective essay by Tax and Durbak (2006) and eluded previously in this book, Eliot Meyerowitz and collaborators (Heisler *et al.*, 2005; Reddy and Meyerowitz, 2005) combined the use of confocal microscopy with the establishment of recombinant probes (that permit the exact identification of specific cell types) to analyze gene-expressing patterns and the instantaneous consequences of perturbation of gene function in living SAMs.

The intercellular movement of two key players that serve as signaling molecules for the formation of new lateral organs (the CLV3 protein and the plant hormone auxin) can now be traced precisely, although one hurdle still exists: the exact quantitation of the hormone (auxin) at the cellular and subcellular levels is not yet available. If such a quantitation would be available, it could be determined at which level of IAA a given peripheral zone of SAM cell will be induced to initiate a leaf. Also, as shall be presented in some detail in this chapter, by the induced activation of transgenes that encode microRNAs with specific mRNA targets, one can eliminate (or strongly reduce) the translation of the respective specific proteins. The original source of the present leaves of angiosperms is not clear. There is good evidence for the extinction of most angiosperm lineages that existed in the very early Mesozoic era (early Cretaceous period, about 150 million years ago). It is also not clear from which seed plants angiosperms evolved. The previous suggestion that angiosperms evolved from Gnetales was not supported by later analyses (see Friis *et al.*, 2005 for review). Did seed ferns with macrophylls that were the precursors of the present angiosperm leaves exist in the very early Cretaceous period, but became extinct after they served as the source of angiosperms? Do the compound leaves of cycads represent the ancestors of angiosperm leaves? There are no sure answers to these questions. In short, the evolutionary source of angiosperm leaves is enigmatic.

While the paleobotanical origin of the leaf is concealed in a mist of extinct seed plants, the structural source of leaves is very clear: it is the stem shoot apical meristem (SAM). Leaves emerge from the periphery of this dome-like structure in an amazingly orderly way, typical for each plant species — this is the phyllotaxis, although phyllotaxis can be changed with the maturation of the SAM in certain plants. Scofield and Murray (2006a), who reviewed the evolution of the concept of meristem, referred to the Swiss botanist Carl Wilhem von Nägeli (1817–1891), who did most of his research in Munich, Germany, and had ample correspondence with Gregor Mendel. He coined the term *meristem*. According to von Nägeli, meristems are one of the two cell division tissues (the other tissue was known as *cambium*). Grossly, the meristem of the SAM can be divided into zones. There is a population of slower dividing cells in the central zone (CZ), and a surrounding peripheral zone (PZ) that constitutes the organogenic zone in which primordia of lateral organs (e.g. leaves) are initiated in a phyllotactic pattern.

We shall see that there is an intricate cross-talk between cellular regions in the SAM that include hormones, macromolecules, and other gene products such as microRNAs, which lead to the emergence of leaf initials and then also participate in the formation of leaf asymmetry and pattern formation of leaves. It is difficult to simultaneously perceive all the known factors that participate in the initiation and shaping of leaves. I shall therefore use "drawers," in each of which I shall focus on one or a few of such factors.

Leaf Development — An Overview of Early Studies

Ample structural information on angiosperm leaves was accumulated during the last 150 years. As our aim in this book is to understand leaf initiation and patterning, we should focus our attention on investigations that were performed since the early 1990s. The detailed review of Neelima Sinha (1999) of the University of California at Davis furnished an update of these investigations (up to 1998). Sinha summarized the then existing knowledge on angiosperm leaves under three main headings: (a) leaf initiation; (b) leaf morphogenesis; and (c) unique features of compound leaves.

For leaf initiation, we should look closely at the SAM, from where all leaves are initiated. While the summit of the SAM may have an indeterminate growth, the PZ projects the leaf initials. The indeterminate growth can proceed almost indefinitely, as in some tree species; but it may also terminate after the projection of a few leaves in a terminal flower, where the terminal floral members are carpels.

Periclinal cell division in the location of the future leaf initial marks the region of the location of the incipial leaf. It was found that an alteration of cell division, rather than an increase in division rate, is the first indication for a new leaf initiation. Moreover, in experiments with wheat in which cell division was suppressed by gamma irradiation, performed more than 40 years ago (when gamma irradiation of plants was fashionable), indicated that the location of the future leaf initials will bulge in the outer (L1) layer of the SAM. Other evidence also indicated that increased cell division is not the morphogenetic trigger of incipial leaf initiation. On the other hand, a loosening of the outer cell wall was implicated with leaf initiation.

By 1998, several genes were already identified that were implicated with leaf initiation. These genes encoded transcription factors or receptor/signal molecules. We shall detail below the homeodomain-containing proteins that are encoded by the *KNOTTED*-like genes (*KNOX* genes). Class I *KNOX* genes such as *Kn1* were found to affect the leaves of maize, and the class I *KNOX* gene *STM1* (*SHOOT MERISTEMLESS1*) was found to be downregulated at the appropriate sites of the future leaf initials. Also, another gene product (of *Rough Sheat1*) was implicated with initiation of a new leaf. Several other genes that were either directly or indirectly involved in leaf initiation were identified by 1989. These include the *CUPSHAPED COTYLEDONS* (*CUC*) and the *CLAVATA* (*CLV*) genes, which were mentioned already in previous chapters. Note that the names of these genes may be misleading; these names do not necessarily indicate the main effects of the respective genes. Another relatively early revealed gene, *PHANTASTICA* (*PHAN*), which encodes MYB domain proteins, was detected in *Antirrhinum* and affected the dorsiventrality of leaves, but was already then proposed to downregulate *KNOX* genes; hence, they also appear to be involved in the initiation of leaf primordia. The gene *LEAFY/FLORICAULA* (*FLO/LFY*) that encodes a protein with a transcriptional activating domain is expressed in newly initiated leaf primordia, but at the time this gene was suggested mainly to affect the determination of floral meristems.

Additional genes were found to affect the shape of leaves that started to grow. Hence, maize plants with a *narrow sheath* mutation will develop narrow leaves.

As for the dorsiventrality of leaves, as noted above, the *PHAN* gene was found to affect the asymmetry in *Antirrhinum* plants. The abaxial surface of emerging leaves grows faster than the adaxial surface, causing arching over the SAM. The abaxial surface also shows earlier cellular differentiation. *Antirrhinum* plants with a *phan* mutation lack the adaxial surface and are almost completely abaxial. This leads to a lack of an abaxial/adaxial boundary and a disruption of the normal flat shape of the leaves. The opposite phenotype was observed in Arabidopsis plants with the dominant mutation *phabulosa-1d*; in this mutant, the leaves lack the abaxial surface.

Less information existed by 1998 with respect to growth and differentiation at the proximal/distal leaf growth and the lateral growth of leaves than on abaxial/adaxial symmetry. The differentiation in the lateral dimension occurs concurrently with the proximodistal differentiation. While quite a number of genes were found to affect leaf shape, many of these had pleiotrophic effects, and thus the mode of the effect of such genes was not ascertained. Examples for the latter genes are *AGAMOUS* and *APETALA3*, which are homeotic genes affecting floral differentiation, but they can also lead to alteration of leaf patterning.

Leaves may be composed of leaflets and thus termed *compound leaves*. Such leaves could also be considered as highly dissected simple leaves, but the terminology is disputed among the various investigators; this discussion is outside the scope of this book. Several types of genes were found to affect the wild-type compound leaf of tomato (see Fig. 55). The dynamics of leaf and leaflet growth varies among dicots. In the tomato example, there is an early basipetal growth in the leaf primordium, which also generates the leaflets. This is followed in the individual leaflets by acropetal growth.

Several mutations, such as the *wiry* mutations in tomato, strongly affect the normal compound (wild-type) leaf. The four types of mutations that affect leaf shape in tomato are listed in Fig. 55. It is plausible that several types of mutations will be revealed in other dicots that have compound leaves. Presently, the molecular genetics of such genes is still obscure. The *KNOX* genes that have orthologs in monocots and dicots may also be involved in the shaping of a form of leaves, although their effect seems to be different in different plants. For example, *KNOX* genes are expressed in leaf and leaflet primordia of tomato; while in compound leaves of the Brassicaceae plants there is a downregulation of *KNOX* genes in leaf initials, and the expression of these genes turns back at a later phase when the leaflets are initiated.

A still different mode of *KNOX* gene expression was observed during the differentiation of the compound pea leaf. One explanation

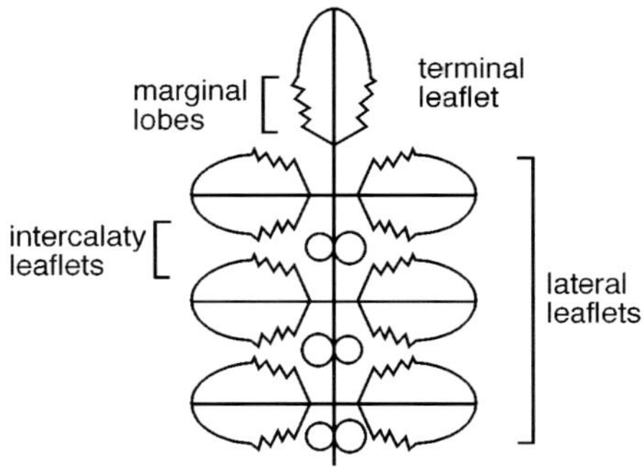

Type 1 mutants	no laminar expansion	wiry wiry-3 wiry-6	
Type 2 mutants	simple or nearly simple leaf, or greatly reduced complexity	Lanceolate entire lyrata	
Type 3 mutants	leaflet margins reduced or unlobed	solanifolia potato leaf	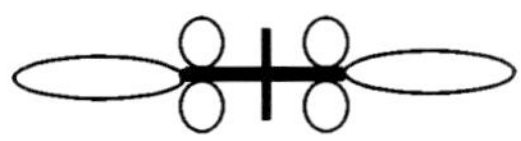
Type 4 mutants	leaves excessively subdivided	Mouse ears Curl Petroselenium bipinnata complicata clausa	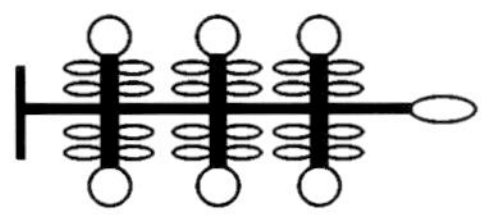

Fig. 55. The four classes of leaf mutations seen in tomato. Mutant phenotypes are described against each class and typical leaves are diagrammed. (From Sinha, 1999.)

was that the difference is only in timing: in tomato, the compound leaf program diverges earlier than in some other compound leaves. On the other hand, it is plausible that the compound leaves of tomato, Brassicaceae plants and pea have arisen by different mechanisms. The same plant genus may contain species with simple leaves and species with compound leaves; moreover, simple and compound leaves may develop on the same shoot. This could mean that the degree of leaf dissection involves only a small number of genes. Indeed, some mutated alleles of the *STAMINA PISTILODA* (*STP*) of pea (a homolog of the *Antirrhinum FIMBRIATA* gene) cause pea leaves to be simpler (rather than the wild-type compound leaves).

Also, another mutant of pea (*unifoliata*) causes a clear phenotype in pea leaves: the pinnate nature of the leaf is lost.

From the above noted mutants, it is clear that genes affecting the shape of leaves have a pleiotrophic effect: the unifoliata is caused by a change in the *PEAFLO* gene, which is the pea homolog of the *FLO/FLY* gene affecting flower differentiation; loss of *FLO/FLY* function causes indeterminacy in the inflorescence and floral meristem. When information is lacking, speculations may flourish. Thus, on the basis of published and personal communications, Sinha (1999) brought up the possibility that the compound leaf of tomato has some stem-like features and may be an intermediate structure between simple leaves and stems.

Focus on *Knotted1*-Like Homeobox Genes

The homeobox gene *knotted1* (*kn1*) was originally isolated from a dominant leaf mutant of maize by transposon tagging (Freeling and Hake, 1984). Since then, due to intensive investigations by Sarah Hake of the USDA in Albany, CA, and others, it became evident that there are *knotted1*-like genes in both monocots and dicots that encode rather similar proteins (and the introns in these genes are fairly conserved), indicating that these are evolutionarily old genes. Moreover, as these are homeobox genes, they share characteristics with other homeobox genes, such as encoding proteins with an evolutionarily conserved DNA-binding domain that is DNA-sequence specific. This strongly suggests that *knotted1*-like genes encode transcriptional regulators that control the expression of specific target genes. Further progress in the research on *knotted*-like genes was made possible after Hake and associates (Lincoln *et al.*, 1994) sequenced two knotted-like genes (*KNOX1* and *KNOX2*) from Arabidopsis. *In situ* hybridization was used to locate the expression of these genes during differentiation. Hence, *KNOX1* transcripts were found to be localized primarily to the SAM of Arabidopsis, and downregulation of this transcript happened as leaf primordia were initiated. Also, the expression of *KNOX1* in the shoot meristem of Arabidopsis decreased during the transition to flowering and became restricted to the cortex of the inflorescence stem.

It is interesting that double mutants of *serrate* (*se*) and either *asymmetric1* (*as1*) or *as2* had lobed leaves, ectopic stipules and sinuses — i.e. phenotypes obtained in transgenic plants, in which *KNA1* (a *knox* gene) was driven by the 35S CaMV promoter (Ori *et al.*, 2000). These results strongly suggested that *KNOX1* (and presumably other *KNOX* genes) plays an important role in morphogenesis. Further studies with *KNOX* genes indicated that these can be divided into two main classes. While class I *KNOX* genes are typically expressed in meristems, the expression of class II genes is in various locations in plants. There is also a difference between these two classes with respect to the location of a conserved intron in the coding region. Phylogenetic analyses that could be used to construct family trees of *KNOX* genes indicated that class I and class II *KNOX* genes are on two distinct branches (Reiser *et al.*, 2000). Further studies were performed mainly with class I *KNOX* genes. However, it should be noted that even within class I there are different genes, and each of these has its unique pattern of expression. For example, *KNAT1* is expressed in a ring around the meristematic dome of Arabidopsis, while *STM* is expressed almost all over the SAM dome of this plant (Reiser *et al.*, 2000). Neelima Sinha and associates (Bharathan *et al.*, 2002) examined the expression of *KNOX1* in various vascular plants, and found that the expression of this gene correlated with complex leaf primordia. However, some of these complex primordia matured into simple (not compound) leaves. Such differences in final complexity were revealed in species of the same genus. Hence, not all simple leaves develop similarly. However, regardless of final complexity or simplicity of the leaves, the *KNOX1* expression was found in angiosperms and gymnosperms to be downregulated at the site of leaf initiation, but no such downregulation was revealed in ferns.

As indicated above, the juxtaposition of abaxial and adaxial cell fates is required for the formation of lamina in simple leaves (e.g. of Arabidopsis). Kin *et al.* (2003) found that the expression of the *LeT6* gene (a tomato *KNOX* gene) and the *LePHAN* gene were reduced during postprimordial leaf development, and this reduction was correlated with a reduction of leaflet formation in tomato *wiry* mutants. The results also suggested that *LeT6* codes for a negative regulator of

LePHAN expression. Hence, it was proposed that the regulatory relationship between *KNOX* genes and *PHAN*-type genes has a role in compound leaf development. However, these results are still far from informing us as to how the typical compound leaf of tomato is patterned, nor is it clear whether or not there are other genes that play an important role in the differentiation of a compound leaf.

Scofield and Murray (2006b) of the University of Cambridge (UK) updated the information of the *KNOX* genes in the shoot and its lateral outgrowth (e.g. leaves, floral members) of plants. The overall effect of *KNOX* (class I) genes is to inhibit cell expansion and differentiation, which lead to organogenesis. Their role in the summit of the SAM is thus simple and clear: the *STM* (a *KNOX* gene) is expressed at the summit, maintains meristematic activity and keeps the indeterminancy of the SAM. We should recall that the *KNOX* genes are related to other homeobox genes of plants, such as *WOX*, *BELL* and *HD-ZIP*, all of which encode proteins that contain a DNA-binding motif. Hence, they can bind to *cis*-regulatory DNA sequences and regulate transcription.

While the first *KNOX* gene (*KNOTTED1* or *ZmKN1*) was revealed in maize (Vollbrecht *et al.*, 1991), many of the more recent studies with *KNOX* genes were performed with Arabidopsis, where there are several *KNOXI* genes such as *SHOOT MERISTEMLESS* (*STM*), *KNAT1* (also termed *BP*), *KNAT2* and *KNAT3*. The phylogenetics and the structural relationship between these genes are schematically shown in Fig. 56. There are additional similarities among the *KNOX* genes: the canonical *ZMKN1* (or *ZmKN1*) is closely related to *OSH1* and *HVKNOX3* (of rice and barley, respectively); and the Arabidopsis *STM* is similar to *LeT6* and *GMSBH1* of tomato and soybean, respectively. Since several of the various class I *KNOX* genes were sequenced, the various homeodomains were located on these genes, but the details are outside the scope of this book (see Scofield and Murray, 2006b for references). What should be remembered is that in some plants different *KNOX1* genes were found to be expressed in typical spatial and temporal manners. For example, in tobacco, several *KNOX1* genes were found to be expressed in different (specific) zones of the shoot apex, *NTH1* and

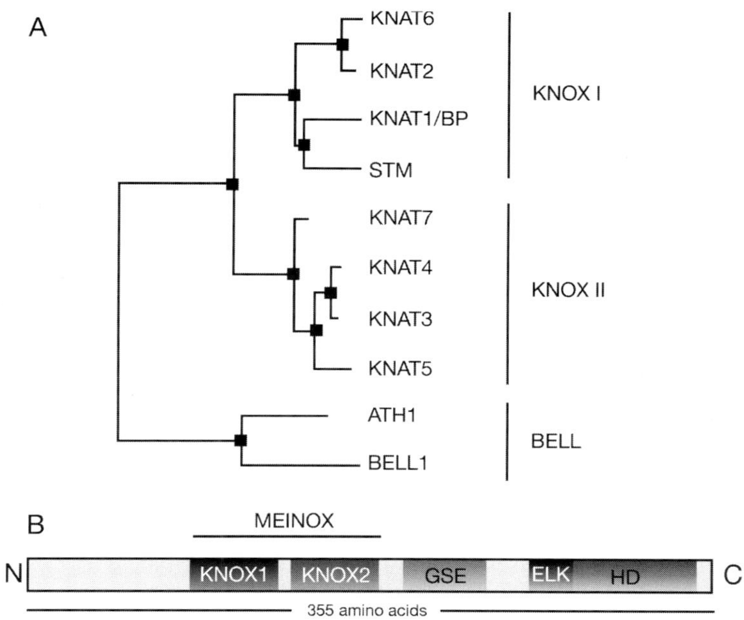

Fig. 56. **(A)** Phylogenetic relationships of eight Arabidopsis KNOX proteins and two Arabidopsis BELL-family proteins based on analysis of amino acid sequences. Distinct KNOXI, KNOXII and BELL clades are evident on the un-rooted tree. **(B)** Schematic representation of a typical KNOX protein (rice OSH15). (From Scofield and Murray, 2006b.)

NTH15 were found to be expressed throughout the corpus of the vegetative SAM (including the CZ and the PZ), the *NTH20* is expressed mainly in the PZ, and *NTH9* was found to be expressed in the rib zone. A specific expression pattern for *KNOX* genes was also revealed in maize, although the expression patterns are not the same in dicots, such as Arabidopsis, and in monocots, such as maize. Thus, for example, the maize genes *ZmRSI* and *ZmKNOX3* are predominantly expressed in a ring of cells at the bases of leaves, auxillary shoot meristems and internodes.

The evolution of SAM, from the embryo phase to the vegetative seedlings, differs between monocots and dicots. In Arabidopsis, the *STM* is activated in the late globular stage of embryogenesis, at the apical domain that will then form the SAM. For the activation of *STM*, the function of *CUC* genes is required. There are transcriptional regulators that control organ separation, and they themselves are regulated by microRNAs during embryogenesis. Then, *STM* has a role in restricting *CUC* expression to the periphery. It should also be noted where the *KNOX* genes are not expressed: they are not expressed during embryogenesis in developing cotyledons.

Since the early studies on *KNOX* genes (Jackson *et al.*, 1994), it became evident that KNOX protein accumulation may occur in different cells from those in which the respective *KNOX* transcription took place. Later studies showed that KNOX proteins can indeed move even between cell layers of the SAM. These proteins are transported through plasmodesmata. Also, in leaves, KNOX proteins were found to be able to traffic from mesophyll cells to epidermal cells (but not in the opposite direction). However, in the SAM, the trafficking (between inner and outer layers) is not unidirectional. Moreover, not only KNOX proteins can move between cells; this ability was also observed for *KNOX* mRNA.

As for the involvement of *KNOX* genes in leaf differentiation, we should first recall that *KNOX* transcription is excluded from the initials of lateral organs: the flanks of the SAM, where lateral primordia are initiated, are devoid of these transcripts. On the other hand, there is no evidence that ectopic expression of *KNOX* genes will inhibit leaf formation. Such ectopic expression (by a 35S CaMV promoter, fused to a *KNOX* code) will cause severe abnormalities in leaves, but not prevent their formation. It is not yet clear how *KNOX* genes affect leaf differentiation because these genes do not seem to directly enhance cell division, but rather have a role in the inhibition of cell expansion and differentiation. In maize leaves, there is an antagonism between the expression of *ROUGH SHEAT2* (*ZmRS2*) and *KNOX*: when *ZmRS2* is expressed, it represses *KNOX* transcription. Hence, the phenotype of *rs2* mutants resembles ectopic expression of *KNOX* genes. Also, in Arabidopsis, mutants of *ASYMMETRIC LEAVES1* (*as1*) resemble the ectopic expression of *KNAT1/BP* and *KNAT2*. A similar competitive interaction was suggested between the shoots *STM* and *AS1*. The actual interactions may be more complicated than presently revealed: it could be that *AS1* and *KNOX* genes competitively regulate common target genes during leaf development. But, which are these target genes?

There is further information on the interaction of *KNOX* genes with other genes. Some members of the *YABBY* (*YAB*) gene family that function in the abaxial cell fate of lateral organs (e.g. leaves) repress the expression of class I *KNOX* genes in leaves. When two *YAB* genes are mutated (*fil* and *yab3*), lobed leaves are formed that express *KNOX* genes and show ectopic meristems on the leave and conversion of abaxial to adaxial cell fates.

The *Antirrhinum PHAN* gene (*AmPHAN*) is closely related to *AS1* of Arabidopsis, and is expressed in primordia of lateral organs. Loss of function of *AmPHAN* results in an adaxial-to-abaxial conversion and leaf radialization. It seems that the adaxial-specific function of *AmPHAN* is required for *AmSTM* function in SAM development of *Antirrhinum*. On the other hand, mutations in the Arabidopsis *PHAN* ortholog *AS1* do not lead to a drastic adaxial/abaxial change. These and, likewise, differences in the detailed activity of orthologous genes in different species, clearly indicated that in spite of the apparent similarity among genes in different species, the end results can differ substantially.

Axillary shoot meristems and *KNOX*-induced ectopic shoot meristems can develop on the adaxial leaf surface. This suggests that factors conductive to meristem functions are located in the adaxial leaf surface, but are normally suppressed by genes such as *YAB*. Also, the homeodomain–leucine zipper proteins (HD-ZIPIII), which have a putative sterol–lipid binding domain, have a role in the abaxial/adaxial leaf symmetry, as shall be detailed in a further section. Two important genes of this group are *PHABULOSA* (*PHB*) and *PHAVOLUTA* (*PHV*). Dominant gain-of-function mutation of these genes causes the conversion from abaxial to adaxial cell types, and thus leaf radialization (see Emery *et al.*, 2003). Loss of function of the closely related gene *REVOLUTA* (*REV*) caused enhanced organ-associated cell proliferation and reduced meristematic activity. As shall be detailed below, the localization of the products of *HD-ZIPIII* genes to the adaxial surface is mediated by microRNA and the products of the *KAN* genes. The latter genes are expressed in the abaxial leaf surface, and loss of function of *KAN* genes results in ectopic expression of *HD-ZIPIII* and consequently in adaxialization (Eshed *et al.*, 2001, 2004).

Ancient Lateral Outgrowth and the Dorsoventrality of Leaves

In order to start understanding the mechanisms that shape the present dorsoventrality of angiosperms, let us look back at the evolution of leaves. The first primitive vascular plants appeared in the middle to late Silurian period (about 420 million years ago). These early vascular plants had no leaves. Later, either of two types of leaves appeared. In the ancestors of certain extant plants, such as lycophytes, microphylous leaves evolved; while macrophylous leaves were formed in enphyllophytes (ferns, gymnosperms and angiosperms). Interestingly, while microphylls and macrophylls apparently evolved independently, it could be assumed that the mechanism of their patterning went in completely different ways. However, this was probably not what happened, as discussed by Harrison *et al.* (2005a) of the University of Oxford (UK). These investigators furnished, by a variety of experimental approaches (e.g. gene expression data from microphyllous lycophytes, phylogenetic analyses and cross-species complementation experiments), results that indicated a common developmental mechanism that underpins both microphyll and megaphyll formation.

We should recall that the indeterminate apical growth is driven by *KNOX* genes, while the determinate leaf growth is marked by the transcriptional or posttranscriptional repression of *KNOX* activity in *Antirrhinum* and Arabidopsis. This repression is manifested by *AS1*, *RS2* and *PHAN*, which are abbreviated as *ARP* genes. The interaction between *KNOX* and *ARP* in the shoot apex regulates the balance between indeterminate (shoot apical meristem) and determinate (leaf initiation) growth. This balance has apparently existed for about 140 million years (initial evolution of angiosperms). Was a similar balance mechanism initiated with the evolvement of microphylls (more than 350 million years ago)? Experimental results of Harrison *et al.* (2005b) suggest that this was so. For example, they isolated a *SkARP1* gene from the diploid *Selaginella kraussiana* lycophyte. The coding region was inserted into a transformation plasmid with a 35S CaMV promoter and introduced the plasmid into *rs2* mutant plants. The resulting transgenic plants varied between an *rs2* phenotype and wild-type plants. When the *SkARP1* transcript was high in the transformed plants, the mutants were rescued. Also, *in situ* hybridizations of *SkARP1* and *SkKNOX* in the lycophyte indicated that the abovementioned balance is rather ancient.

The investigators proposed that this ancient mechanism might have originally operated in the context of primitive plant apices to facilitate bifurcation of shoots and later led to determinate lateral outgrowth. This proposal hints at the idea that leaves originated from metamorphosed axillary (bifurcated) shoots. The 18th-century claim of Caspar Friedrich Wolff and Wolfgang von Goethe that all the organs of plants except the shoot and the roots are modified leaves may have to be changed: even the leaves themselves could be regarded as modified shoots. Such speculations are not directly helpful in understanding the dorsoventrality (or the abaxial/adaxial asymmetry) of present angiosperm leaves. An early step in such an understanding was a study with *Antirrhinum* leaves by Waites and Hudson (1995) of the University of Edinburgh (UK). These investigators isolated and characterized mutations in the *phan* locus of *Antirrhinum*. They found that the leaves, bracts and petal lobes of *phan* mutants had varying degrees of reduction in dorsal (adaxial) tissues. This indicated that wild-type *PHAN* is required for the establishment of dorsal (adaxial) cell identity. In some mutants, the boundaries between abaxial and adaxial cell types were modified. This indicated that *PHAN*-dependent adaxial cell identity is required for lateral growth of wild-type leaf and petal lobes. It was also found that this abaxial/adaxial characterization happens very early in leaf initiation. Interestingly, temperature affected the rate of severity of the suppression of adaxialization: greater malformation of leaves occurred at 17°C than at 25°C.

In a further study of the same team of the University of Edinburgh (Waites *et al.*, 1998), the investigators isolated the *PHANTASTICA*

(*PHAN*) gene from *Antirrhinum majus*, sequenced the cDNA of this gene, and provided the derived amino acid sequence. The sequencing showed that *PHAN* encodes a MYB transcription factor homolog. The derived amino acid sequence was thus compared with that of other MYB proteins from Arabidopsis and maize. Moreover, the *PHAN* gene of *Antirrhinum* could be used for *in situ* hybridization to exactly locate the appearance of the transcript in lateral organs. It was found that *PHAN* mRNA is confined to lateral organ initials and primordia. It is first detected before primordia emergence (P_0 initials) and persists in primordia until stages P_3–P_4. It was also revealed by these investigators that in nonpermissive temperatures (17°C) the leave of *phan* mutants are not only abaxialated, but their proximodistal axis is shorter (and no flat blades are formed). These malformations are reduced with the increase in temperature. It is also noteworthy that in *Antirrhinum* the expression of the STM homolog disappears from the leaf initials at about the same time they begin to express *PHAN*. This hinted at the possibility that the *STM*-encoded product acts as a negative regulator of *PHAN* expression and organ identity.

A new player in the adaxial/abaxial asymmetry of leaves was handled in a study of McConnell and Barton (1998) from the University of Wisconsin in Madison. Actually, these investigators handled not only leaf asymmetry, but also looked at the normal development of branches from the axils to the leaves. In Arabidopsis, axillary SAMs develop in close association with the adaxial leaf base, at the junction of leaf and stem (meaning the leaf axil). They found that a dominant mutation of the gene *PHABULOSA* (*PHB*), termed *phb-1d*, caused altered leaf polarity and adaxial characters to develop in place of abaxial leaf characters. Such adaxial leaves fail to develop leaf blades (they lack the required juxtaposition of adaxial and abaxial cell fates; see Waites and Hudson, 1995). It should be noted that in Arabidopsis (and other dicots), there are clear differences between the normal abaxial side and the normal adaxial side. The adaxial epidermis is glossy, dark green and rich in trichomes; the abaxial epidermis is matte, grey-green and, especially in early leaves, poor in trichomes. There are also internal differences in the inner tissue of the leaves, such as palisade cells which underlay the adaxial epidermis. The investigators also found that the *phb-1d* mutants may develop ectopic SAMs on the underside (ventral side) of the leaves and weakly suppress the phenotype of the *stm (shoot meristemless)* mutant. Hence, there is probably a role for the abaxial cell fate in promoting the development of axillary SAMs.

Another team, Siegfried *et al.* (1999) which included investigators from UC Davis (e.g. S.F. Baum, J.L. Bowman, Y. Eshed) and from the University of Utah, Salt Lake City (e.g. G.N. Drews and D. Otsuga), started to investigate the abaxial–adaxial axes of symmetry from another angle. As noted above, genetic studies indicated that radially symmetric leaves developed in *phan* mutants of Arabidopsis. The leaves were interpreted to be abaxidized. The wild-type *PHAN* was thus considered to promote adaxial cell identity. It should also be noted that *phb-1d* mutants, in addition to causing adaxilation, caused an enlargement of the SAM. It was therefore suggested that there is a positive influence of adaxial cells on the meristem formation, and the SAM was assumed to provide a signal that promotes adaxial cell fate. In the absence of such a signal, the default is abaxilation. To further test the abovementioned indications, Siegfried *et al.* (1999) isolated three genes of the *YABBY* gene family. The analyses of the sequences revealed similarity between the genes. They had six or seven introns, a zinc finger domain and a *YABBY* domain. The investigators used the coding sequence to locate the expression of *FILAMENTOS FLOWER* (*FIL*), *YABBY2* (*YAB2*), and *YABBY3* (*YAB3*). Each of these three was found to be expressed in a polar manner in all lateral organ primordia derived from the apical and flower meristems. These expressions were precisely correlated with the abaxial cell fate of mutants in which the abaxial cell fates were found ectopically, and were reduced or eliminated correspondingly. Ectopic expression of either *FIL* or *YAB3* was sufficient to specify the development of ectopic abaxial tissues in lateral organs, while loss of polar expression of these two genes resulted in a loss of polar differentiation of tissues in lateral organs. These results therefore indicated that members of the *YAB* gene family are responsible for abaxial cell fate in lateral organs of Arabidopsis, and that ubiquitous abaxial cell fate and maintenance of a functional apical meristem are incompatible.

It should be noted that the founding member of the *YABBY* gene family is *CRABS CLAW* (*CRC*); but since expression of *CRC* is largely limited to carpels and nectaries, the *CRC* gene will be mentioned in Chapter 12. The *FIL* gene of Arabidopsis was investigated in studies by Sawa *et al.* (1999a,b) in parallel to the investigation of Siegfried *et al.* (1999), but the former Japanese investigators focused on *FIL*'s role in inflorescence and floral meristems. The various *YABBY* gene family members all have a similarity in their *YABBY* domains and zinc-finger regions. This is true for the *FIL*, *YAB2*, *YAB3* and *INNER NO OUTER* (*INO*) genes; thus, their function to promote the abaxial identities of the organs in which they are expressed (e.g. leaf initials, floral members) as well as their (at least partial) redundancy (as among *FIL* and *YAB3*) can be explained by their similarities.

The dominant mutation (*phb-1d*) of *PHB* was mentioned above (McConnell and Barton, 1998). Like the *phb-1d*, a dominant mutation of *PHAVOLUTA* (*PHV*) also causes a dramatic transformation of abaxial leaf fates into adaxial leaf fates (McConnell *et al.*, 2001). The latter authors found evidence that this transformation could be traced to an altering of a sterol/lipid-binding domain. These changes were found to render the protein constitutively active, and the authors hypothesized that the proteins encoded by *phb* and *phv* mutants have an altered START domain; therefore, a hypothetic ligand is no longer required to activate the protein, which normally happens only in defined regions of the leaf primordium. Consequently, the PHB (and PHV) is (are) active all over the young primordium rendering it adaxial.

Based on previously existing information on the involvement of several genes in the polarity of leaves and floral organs and on additional experimental results, Eshed *et al.* (2001) formulated a model for polarity establishment in lateral organs. This model is summarized in Fig. 57, and its main message is that the establishment of polarity in lateral plant organs (e.g. leaves and floral members) occurs via mutual regressions/interactions between abaxial and adaxial factors after primordium emergence. The basic features of this model were also accepted by Gotz and Hudson (2002), who reviewed the literature on signaling in plant lateral organ development up to the end of

2001. The latter authors also handled the discussion of two apparently contrasting theories on the mechanisms of patterning and cell identity: is it an intrinsic function of the respective cells that have inherited information and behave accordingly, or do cells sense their position and respond to it accordingly? The majority of experimental evidence favors the position theory; but there are also observations that cell displacement does not affect the behavior of cells, meaning that cells follow their dictated behavior, even if displaced (e.g. in maize leaves; Kessler *et al.*, 2002). Although this is an important issue, I shall not go further into details, but indicate that in most cases the position is decisive.

From our point of view (which may not represent what exactly happens in plant tissues), the position determination has an advantage because, if a mistake happens in the behavior of a cell, which is controlled only by its cell line origin, this will lead to an aberrant tissue that cannot be corrected. The position determination is more flexible and can correct itself; it is thus reasonable that the position determination had a selective advantage during evolution. It should be noted that the two theories are not necessarily mutually exclusive. As for the specification of lateral organs (e.g. leaves) in Arabidopsis, Gotz and Hudson (2002) added the roles of *CUP-SHAPED COTYLEDONS* (*CUC1*) and *CUC2* as players in lateral organ initiation. These genes were proposed to limit the growth of nonlateral organ regions, possibly by promoting *STM* expression. The latter authors also discussed the role of auxin in lateral organ development. It appears that auxin has a decisive role in the outgrowth of leaves, but not in the very early stage of organ fate specification. Accepting the old hypothesis that floral members are merely metamorphosed leaves, it should not surprise us that the same genes that play a role in dorsoventrality of leaves have a role in shaping floral members such as the gynoecium.

An additional study of the Davis investigators (Emery *et al.*, 2003) furnished important additional information on the regulation of abaxial/adaxial symmetry. Another member of the class III HD-ZIP gene family was found to take part in this regulation: *REVOLUTA* (*REV*). Gain of function of a *rev* allele caused adaxilation of lateral organs and alteration in the radial patterning of vascular bundles

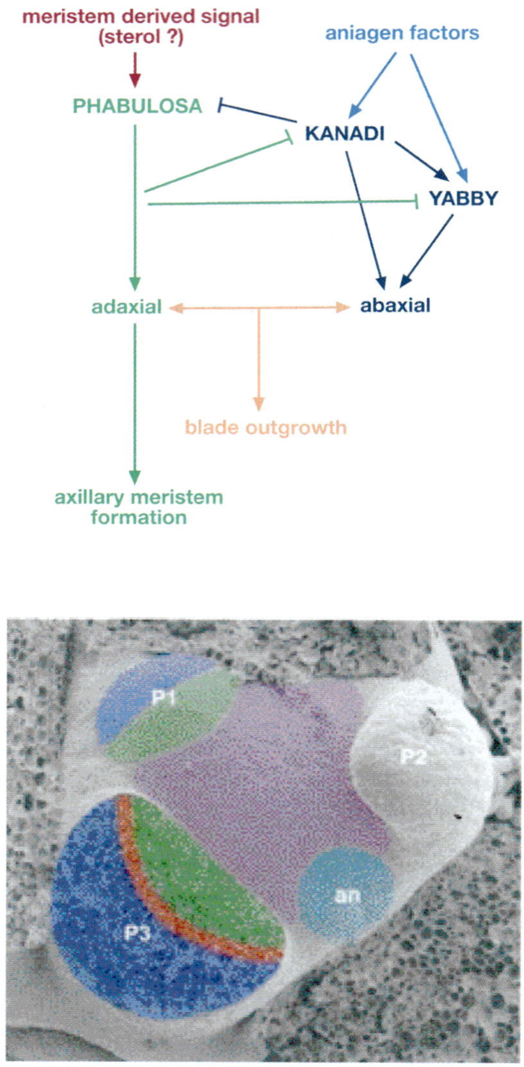

Fig. 57. Model of polarity establishment in lateral organs. Top upper panel details a genetic model of lateral organ polarity establishment, with the spatial and temporal aspects mapped onto a potato apical meristem in the lower panel. An emerging picture from classical and molecular genetic analyses is that as incipient lateral organ primordia develop from the flanks of the shoot apical meristem, factors both intrinsic and extrinsic to the organ primordia contribute to the specification of cells as adaxial (green) or abaxial (blue). The apical meristem (purple) itself likely provides a signal(s)

in the stem of Arabidopsis. These investigators also revealed a novel patterning effect: the effect of microRNA. They found that the gain of function can result not from the change of amino acids encoded by *REV*, but changes in nucleotides that do not change the amino acid sequence can have this effect. Such a change in nucleotides, although it does not affect the encoded protein, does change the ability of a specific microRNA to bind to the *REV* gene and to suppress the formation of this protein. The model previously suggested by this team (see Fig. 57) should thus be amended to include the role of microRNA 165/166 that normally negatively regulates the class III HD-ZIp genes (e.g. *REV*, *PHB*, *PHV*), but this microRNA is not effective when certain nucleotides are changed.

In a further study, the Davis team returned to the roles of *KANDI* and *YABBY* gene families as players in the asymmetric leaf development and blade expansion in Arabidopsis (Eshed *et al.*, 2004). Readers will recall that the abaxial fates were grossly found to be promoted by the *KANADI* and the *YABBY*. When the leaf initial is totally abaxilized or adaxilized, the leaf lamina are not formed. In this investigation, Eshed *et al.* (2004) found evidence indicating that the *YABBY* gene activity is associated with lamina expansion; and hence, it was proposed that boundaries of *YABBY* gene expression, which mark the abaxial–adaxial boundary, are intimately linked to the

---

that promotes adaxial cell fate, whose perception may be mediated through *PHB/PHV/REV* (*PHABULOSA* in figure). The ultimate source and biochemical nature of the ligand is unknown. *PHB*, *FIL*, and *KAN* are all expressed in the leaf anlagen, but their expression becomes confined to mutually exclusive domains as the primordia form. Abaxial cell fate may be a "default" in the absence of signal, for instance, if the lateral organ primordia are separated from the apical meristem. This default state could be the result of the failure to repress genes promoting abaxial identity) e.g. *YABBY* and *KANADI* genes), which are initially activated throughout the anlagen (aqua). Surgical experiments indicate that while polarity is labile in P1, it is irreversibly established by P2. KANADI activity may mediate between PHABULOSA and YABBY activities; however, the precise relationships between these pathways remain to be elucidated. Subsequent interactions between the juxtaposed adaxial and abaxial domains, perhaps mediated by relative levels of KANADI and YABBI activity, are required for lamina outgrowth (red). (From Eshed *et al.*, 2001.)

proposed communication between the abaxial and adaxial domains during leaf development.

The authors suggested that the boundaries of *YABBY* gene activity appear to promote organ growth in adjacent cells; thus, in leaves, this activity acts as a focus for lamina formation. Specifically, the genes *KAN1*, *KAN2* and *KAN3* (all of the *GARP/KANADI* gene family) are already expressed abaxially at the heart stage of embryogenesis (see Chapter 7), while *KAN4* has a limited expression pattern in the ovules. Double or triple mutants (*kan1 kan2*, or *kan1 kan2 kan3*) had a much stronger phenotype than single mutants. For example, *kan1 kan2 kan3* triple mutants had nearly radial leaves at initiation and the mature blade expanded in various planes, giving rise to long narrow leaves with a fan-like blade. The results clearly showed that the simultaneous inactivation of three *KANADI* genes in Arabidopsis is enough to change most abaxial features of leaves into adaxial features. Moreover, such transformation in leaf patterning could include abaxilation of vasculature and mesophyll. When a *YABBY* mutation was added, there was further reduction in the polarity of lateral leaves. Obviously, the *AS1* gene (see Ori *et al.*, 2000, mentioned above) and its *Antirrhinum* ortholog *PHAN*, which were not specifically handled in the studies of the Davis team, are also players in the patterning of leaves.

In a further study of lateral organ symmetry by the team of Yuval Eshed (now at the Weizmann Institute of Science in Rehovot), *ETTIN* [*ETT*, also termed *Auxin Response Factor3* (*ARF3*)] and a closely related gene *ARF4* were analyzed with respect to their role in asymmetry and their interaction with previously mentioned genes (*KANADI*). The results of this study (Pekker *et al.*, 2005) suggested a model for the role of auxin in mediating abaxial/adaxial partitioning in lateral organ primordia, shown in Fig. 58. According to this model, it was suggested that with the rapid expansion of the growing primordium, auxin concentrations form a slight gradient via asymmetric auxin influx carrier distribution from being a sink to a new source of auxin synthesis. This gradient of auxin is translated into differential action of specific subsets of ARFs (as ETT, ARF4 and others), enabling KAN to override PHB-like activities at the abaxial

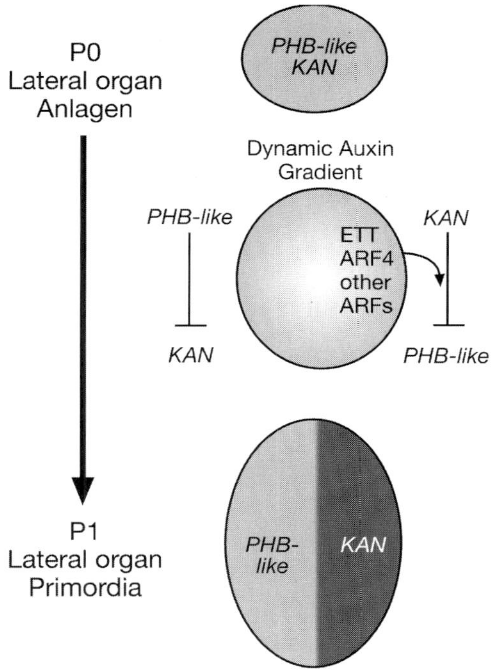

Fig. 58. Proposed role for auxin in mediating abaxial/adaxial partitioning of organ primordia. Initiating organ primordia cells coexpress abaxial and adaxial factors. The partitioning of lateral organs into the abaxial *KAN*-expressing domain and the adaxial *PHB-like*-expressing domain is gradual and evolves by mutual antagonism between the two types of factors and external morphogenic input. With the rapid expansion of the growing primordium, auxin concentrations form a slight gradient via asymmetric auxin influx carrier distribution and due to conversion from being a sink to a new source of auxin synthesis. This gradient of auxin is translated into differential action of specific subsets of ARFs (ETT, ARF4, and others), enabling KAN to override PHB-like activities at the abaxial domain. Subsequently, the gradients of these ARFs help differentially translate auxin presence to maintain abaxial fate, leading to stable abaxial/adaxial partitioning. (From Pekker *et al.*, 2005.)

domain. Thereafter, gradients of ARF3 help to differentially translate auxin presence to maintain the abaxial fate, thus leading to a stale abaxial/adaxial partitioning.

Two recent reviews updated the differentiation of lateral organs in plants: Golz (2006) and Aida and Tasaka (2006). I shall not summarize these rather authoritative reviews because they actually provided overviews on controlling systems that were already mentioned above. Moreover, this is an open subject, and further information on it is expected to be provided continuously in the next few years. I shall only note that Golz (2006) devoted his discussion to the role of the PHB-like transcription factors in signaling adaxial cell identity and to the involvement of specific microRNAs in retention of these transcription factors in defined domains of the emerging lateral outgrowth. Golz (2006) also summarized the role of auxin flow in the patterning of leaves. The review of Aida and Tasaka (2006) stressed the boundary issue and its relevance to leaf blade growth. They claimed that cells at the boundary display reduced growth activity that results in separating adjacent organs or tissues, and that the morphological boundary coincides with the border of different cell identities. The boundary was also suggested to act as a reference point for the generation of new meristems, such as auxillary meristems. The initiation of auxillary meristem is another emerging issue in lateral organ differentiation, and this issue was recently reviewed by Bennet and Leyser (2006).

Vein Differentiation in Leaves

The *vascular patterning*, or *vein patterning*, in angiosperm plants has attracted the interest of many generations of botanists. In the distant past, the investigators were mainly engaged in the description of this patterning. Since this patterning is species-specific, it also served in the taxonomy of angiosperms, although within the same plant the pattern of veins in leaves of juvenile plants may differ considerably from the pattern of veins in adult plants. I shall not review the various patterns; readers who are interested in this subject will find references in books and reviews (e.g. Esau, 1965; Sachs, 1989). This section mainly deals with the known mechanisms of vein formation and the available information on vein pattern establishment. The role of leaf veins in the transport of water, minerals, metabolites, specific signals and viruses is excluded from this section.

The main components of veins are the phloem and the xylem that usually take abaxial and adaxial positions, respectively. As with the studies on other patternings in angiosperms, the majority of the studies on vein patterning during the last 10 years was performed with Arabidopsis (representing dicot veins) and maize (representing monocot veins). The vascular patterning of mature Arabidopsis and maize leaves is compared in Fig. 59. The scheme for vascular patterns in the rosette leaf of Arabidopsis (see Fig. 59A) shows the midvein, secondary veins, tertiary veins and quaternary veins. In leaves of the plants, there may be additional orders of veins. A different architecture of veins exists in the maize leaf (see Fig. 59B), which consists of a midvein, large veins, intermediate veins, small veins and transverse veins. In both species, as in all other leaves, these architectures assure that all leaf cells are only a few cells away from a vein. Readers inclined to better understand vein patterning by mathematical modeling will appreciate the publication of Runions *et al.* (2005), where vein patterning in leaves of diverse shapes is treated by modeling.

The overall ontogeny of venation in dicots (see Figs. 60A–D) proceeds in the following manner. First, the midvein provascular strains extend, acropetally, from the stem into the leaf primordium. Then, the secondary vein provascular strands extend from the midvein towards the margin of the leaf lamina. Thereafter, the reticulum of tertiary and higher-order veins is established during the intercalary expansion growth of the leaf (see Figs. 60C and 60D). The formation of minor veins usually proceeds in a basipetal direction. In dicots with pinnate (or compound) leaves, the formation of secondary veins is coordinated with the differentiation of the lamina. As the leaf lamina expands, areoles formed by tertiary veins are subdivided by the next-order veins, and this may go on further, causing the initial coarse reticulum to become finer and finer.

The regular parallel (striate) venation in (most) monocots also arises in a hierarchical sequence, as schematically featured in Figs. 60F–H. First, the midvein provascular strand extends acropetally from its initiation at the disk of leaf insertion, which in most monocots occurs without initial attachment to the stem vasculature. Then, the major lateral

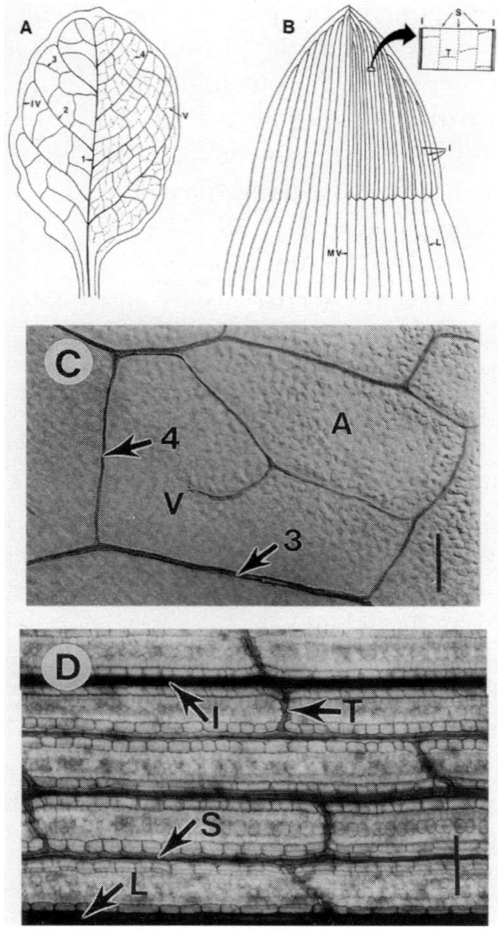

Fig. 59. Comparison of vascular pattern in mature leaves of Arabidopsis (ecotype Columbia) and *Zea mays*. **(A)** Diagram of vascular pattern in a rosette of Arabidopsis. Note midvein (1), secondary veins (2), which are joined by an intramarginal vein (IV), tertiary (3), and quaternary (4) minor veins and freely ending veinlets (V). **(B)** Diagram of vascular pattern in seedling leaf of maize. Note midvein (MV), large (L), intermediate (I), and small (S) longitudinal veins and transverse (T) veins. **(C)** Photograph of cleared Arabidopsis leaf. The image shows tertiary (3) and quaternary (4) minor veins surrounding an areole (A), and freely ending veinlets (V). Bar = 100 μm. **(D)** Photograph of cleared maize leaf blade. The image shows large (L), intermediate (I), and small (S) longitudinal veins and transverse (T) veins. Bar = 100 μm. (From Nelson and Dengler, 1997.)

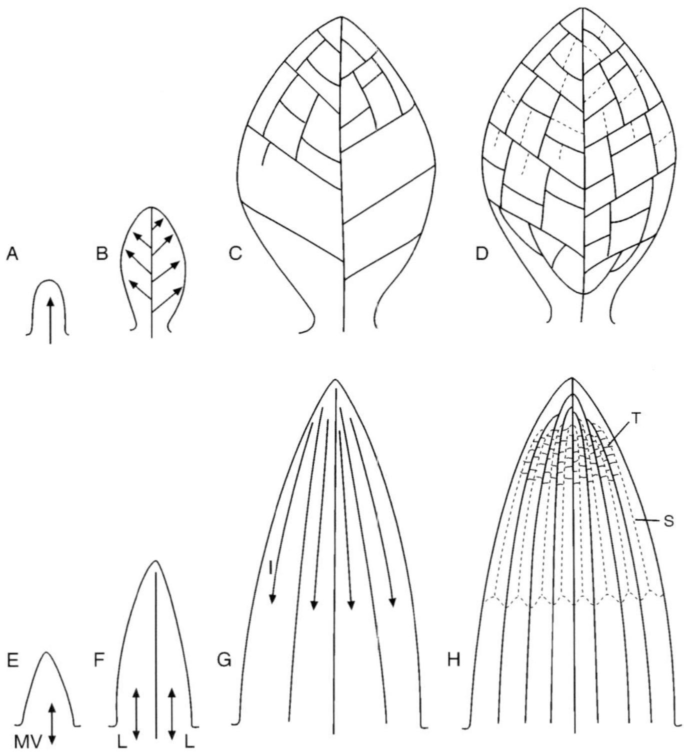

Fig. 60. Vascular pattern ontogeny in dicots and monocots. **(A–D)** Arabidopsis (a dicot); **(E–H)** maize (a monocot). **(A)** Acropetal development of midvein provascular strand from stem vasculature (*arrow*). **(B)** Progressive formation of secondary vein provascular strand (*arrows*). **(C)** Simultaneous formation of tertiary vein network. Tertiary vein formation begins near the leaf apex and proceeds in a basipetal direction. **(D)** Formation of quaternary veins and freely ending veinlets. The formation of minor-order veins (dashed lines) also proceeds in a basipetal direction from the apex of the leaf towards the petiole. **(E)** Formation of midvein (MV) provascular strand in the disk of insertion. The midvein extends acropetally into the leaf primordium and basipetally to connect to the stem vasculature (two-headed arrow). **(F)** Formation of large lateral vein provascular strands (L) in the disk of insertion. Large lateral veins develop acropetally into the leaf primordium and later basipetally to connect to the stem vasculature. **(G)** Formation of intermediate longitudinal vein provascular strands (I) in the distal portion of the leaf. Only some of the intermediate longitudinal veins connect basipetally with the stem vasculature. **(H)** Formation of small longitudinal (S) and transverse (T) veins in leaf blade region. Note the basipetal pattern of transverse vein formation. (From Nelson and Dengler, 1997.)

provascular strands, initiated at the disk of insertion, extend acropetally. Venation then proceeds as schematically featured in Figs. 60G–H. It is noteworthy that in leaves of Gramineae (or Poaceae, grasses), the initiation of the midvein and the major lateral veins is independent from the stem vasculature. Only later is the association with the stem vasculature established: there are early gaps between the veins of very young leaves and the nearest stem vasculature. The gaps are then basipetally eliminated by the extension of provascular leaf veins into the stem vasculature. More details on this process in maize and other monocots are reviewed by Nelson and Dengler (1997).

One of the theories for vascular pattern formation is the canalization of signal flow that involves auxin flow. It resulted from intensive investigations of the late T. Sachs of the Hebrew University in Jerusalem (see Sachs, 1991a, 1991b). These investigations and the respective theory for vascular patterning were supported by a study of Carland and McHale (1996), which identified and characterized the *LOP1* gene from Arabidopsis and found that the *lop1* mutants were defective in basipetal transport of IAA and that their leaf midvein was aberrated: it was bifurcated and the veins were rotated. Interestingly, while the basipetal transport of IAA was strongly inhibited in *lop1* mutants, the acropetal transport was not affected and possibly even enhanced. The role of auxin flow in the canalization that leads to vein formation was substantiated by studies of Aloni *et al.* (2003) and Mattsson *et al.* (2003).

A further search for Arabidopsis mutants defective in vascular patterning (mainly in cotyledons) revealed two genes, *COTYLEDON VASCULAR PATTERN1* (*CVP1*) and *CVP2*, which had disrupted patterns of vascular bundles when mutated (Carland *et al.*, 1999). These genes are probably not involved in auxin perception, biosynthesis or transport. On the other hand, when Sieburth (1999) grew Arabidopsis seedlings in a medium that contained polar auxin transport inhibitors, the veins of mature leaves were aberrated. When Berleth *et al.* (2000) reviewed the then known genes involved in vascular differentiation, they could list 10 genes involved in this differentiation, half of which affect auxin physiology. These authors also proposed a model to explain how inhibition of auxin flow can distort the venation of Arabidopsis leaves. For example, severe auxin transport inhibition causes the restriction of venation to the leaf margin.

In his book *The Art of the Genes*, Coen (1999) mated science with art and described development of patterns as a work of art, where there is a continuous interaction of the "master" and his "creation," gradually increasing the resolution of his work of art. Such a gradual resolution of the picture of vein patterning seems to operate in vein differentiation. Hence, by 2002, Ye (2002) could provide an advanced picture of vascular differentiation. The advanced understanding was based on the characterizaton of many genes and molecular markers (e.g. genes expressed specifically in xylem cells, whose promoters can serve in *in situ* hybridization to trace xylem differentiation). It was revealed that procambial cells in leaves are formed at a very early stage of leaf development. The sites of procambial cell initiation then determine the pattern of vascular organization. It gradually became clear that auxin induces the formation of procambial cells. However, cytokinin also has a role because it is essential for promoting the division of these cells. The molecular mechanisms of auxin and cytokinin effects were still not clear by 2002. Also, Kuriyama and Fukuda (2001) reviewed the possible involvement of additional plant hormones in the formation of tracheary elements, such as brassinosteroids, gibberellins, ethylene and jasmonic acid; but again, the molecular levels of these involvements were still obscure.

The acquisition of biological knowledge is promoted mainly by genetic information, by molecular biology tools, and by wise investigators who know how to make the best use of this information and the tools. This declaration brings us to the studies of Thomas Berleth and associates from Toronto, Canada, (Mattsson *et al.*, 2003; Scarpella *et al.*, 2004, 2006). As previous information indicated that the procambial cells are the precursor or vascular cells, they first set out to investigate the early procambial formation. They thus identified three genes that could serve as expression markers of three distinct procambial stages. They also used one marker that appeared to be perfectly congruent with the appearance of procambial cells. The expression profiles revealed a common spatiotemporal pattern of early vein formation. The promoters of the respective genes were fused to

GUS, and the GUS staining of leaves was then analyzed microscopically. Of these marker genes, *Athb8* is the earliest one, and its expression is started in the ground meristem cells at the preprocambial cells. Thereafter, starts the expression of *553–643*, *ET1335*, and *GT5211*. The last gene (*GT5211*) is initially weakly expressed in all ground meristem cells in addition to its strong expression in procambial cells. In all vein orders, the expression of *Athb8* is initiated next to a pre-existing vasculature, and then the expression progresses polarly away from its point of origin.

Initiation of *Athb8* expression identified a critical stage of preprocambial cell specification. Expression of *Athb8* and the emergence of mesophyll cells seem to mark two mutually exclusive and typically irreversible stages leading either to veins or to mature mesophyll. Procambium formation was marked by *ET1335* expression. The polar progression in the acquisition of a preprocambial cell identity was found to be the first vascular patterning event, and it was suggested that the vein network is formed through the fusion of initially free branches. In a recent study of this team, Scarpella *et al.* (2006), in which J. Friml of Tübingen also participated, the focus was on the role of auxin flow in vein patterning. These authors introduced their publication by making an important and honest statement, saying that a basic component of the mechanisms of vein patterning is still enigmatic; the molecular nature underlining the positional information is unknown. This enigma also exists for other systems of plant patterning and was indicated in the Introduction of this book. However, once a cell (or a group of cells) perceives where it is located and consequently which changes in gene activities are programmed, these changes can be followed by the investigators and important knowledge can be obtained. The study of Scarpella *et al.* (2006) focused on the role of polar auxin transport in leaf vascular patterning.

As previously indicated in this book, the quantification of auxin in individual cells and, even more, the intracellular levels and the intercellular flow of auxin are not yet possible. Therefore, the investigators used the localization of the auxin efflux–associated protein AtPIN1 to follow auxin flow. The PIN1 was visualized histologically by fusion with green fluorescence protein (*PIN1:GFP*). They also traced the expression of the early gene associated with vein formation and fused its promoter to the GUS reporter (*Athb8:GUS*). As indicated in their previous study (Scarpella *et al.*, 2004), the expression of *Athb8* marks the pre-procambial stage of vascular differentiation. To simulate the natural auxin source, very small drops of an IAA-containing paste were surgically applied to the required locations of emerging Arabidopsis leaves. It was found that the early expression of *PIN1* preceded the expression of *Athb8*, which starts at the preprocambial stage. AtPIN1 was found to be expressed in restricted domains of future procambium formation. The subcellular *AtPIN1* polarity indicated that auxin is directed to distinct convergence points in the epidermis, from where it defined the position of major veins. The integrated polarities in all emerging veins indicated that there is auxin drainage towards pre-existing veins. Auxin application and auxin transport inhibitors revealed that convergence-point positioning and *AtPIN1* expression domain dynamics are auxin-dependent processes. In general, the venation proceeds towards the auxin source. However, what mechanism causes the formation of an auxin "fountain" in specific (e.g. epidermal) cells is still enigmatic. While it is now known that *PIN1* has a major role in auxin efflux from plant cells, there are also plant genes that have a role in the influx. Clearly, for intercellular flow of auxin, there should be both influx and efflux of this hormone. It is expected that future studies will handle the role of influx genes in leaf venation.

In the very same Department of Botany of the University of Toronto, where T. Berleth and associates conducted their research on leaf vein patterning, there was another research team, the team of Nancy Dengler, working on a very similar subject (see Kang and Dengler, 2002 for references). The latter team actually used methodologies that were similar to those used by the Berleth team (e.g. using *Athb8:GUS* to trace early changes in undifferentiated cells that lead to procambium cells). While the Dengler team dealt with some details of the venation of adult Arabidopsis leaves and related the initiation of files of procambium to the expansion of these cells, they did not study the auxin role in leaf venation. The results of the works of the two teams complemented each other, and the two teams discussed their manuscripts before publication; however, there are no joint publications of the two teams: spatial proximity does not necessarily avoid independence.

A recent review by Hiroo Fukuda of the University of Tokyo dealt with a similar subject: signals that control plant vascular cell differentiation. However, this review focused on the differentiation of cells, such as the process of xylem cell formation, rather than on the patterning of veins. One of the experimental tools of Fukuda and associates was to follow the formation of xylem cells by *in vitro* cultured undifferentiated *Zinnia* cells. The details of the experimental results of Fukuda and associates are outside the scope of this book. Readers interested in these studies will find a comprehensive report in the review of Fukuda (2004).

A team of nine investigators from the University of Utah in Salt Lake City and Wake Forest University in North Carolina (Sieburth *et al.*, 2006) looked at the venation in leaves and the role of auxin efflux from another angle. They analyzed the *SCARFACE* (*SFC*) gene and its mutants in Arabidopsis. The cotyledon and leaf veins of the *sfc* mutants are largely fragmented, unlike the interconnected vein network in wild-type *SFC*. *SFC* encodes an ADP ribosylation factor GTPase activating protein (ARF-GAP), which belongs to a class of proteins with known roles in the regulation of vesicle trafficking. Quadruple mutants of *SFC* and three homologs (ARF-GAP DOAMIN1, 2 and 4) had modestly enhanced vascular phenotypes. The involvement of *SFC* and *PIN1* is an elaborated issue, but we should recall that *PIN1* recycles intracellularily from the membrane to the endosome (and *vice versa*); thus, recycling is probably essential for its impact on auxin efflux. The cycling is regulated by ARF. ARF-GTP is the active form that participates in the formation of transport vesicles (ARF-GDP is the inactive form). The conversion of ARF-GTP to ARF-GDP is again regulated by two additional classes of proteins. The GN/EMB30 is a positive regulator and the latter is inhibited by brefeldin A (BFA).

The authors studied the results of genetic interactions between *sfc* and *pinoid* (the latter, *PID*, encodes a serine–threonine protein kinase, implicated in the control of auxin efflux). The *sfc* was also analyzed in Arabidopsis roots, and the investigators found an exaggerated response to exogenous auxin application and a higher auxin transport capability in the respective mutants. Tracing *PIN1:GFP* (with confocal microscopy) showed normal *PIN1:GFP* localization at the apical membrane of root cells, but treatment with BFA (which inhibits the positive regulator, GN/EMB30, of ARF) resulted in PIN1 accumulation in smaller and more numerous compartments than in wild-type. The sum of the experimental results of Sieburth *et al.* (2006) suggested that SFC is required for normal intracellular transport of PIN1 from the plasma membrane to the endosome. Since such a transport is required for auxin efflux from cells and such an efflux is required for the auxin-mediated venation, SFC seems to be required for vein patterning in Arabidopsis.

Remarks on the patterning of monocot leaves

There is ample morphological information on the shapes of leaves in diverse monocot species, but information on the molecular genetic mechanisms that control leaf patterning in monocots is restricted to very few species of the Gramineae (or Poaceae, the grass family). The bulk of this information comes from studies with maize. This should not surprise us because maize ("Indian corn," or sometimes termed "corn" in the literature on agriculture in the USA) was the subject of intense research since the early years of genetics in several laboratories in the USA; for several years the corn breeders of R.A. Emerson at Cornell University took the lead in these investigations and updated the genetics of maize (Emerson *et al.*, 1935). Since then, a plethora of genetic information has accumulated on maize. Although the genome of maize is large and is not yet fully sequenced, the known genome sequence of rice can be helpful in the location of maize genes. As for the genetics of maize leaf patterning, we can start with the review of Michael Freeling (1992), who opened his article with the following sentence: "What is and is not known about the maize leaf is reviewed." This statement, while not overly modest, is however, probably correct. This review can therefore serve as a baseline for our deliberations on the patterning of maize leaves.

As in dicots, one can look at the shoot of maize as composed of *phytomers.* Each phytomer is composed of leaf, prophyll, axillary bud and internode. Genetic lines of maize differ considerably in the number of leaves up to the terminal (male) inflorescence (tassel); commonly, there are about 20 leaves. Of these, five are termed *fetal leaves*, because they are formed during embryogenesis and already exist in the maize kernel. These leaves are sometimes termed *juvenile* leaves; that is an ambiguous term. An early indication that a new leaf is differentiating from the SAM is a ring of ~250 cells with overlapping ends (margins), and in the middle of this ring is the location of the future midrib, flanked by cells that will form the blade. These are the founder cells of the future leaf. After the ring of founder cells is established, there are two further main phases: primordium development and postprimordial cell division and differentiation.

Contrary to the initiation of leaves from the periphery of the dicot shoot apical meristem, where the initial appears as a restricted protrusion, the leaf initial of maize has the shape of a hood encircling this SAM. The main vein is established with the initiation of the leaf and, soon after, the parallel secondary veins are formed. The ligule marks the border between the blade of the maize leaf and the sheath that encircles the shoot. Freeling (1992) studied the various mutants of maize that had defective ligules. Such mutants may shift the sheath/blade border towards the sheath, but not *vice versa*. Freeling also rationalized that, as in animal systems, there should be a morphogenetic induction from layer to layer of cells; in the leaf, he assumed that the middle layer, where veins are formed, will affect the outer epidermis layer. However, there was no experimental evidence for this. We saw that in dicots the flow of auxin in the epidermis cells can affect venation. Hence, the conclusion of Freeling (1992) that "any hypothesis that places primary morphological control in the epidermis should be detracted" may require amendment. One maize gene was defined, *Rld* (*Rolled*), which can invert adaxial and abaxial epidermis identities. Is *Rld* orthologous to genes such as *PHV* in Arabidopsis? In his review, Freeling also presented arguments that specify the sheath and the blade as the juvenile and the mature components, respectively, of maize leaves. Thus, mutations that retain juvenility and/or retard maturation will shift the sheath/blade border causing longer sheaths.

It was noted above that two nonexclusive theories exist with respect to patterning: fixed by the cell lineage or affected by the position of cells, relative to other cells. A study by Smith *et al.* (1996) with the *tangled-1* (*tan-1*) mutant of maize lent support for the position theory. This (recessive) mutation causes cells to divide in abnormal orientations throughout leaf development. However, this *does not* alter the overall shape of the maize leaf, although leaf growth in the mutant is slower than in wild-type maize. This suggested that the generation of leaf shape does not depend on the precise spatial control of cell division.

In a previous section of this chapter, the *KNOX* genes were discussed in some detail. Already in the early years of *KNOX* gene study, it was suggested that, in general, the downregulation of *KNOX* gene expression is essential for normal leaf initiation and development. Indeed, Schneeberger *et al.* (1998), a team of investigators from UC Berkeley and the University of Oxford (UK), found a maize gene, *ROUGH SHEAT2* (*RS2*), which affects *KNOX* gene expression and leaf initiation when mutated (*rs2*). They found that this gene defined a new class of genes that function to either directly or indirectly regulate *KNOX* homeobox genes. The *rs2* mutants displayed a wide range of leaf phenotypes that resemble those caused by dominant mutations in several maize homeobox genes (e.g. *Kn1* and *Rs1*). In some cases, the leaf blade adopted a sheat-like identity. They suggested that the *rs2* phenotype resulted from the elimination of a negative regulator, the *rs2* gene product, which normally suppresses *KNOX* gene expression in developing leaf primordia. However, they guessed (prophetically!) that a posttranscriptional downregulation could also be involved.

A much more recent study of maize leaf development was performed by teams from NY State, Cold Spring Harbor and Stony Brook University (Juarez *et al.*, 2004a, 2004b). This study concerned the adaxial/abaxial polarity in maize leaves and the three genes that are involved in this polarity. Their results suggested the following scheme for the involvement of *rolled leaf1* (*rld1*), *leaf bladeless1* (*lbl1*) and *YABBY* in adaxial/abaxial specification:

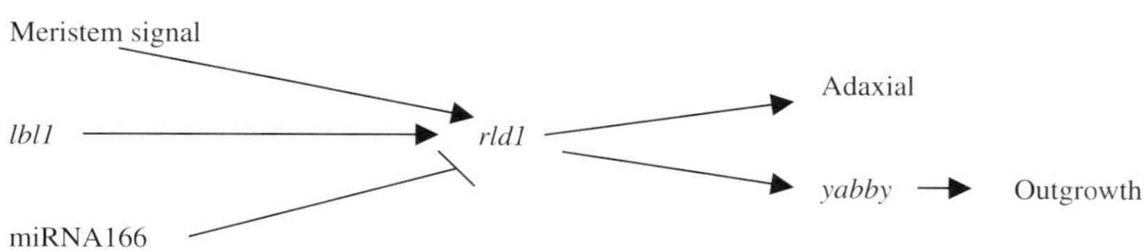

This means that *lbl1* and *miRNA166* have an opposing effect on the accumulation of *rld1* transcripts, and together lead to adaxial-specific expression of *rld1.* The RLD1 (protein) specifies adaxial cell fate, possibly upon activation by the proposed meristem-derived signal. The RLD1 protein also induces *yabby* expression in the adaxial domain, and the juxtaposition of *yabby*-expressing and yabby-nonexpressing cells mediates mediolateral outgrowth. Support for such a scheme came from a mutation, *Rld1-O*, which is *not* repressed by miRNA166 due to small changes in the coding sequence that avoid *rld1* mRNA cleavage by miRNA166.

The team of Jane Langsdale of Oxford University (Alexander *et al.*, 2005) studied the mechanism of domain specification in the SAM of maize and its impact on leaf patterning. They used recessive *corkscrew1* (*cks1*) mutants that disrupt both leaf initiation patterns in the SAM and domain specification within the mediolateral and proximodistal leaf axes. They introduced their study by noting that leaf initiation in Arabidopsis, *Antirrhinum* and maize is correlated with downregulation of *knotted1*-like (*knox*) genes on the flanks of the SAM. If *knox* gene expression is *not* suppressed in leaf primordia, domains within the leaf are misspecified. Also, ectopic expression of *knox* in eudicots appears to delay or prevent the switch from indeterminate to determinate growth, which is normally associated with meristem-to-leaf transition. This is not so in monocot leaves; there, ectopic *knox* expression leads to specific perturbations in the proximodistal axis, such as sheath, auricle and ligule tissue displacements in the blade domain. In most cases, in monocots, *knox* expression does not lead to ectopic meristem formation in leaf blades. This suggested to Alexander *et al.* (2005) that the monocot leaf represents a distinct developmental context from that found in eudicot leaves. The latter investigators used the *cks1* mutants; these mutants are dwarfed, their leaves have multiple midribs and their leaf-sheath tissue differentiates in the blade domain. Such perturbations are a common feature of maize mutants that ectopically accumulate KNOX proteins in leaf tissue. The investigators found that *knox* genes were ectopically expressed in the leaves of *cks1* mutants. However, ectopic KNOX proteins could not be detected.

Consequently, Alexander *et al.* (2005) suggested that CKS primarily functions within the SAM to establish boundaries between meristem and leaf zones. The loss of gene function will thus disrupt boundary formation, impact phyllotactic patterns and lead to aspects of indeterminate growth within the leaf primordium. Moreover, because these perturbations arise independently of ectopic *KNOX* activity, the *cks1* mutation seems to define a novel component of the developmental machinery that facilitates leaf-versus-shoot development in maize.

Trichome Development

Trichomes are epidermal hairs that develop on above-ground organs of plants. In this section, the discussion shall be limited to leaf trichomes, with a focus on Arabidopsis leaves, because most of the genetic and molecular studies on trichomes were conducted on these leaves and consequently the bulk of information on leaf trichomes concerns Arabidopsis. Clearly, trichomes are favorable organs for genetic studies because they are not essential for the vitality of plants and homozygous mutants that are defective in trichomes can be sexually reproduced. Moreover, they are relatively simple organs and can easily be observed during all stages of development. Trichomes are commonly absent from roots, hypocotyls, cotyledons, petals, stamens and carpels of Arabidopsis; but do develop on leaves, sepals and stems of this plant. On leaves, the first trichomes are initiated at the tip of the adaxial surface of the first leaf primordium, when the latter reaches a length of about 100 μm. With the maturation of the first trichomes at the tip of the leaf, additional trichomes emerge progressively towards the base of the leaf. Interestingly, trichomes are initiated in regions where epidermal cells divide; but the first detectable step in the commitment to the trichome fate is the cessation of further cell division with a continuation of DNA synthesis, with at least two rounds of endoduplication (up to 8 N). The epidermal cells surrounding the committed trichome cell continue their division. The committed cell then grows and the nucleus migrates into the newly formed stalk. Then, the branching is started, and the nucleus undergoes further DNA endoduplication and migrates to the base of the last branch. The cells around the trichome attain a socket form. There is an unequal distribution of trichomes on leaf and leaf-like organs. For example, in the first bracts of Arabidopsis, the trichomes develop on the abaxial and the adaxial surfaces; later bracts have fewer trichomes on their adaxial surface. It was suggested that the spacing between trichomes involves a cell-to-cell communication rather than being controlled by cell lineage.

A thorough review of the molecular genetics of trichomes in Arabidopsis up to about 10 years ago was provided by David Marks of the University of Minnesota in St. Paul (Marks, 1997). Marks reported already on more than 20 mutations that had phenotypes of changes in trichome development. The genes concerned with some of these mutants were cloned and sequenced. Of these, the *gl1* mutation affected trichomes specifically, while others such as *ttg* affected other features of the plant. The *GL1* gene, which is required for trichome development, is probably cell-autonomous. The mutants of *GL2* (which is a very different gene from *GL1*, but also acts cell-autonomously) have various defects on trichome development. Mutations that affect branching, expansion and maturation were also detected, and classical genetic studies were conducted to look for interactions between the various genes. The genes *GL1* and *GL2* were sequenced, and *TTG* was inferred to be a homolog of the *R* gene of maize. After the identification of these genes, molecular studies on trichome development could be performed.

There are several mutated alleles of *GL1* with different severity of phenotypes (e.g. *gl1-43* lacks trichomes only on the stem, but not on the leaves). The *GL1* encodes a member of the myb class of transcription factors that contain helix-turn-helix structures and can bind specific DNA sequences. Two helix-turn-helix structures and a typical 120-amino-acid sequence at the carboxyl end were recognized in the GL1 protein.

In situ hybridization of *GL1* mRNA showed that the transcript was localized in fields of epidermal cells from which trichomes are initiated and the cells are committed to become trichomes. The latter cells contained more transcripts than surrounding cells. The promoter of *GL1* was linked to a GUS construct and transgenic plants were obtained, where the activity of *GL1* could be traced by histological staining of the GUS. Other studies indicated that the level of *GL1* has an effect on trichome development and that ectopic expression of *GL1* in *gl1* mutants restores trichome formation. The interaction between *GL1* and *TTG* was also followed. Clearly, even constitutive expression of *TTG* cannot bypass the requirement for *GL1*. However, when 35S:*GL1* plants were crossed with 35S:R plants (*R*-representing TTG), plants were obtained that formed abundant trichomes on the hypocotyl as well as on adaxial and abaxial surfaces of cotyledons. It was suggested that *GL1* and *TTG* expressions have roles in controlling lateral inhibitions (spacing of trichomes).

The sequencing of *GL2* suggested that the encoded protein belongs to a class of proteins that has a leucine zipper domain on the carboxyl side of the homeodomain. This gene was found to be highly expressed in developing trichomes. *GL2:GUS* plants indicated that the expression of *GL2* is more complicated than the expression of *GL1*: the former is also expressed in mesophyll cells. In developing trichomes, the GL2 protein was found primarily in the nucleus. A third sequenced gene was *ZWICHL* (*ZWI*). Contrary to w.t. trichomes of Arabidopsis leaves that have a stalk and three to four branches (see Fig. 61), the *zwi* mutant trichomes have a shorter stalk and only two branches. It was found that *ZWI* encodes a large kinesin-like protein. As detailed in Chapter 3, kinesins are microtubule motor proteins. In the ZWI protein, there is a motor domain located towards the carboxyl terminus. The 5′ part of *ZWI* encodes a region with similarity to class IV myosins. The sequence of the ZWI protein has similarity to a calmodulin-binding protein, and it is therefore possible that Ca^{2+} and calmodulin are involved in regulating *ZWI* activity. Clearly, because ZWI has a similarity to kinesin, the cytoskeleton should play a role in trichome morphogenesis.

In summary, already in 1997, it was found that many genes affect the discrete aspects of trichome development. Genetic analyses indicated that some of these genes act in a linear pathway, while others probably work through unrelated parallel pathways. *GL1*, *GL2 TTG* and *TTG2* seemed to encode transcription factors, while *ZWI* seemed to encode a protein of the cytoskeleton. Indeed, evidence for the involvement of the cytoskeleton in the trichome morphogenesis came from the study of D.B. Szymanski and collaborators (Szymanski *et al.*, 1999). They found that F-actin plays an important

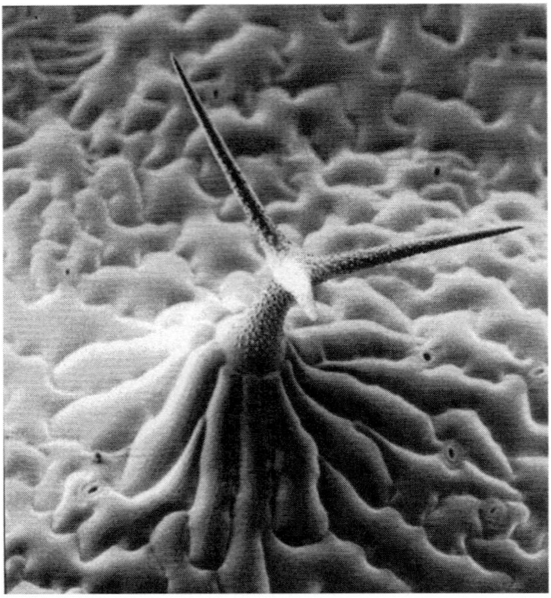

Fig. 61. A trichome on the surface of an Arabidopsis leaf. (From Johnson *et al.*, 2002.)

role during trichome morphogenesis in Arabidopsis. By pharmacological experiments, they revealed that there are distinct requirements for actin- and microtubule-dependent functions during trichome morphogenesis.

Additional evidence for the involvement of cytoskeleton components in the morphogenesis of trichomes in Arabidopsis resulted from studies of J. Mathur, N.-H. Chua and associates from the National University of Singapore and Rockefeller University (Mathur *et al.*, 1999; Mathur and Chua, 2000). These investigators first produced transgenic plants carrying a *GFP-talin* chimeric gene, and found that during trichome cell development, the actin microfilaments assumed an increasing degree of complexity. Disruption of the F-actin cytoskeleton by actin antagonists distorted the trichome in a similar manner to that in specific trichome mutants. On the other hand, *zwi* mutants, which are defective in a kinesin-like protein, as noted above, had normal actin cytoskeleton. Mathur *et al.* (1999) concluded that microtubules establish the spatial patterning of trichome branches, while actin microfilaments elaborate and maintain the overall trichome pattern during development.

In a further study (Mathur and Chua, 2000), these investigators found that, during trichome branching, microtubules reorient with respect to the longitudinal growth axis. The use of the mutations *stichel* and *zwi*, as well as drugs that affect the cytoskeleton, provided results suggesting that trichome cell branching in Arabidopsis may be mediated but transiently stabilized microtubular structures, which may form a component of a multiprotein complex required to reorient freshly polymerizing microtubules into new growth directions.

A series of additional studies by several research teams (e.g. Mathur *et al.*, 2003; Reddy *et al.*, 2004b; El-Assad *et al.*, 2004; Deeks *et al.*, 2004; Fu *et al.*, 2005; and Djakovic *et al.*, 2006) investigated the role of the cytoskeleton in trichome patterning in Arabidopsis.

J. Mathur, M. Hülskamp and associates of the University of Cologne (Mathur *et al.*, 2003) used the mutants of the *CROOKED* gene to study the evolution of shape in epidermal cells of Arabidopsis. It was previously known that the *crooked* mutants had aberrant shapes because of misdirected expansion. The investigators found that localized expansion is promoted in cell regions with fine F-actin, but is restricted in regions of dense F-actin. The *CROOKED* gene was molecularly identified as the plant homolog of ARPC5, the smallest subunit of the ARP2/3 complex. The latter plays a role in forming arrays of fine F-actin in animals. The *crooked* phenotype could be rescued by the human ortholog. Thus, Mathur *et al.* concluded that (1) fine F-actin, generated through the mediation of the ARP2/3 complex, is required for cell expansion; (2) a localized increase in the presence of organelles (vesicles) at a spot does not necessarily cause local expansion, but expression can proceed only when vesicles carrying cell-binding material can be released and assimilated; and (3) the spatiotemporal regulation of F-actin density within a cell can create a loose sieve to allow vesicle release or a dense barrier to trap vesicles, and thus determine where and when vesicles will be released for effective local expansion.

An additional study handled ARP2/3 and trichome differentiation. The Szymanski lab (El-Assal *et al.*, 2004) focused on the *GNARLED* (*GRL*) gene of Arabidopsis. In animal systems, WAVE protein leads to ARP2/3 activation. The *GRL* gene encodes a subunit of WAVE *grl* mutants of Arabidopsis that have distorted trichomes, but the animal ortholog of *GRL* could restore the mutant phenotype of *gnl*.

A further step in clarifying the role of the cytoskeleton in trichome development came from investigators of the Colorado State University (Reddy *et al.*, 2004b). It was indicated above that the *zwi* mutants have aberrated trichomes resulting from a defect in kinesin, a motor protein that interacts with microtubules. One newly discovered protein in Arabidopsis is the kinesin-like calmodulin-binding protein (KCBP). The latter investigators looked for a protein that interacts with KCBP and isolated a novel KCBP-interacting Ca^{2+}-binding protein, and termed it KIC. KIC was found to interact with KCBP in a Ca^{2+}-dependent manner, and by that inhibited the microtubule binding of KCBP. The Ca^{2+} concentration required for this inhibition was much lower than that required for calmodulin binding. The overexpression of KIC in Arabidopsis resulted in trichomes with reduced branch number (as the *zwi* phenotype). The authors thus suggested that KIC modulates the activity of KCBP in response to changes in the concentration of cytsolic Ca^{2+} and regulates trichome morphogenesis.

Further progress in understanding the roles of the cytoskeleton in trichome differentiation resulted from the studies of Ying Fu and collaborators at UC Riverside (see Fu *et al.*, 2005 for references and details). These investigators were interested to gain information on the mechanisms that lead to the interlocking jigsaw puzzle–shaped pavement cells in the leaf epidermis. They found that the Rho-related GTP of plants activates RIC4 (rho-of-plants interactive CRIB motif–containing protein). The activated RIC4 promoted the assembly of cortical actin microfilaments (MFs) that are required for localized outgrowth. However, ROP2 also activates another target: *RIC1*. *RIC1* activity then promotes the well-ordered cortical microtubules (MTs). The RIC1-dependent MT organization locally inhibits outgrowth and *suppresses ROP2* activation in the indentation zone. The authors thus suggested that the counteractivity of these two pathways demarcates outgrowth and indenting cortical domains, and by that coordinates a process that gives rise to interdigitations between adjacent

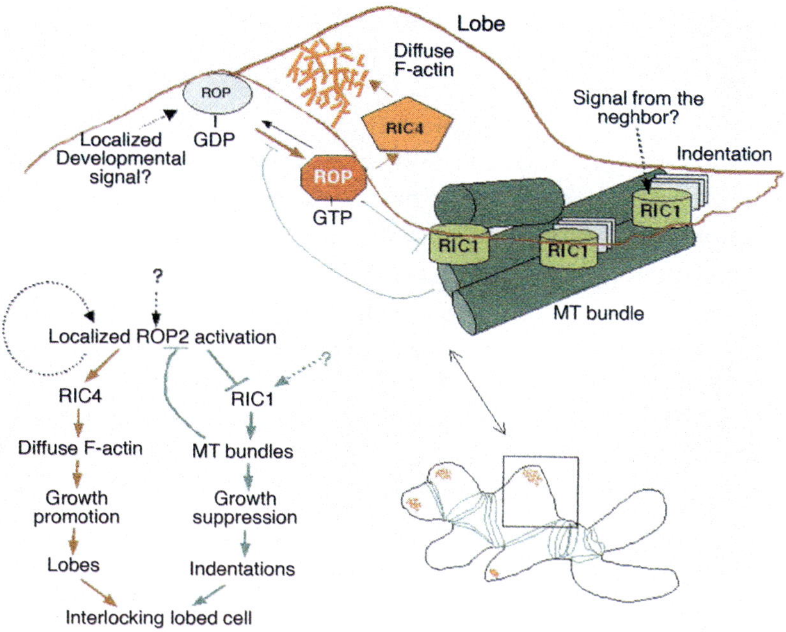

Fig. 62. A model for ROP GTPase control of pavement cell morphogenesis. Interdigitating growth in Arabidopsis leaf pavement cells is controlled by a ROP GTPase signaling network containing two counteracting pathways with opposing actions on localized cell expansion: the ROP-RIC4 pathway promotes lobe outgrowth via a fine cortical MF network, while the RIC1 pathway suppresses outgrowth via well-organized cortical MTs. It is postulated that, during the initiation of lobes and indentations, these two pathways are locally activated and spatially separated, but they may subsequently counteract each other as a result of their respective self-amplification along the cell cortex. The countersignaling establishes borders that separate outgrowth-promoting from outgrowth-suppressing domains of the cell cortex. Solid arrows indicate steps supported by direct experimental evidence, and dotted arrows show hypothetical steps with some direct evidence. (From Fu *et al.*, 2005; see text for meaning of acronyms.)

pavement cells. A model for ROP GTPase control of pavement cell morphogenesis was thus suggested by Fu *et al.* (2005), as shown in Fig. 62. Reviews by Szymanski (2005), Bannigan and Baskin (2005) and Mathur (2006) attempted to connect the cytoskeleton involvement in the patterning of pavement cells to the differentiation of trichome, taking into account what was learned with respect to microfilaments and microtubules in epidermal cells.

In summary, while a great number of genes and their encoded proteins involved in trichome differentiation have already been revealed (see Djakovic *et al.*, 2006 for an update), more genes and their respective encoded proteins are expected to be found. Their interactions will probably be revealed, but a full picture of trichome patterning is still far away.

Finally, because trichomes can be easily defined morphologically, the evolution of trichome morphology can be combined with the evolution of other features of a taxonomic family. This was actually performed by Beilstein *et al.* (2006) of the University of Missouri, St. Louis, at a lab (Kellogg) where angiosperm taxonomy is the focus of investigations. The results of Beilstein *et al.* (2006) showed (among other findings) that branched trichomes likely evolved independently several times in Bassicaceae (also termed Cruciferae).

Stomatal Patterning

Stomata have a vital role in terrestial plants: they serve as the gate to the environment with respect to gasses and vapors. In this section, we shall not deal with the physiology of stomata nor with the mechanisms that control the opening and closure of stomata. This section will rather focus on the differentiation and patterning of stomata. Clearly, stomata are well suited for investigations because they occur on organ (e.g. leaf) surfaces and are readily accessible. Also, stomatal patterns are two-dimensional, rendering pattern assessment a relatively simple task. Clearly, stomata are not restricted to leaf surfaces. They exist on other plant organs, but our deliberations will be focused on leaf stomata on which there is adequate information.

Judith Croxdale of the University of Wisconsin in Madison (Croxdale, 2000) updated the knowledge about stomatal patterning in angiosperms up to the end of 1999. However, her definition of "stomatal patterning" is not how an epidermal cell differentiates into a stoma, but rather the distribution and density of stomata on the organ's (e.g. leaf) surface — i.e. the degree of order and the number of stomata per unit surface of a given organ. There are some bits of information on stomata that we should take into account. One of them is that not all initiated stomata reach maturity. They may be arrested before maturity. This arrest happens in dicots and monocots, and ranges between 10% and 50% of the initiated stomata. Hence, there can be a commitment to produce a guard mother cell (GMC), but full differentiation will never take place. The question thus is whether, for distribution, the arrested stomata are taken into account when distribution is calculated. There is no clear answer for that, but there is a tendency to ignore arrested stomata.

Since cell division and maturation are cellular events that differ considerably among angiosperm leaves, the stomatal patterning (according to Croxdale, 2000) is problematic. For dicots, Bünning (1956) suggested that mature stomata release an inhibitor that prevents the formation of new stomata in their vicinity; but when there is additional cell division between existing stomata, this inhibition could be removed. However, such inhibitory substances were not yet found. There is also information on the relevance of the phase of the cell cycle, meaning the commitment of an epidermal cell to become a GMC is restricted to a certain phase of this cycle.

Another point that should be mentioned is endoduplication. As noted above, there is a clear endoduplication of DNA in the differentiation of trichomes from epidermal cells. There is no such endoduplication during the differentiation of stomata. There is a difference between dicots and monocots with respect to "stomatal patterning", and models for these patternings were suggested by Croxdale (2000), as shown in Fig. 63. Briefly, while in the studied monocots, stomata develop serially, in cell files, stomatal formation in the studied dicots is dispersed in time and space during the mosaic growth of the leaf.

Following the mostly structural information on stomatal distribution as reviewed by Croxdale (2000), there were several years of research that handled the molecular genetics of stomatal differentiation. These were reviewed briefly by Nadeau and Sack (2003) of Ohio State University in Columbus. These authors first outlined the epidermal cell divisions (symmetric and asymmetric) that lead to mature stomata in Arabidopsis. In total, there is at least one asymmetric and

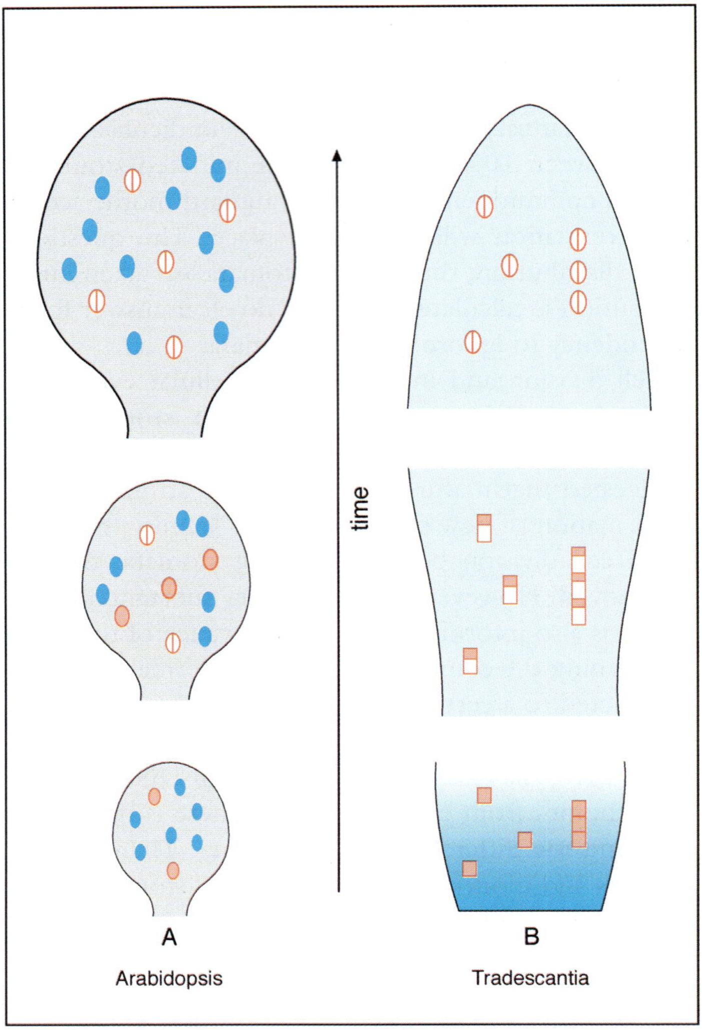

Fig. 63. An unified theory of stomatal patterning in angiosperms is predicated on the cell cycle theory of Charlton. In *Arabidopsis*, the dicot example, leaves have scattered, clonal growth, and only proliferating cells in the appropriate positions of the cell division cycle (red) are capable of producing a stoma; other groups of proliferating cells (blue) are not capable of being specified to the stomatal fate. As the leaf grows, new groups of proliferating cells arise and some of the existing proliferating cells are now at the appropriate position in the cycle to become stomata. In *Tradescantia*, the monocot example, leaves have polarized growth and the proliferating

one symmetric cell division. Schemes for the stomatal development as well as a diagrammatic representation of two genes that are involved in this development are presented in Figs. 64A and B, respectively. The first epidermal cell division that takes place in a presumed stem cell (the meristomoid mother cell), which is to be committed to the stomatal pathway, is an asymmetric division that results in a small precursor cell, meristemoid (M), that is eventually converted into a guard mother cell (GMC), and a neighbor cell (NC). The GMC divides symmetrically to produce two guard cells that constitute the stomata.

In this series of division, there are two developmental plasticities. The smaller M cell may or may not divide again asymmetrically before assuming the GMC fate. If M divides after its initial formation, the final stoma will be surrounded by these NCs that share the same lineage. There is another possibility for plasticity: the larger sister cell to M (the first NC) can also follow several fates. It may or may not divide. When it does not divide, it differentiates into a pavement cell. Cell division in such NCs may affect the density of stomata. Abundance of NCs around the stomata will clearly assure that the stomata will not be adjacent to each other. Obviously, the orientations of divisions on NCs also have a role in stomatal distribution. These division orientations appear to be controlled by intercellular signaling. Such controls may locate the new precursors apart from pre-existing stomata.

Stomatal distribution was found to be affected by two genes: *TOO MANY MOUTHS* (*TMM*) and *STOMATA DENSITY AND DISTRIBUTION* (*SDD1*). Mutations in *TMM* (i.e. *tmm*) were found to randomize the plane and alter the number of asymmetric divisions in NCs (Geisler *et al.*, 2000). Wild-type stomatal distribution is largely affected when divisions next to a pre-existing stomata or precursor are oriented so that a new M is placed away. The mutation *tmm* randomizes

---

cells are massed at the leaf base (blue). Some of these cells are at the appropriate position in the cell division cycle and become stomatal precursors (red). As the proliferating cells produce additional derivatives, the stomatal precursors divide unequally to produce a small stomatal initial (red) and a large epidermal cell. With further growth, the stomatal initials divide and differentiate as guard cells. Mature stomata first appear at the leaf tip and advance basipetally. (From Croxdale, 2000.)

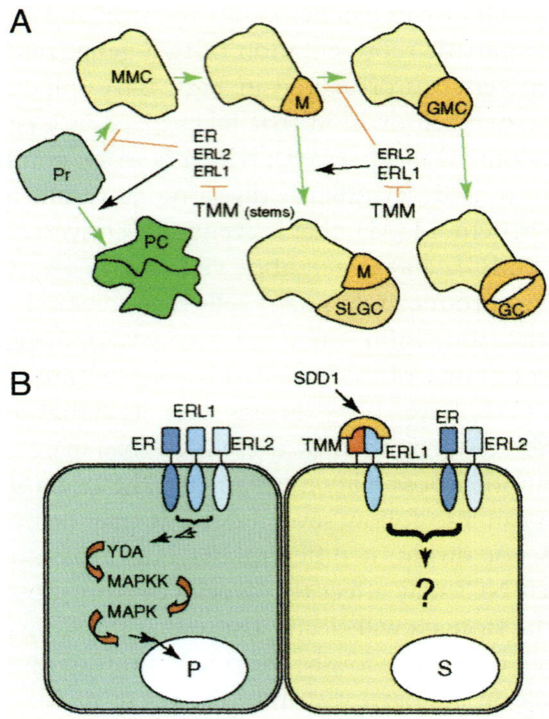

Fig. 64. Stomatal development. **(A)** A simplified schematic representation of stomatal development. In wild-type development most protoderm cells (Pr) undergo proliferative cell divisions to form pavement cells (PC). A subset makes a developmental switch and enters the stomatal lineage, becoming meristemoid mother cells (MMC). These divide asymmetrically to give meristemoids (M). Meristemoids can either undergo proliferative cell divisions to give a stomatal lineage ground cell (SLGC) and another meristemoid, or become guard mother cells (GMC). Guard mother cells divide once, symmetrically to give the guard cells (GC) of the stomatal pore. The points at which *ER*-class genes and *TMM* control these processes are indicated. Red T-bars show inhibitory effects and black arrows show positive regulation. **(B)** A possible model for the molecular basis of interaction between *TMM* and the *ER*-class genes. In the green cell (left) TMM is either not present, or is unable to associate with ERL1, possibly due to lack of an SDD1-generated ligand. Synergistic interactions between ER, ERL1 and ERL2 promote proliferative cell division (P), possibly via the YDA MAP-kinase cascade, although this remains to be investigated. During the switch to meristemoid mother cell or guard mother cell fate (orange), TMM associates with ERL1, possibly due to binding of the SDD-1 generated ligand, interrupting the synergistic interactions of the ER-class proteins and promoting stomatal development (S). (From Ingram, 2005.)

this orientation and releases a prohibition on asymmetric division in cells at specific locations, causing stomatal clusters. The wild-type gene thus seems to be necessary for two position-dependent events: the orientation of asymmetric divisions that pattern stomata distribution, and the control of which a cell will enter the stomatal pathway. The *TMM* encodes a leucine-rich repeat receptor-like protein. This protein has no cytoplasmic kinase domain, and thus would need to interact with other factors to participate in the intracellular signaling pathway. *TMM* is expressed in stomatal precursors, in NCs and in some nonneighboring cells. Apparently, according to the cells where *TMM* is expressed, it can either increase or decrease cell division frequency. Interestingly, stomata are absent from stems of *tmm* plants. The *sdd1* mutants, as the *tmm* mutants, have many more stomata than the wild-type, suggesting that *SDD1* acts as a negative regulator of asymmetric divisions in NCs, but there is a phenotypic difference: the *sdd1* mutants have fewer and smaller stomata than *tmm* plants. The *SDD1* encodes a subtilisin-like serine protease (Berger and Altmann, 2000), and may signal its effect through the cleaving or modifying of developmental signals. The *SDD1* is mainly expressed in M and GMCs, but not in NCs. Overexpression of *SDD1* in wild-type plants (by a 35S:SDD1 transgene) reduced the number of stomata and caused many GMCs to arrest.

A team from the Flanders Interuniversity Institute for Biotechnology, Ghent University (Boudolf *et al.*, 2004) found that stomatal cell differentiation is independent of cellular and nuclear division: even when the stomatal division was inhibited (in Arabidopsis), the cells still acquired stomatal identity. This conclusion followed a study of a cyclin-dependent kinase (CDK) of Arabidopsis, *CDKB1;1*. This gene is highly expressed in guard cells and stomatal precursors. Plants with reduced B-type of CDK had a decreased stomatal index because of an early block in M division and inhibition of satellite meristemoid formation. Aberrant stomatal cells were observed (all blocked at the G2 phase of the cell cycle); but in spite of this situation, stomatal identity was not completely abolished.

Another gene of Arabidopsis involved in stomatal development was studied by Chris Somerville and associates of the Carnegie Institution's Department of Plant Biology in Stanford, California (Bergmann *et al.*,

2004). This gene was termed *YODA* (*YDA*), and it encodes a mitogen-activated protein kinase. Bergmann *et al.* (2004) based their studies on the two previously mentioned genes that are involved in stomatal development, *TMM* and *SDD1*, which encode a predicted cell surface receptor and a serine protease, respectively. As also noted above, mutations in these genes frequently violate the one-cell spacing rule between stomata, while mutations in neither of these genes completely disrupt stomatal patterning. The latter observation caused the authors to assume that there should be other genes that act in concert to establish the wild-type patterning. They found such a gene, *YDA*, which encodes a mitogen-activated kinase protein (MAPK) kinase kinase (MAPKK). The *YDA* gene was identified by the same team (Lukowitz *et al.*, 2004) in a study of asymmetric cell division in Arabidopsis zygotes. Bergmann *et al.* studied the phenotypes of six mutants of *yda*. The respective plants showed an extreme overproduction and clustering of guard cells in the epidermis of the cotyledons and hypocotyls. The early development of epidermis in the *yda* mutants was similar to that of wild-type plants up to 72 hours after germination, but then it appeared that more epidermal cells entered the stomatal pathway and that the one-cell spacing between stomata was not maintained. In *yda* mutants, the M cells divided to produce two GMC cells rather than one pavement cell and one GMC — there was no asymmetric cell division in the M cells.

One of the mutants that was used by this team, with the *yda-Y295* allele, was predicted to be a protein null allele. In many MAPKK kinases (the YDA included), the N-terminus serves as a negative regulatory domain. Hence, the removal of this domain will result in a constitutive activation of the kinase; in a way, it will result in an opposite effect to loss of function. The modified gene was termed *ΔN-YDA*. The authors found that guard cells were abolished in Arabidopsis plants homozygous for *ΔN-YDA*. This was interpreted by the authors to mean that in wild-type plants *YDA* must be downregulated in order to allow epidermal cells to enter the stomatal lineage. They suggested that *YDA* is part of a cell fate switch. The switch probably occurs very early because, in *ΔN-YDA* plants, the epidermal cells did not even start the stomatal pathway; they developed into typical pavement cells. The authors further suggested that TMM, DD1 and YDA act in a common signaling pathway, with SDD1 cleaving a ligand that is perceived by TMM and a coreceptor kinase; a signal transduced through this receptor pair could trigger the phosphorylation and activation of YDA MAPKK kinase and its downstream signaling cascade. Genetic studies supported the authors' suggestion. This suggestion was also supported by analyzing the RNAs from the various mutant lines by probing (with Affymetrix) full-genome microarrays.

The roles of receptor kinases in stomatal patterning were investigated by a trans-Pacific team that consisted of Elena Shpak of the University of Washington, Keiko Torii of the Japan Science and Technology Agency in Saitama and their associates (Shpak *et al.*, 2005). This team previously collaborated in studies on the ERECTA family receptor-like kinases that control development in Arabidopsis (e.g. Shpak *et al.*, 2003, 2004). In their later study, Shpak *et al.* (2005) focused on the ERECTA (ER) family leucine-rich repeat–receptor-like kinases (LRR-RLKs) to elucidate their role in stomatal patterning. For that they also analyzed the interactions of mutants (*er*, *erl1* and *erl2*) with each other and with mutations in *tmm*. The phenotypes of various combinations of these mutants as well as the pattern of their expressions in epidermal cells were analyzed (e.g. by the use of the chimeral transgenes ER::GUS, ERL1::GUS and ERL2::GUS). Interestingly, the impact of the mutation in the ER family of LRR-RLK members and their combinations with either *TMM* or *tmm* on the phenotypes of stomatal development are not the same in the various plant organs (e.g. leaves, stems and siliques).

A later study of the team of Fred Sack (Lai *et al.*, 2005) handled the last stage of stomata development: the differentiation of the guard cells from the GMCs. In the normal differentiation, the GMC undergoes one symmetric cell division to yield two daughter cells that develop into the respective two guard cells. Recessive mutations in the gene *FOUR LIPS* (*FLP*) cause the abnormality of four guard cells that are in contact with each other. This phenotye probably results from a presistance of a precursor cell identity, leading to an extra symmetric division at the end of the cell lineage. There are several mutant alleles of *flp*. Alleles causing weaker phenotypes (e.g. *flp-1*, *flp-2*) produce plants that typically contain two laterally aligned stomata. More severe phenotypes are caused by *flp-7* and *flp-8*. In the respective plants, the stomata appear in clusters that have, in addition to malformed stomata, other cell types.

The investigators found (by map-based cloning) that FLP is an R2R3 MYB protein, meaning it possesses two tandem imperfect MYB repeats of about 50 amino acids each. The full-length FLP is encoded by 11 exons and contains 471 amino acids. It contains a predicted nuclear localization signal, a coiled–coiled domain and a C-terminal region of 163 amino acids that is rich in serine and proline. This MYB protein has conspicuous changes in plants with specific mutated alleles. For example, in plants homozygous for the *flp-8* allele, there is a missense mutation in the first R3 helix that replaces the highly conserved glutamic acid with the basic lysine. The transcript of *FLP* accumulates just before the (last) symmetric division.

There is a paralogous gene to *FLP* in Arabidopsis, termed *MYB88*. The protein encoded by the latter gene has all the main domains of FLP (e.g. nuclear localization signal, MYBR2, MYBR3, coiled–coiled domain and serine- and proline-rich domain), and there is a great similarity in the amino acid sequence of both proteins. The phenotypes of mutated *flp* alleles could be corrected by the *MYB88* transgene, while double mutants of *flp* and *myb88* have severely malformed stomata. It should be noted that guard cells are probably arrested in the G_1 phase of the cell division cycle or in the exit of this cycle (G_0), because they do not divide. They thus retain the 2C level of nuclear DNA. The GMC division appears normal in *flp* mutants, but then the daughter cells divide symmetrically one or more times. Because these extra divisions occur in cells that exhibit GMC traits and lack guard cell traits, the loss-of-function mutations in *FLP* appear to prolong GMC identity. *FLP* expression, as noted above, apparently starts before GMC mitosis and ends after the initial specification of guard cells. Lai *et al.* (2005) thus reasoned that *FLP* functions normally do limit division during this period directly by regulating cell cycle genes, or indirectly by promoting a transition in cell fate.

Chapter 12

PATTERNING OF FLOWERS AND THEIR MEMBERS

Although we are still far away from a full understanding of patterning of the angiosperm flower, the amount of information that is available on this subject is huge. While up to 1988 a wealth of knowledge was accumulated, especially in the structural aspects of flower differentiation, the "explosion" of molecular genetic information on this subject started in the late 1980s. Since then, dozens of research groups have become engaged in various aspects of flower patterning and have furnished hundreds of meaningful publications. Therefore, a comprehensive report on our present understanding of flower patterning would require a book rather than a chapter. This chapter will thus not fully cover this subject, but rather deal with a choice of aspects of this patterning. Accordingly, I shall briefly handle the transition from vegetative growth to floral induction, a subject that was already handled in some detail in Chapter 10. After indicating the structural variation of inflorescences and flowers in angiosperms, I shall devote a section to the ABC model that was suggested for the determination of the identity of floral organs (floral members) on the basis of investigations with the model plants, snapdragon (*Antirrhinum majus*) and Arabidopsis, and the recognition of the roles of MADS proteins in floral/organ identity. I shall also note how this model fits other dicot plants (e.g. Petunia) and devote a special section to the patterning of monocot flowers (mainly represented by flowers of the Gramineae family).

The initially suggested ABC model was then further elaborated due to the recognition of additional genes that participate in flower patterning, leading to a better understanding of the patterning process. Some of these genes encode specific regulatory proteins (MADS as well as others) that interact to control the activation of specific patterning genes. These interactions will be summarized. While the organ identity genes lead towards specific floral organs, the former genes do not shape the respective organs. The structural differentiation is performed by shaping genes that are mostly still unknown. These shaping genes constitute the targets of the organ identity genes. A special section of this chapter will deal with the present knowledge on these targets. In Chapter 4, I provided basic information on gene duplication. It is evident that gene duplication had major roles in the evolution of genes that participate in the patterning of the angiosperm flowers. Therefore, these roles will be summarized below. In most angiosperms, the very young flower bud has initials of stamens and carpels. Both of these organs may reach maturity to establish the bisexual flowers; but in monoecious and dioecious species, the further differentiation of the initial stamens or the initial carpels is arrested, leading to female (pistillate) or male (staminate) flowers, respectively. This sex determination of angiosperm flowers will be briefly handled.

The term "gene expression" is ambiguous. It can mean the transcription of a gene to mRNA or the translation into a protein that is encoded in a gene. In recent years, it became evident (see Chapter 5 of this book) that eukaryotic cells have a plethora of means to regulate the level of specific proteins; among these are the ubiquination of proteins and the reduction of specific mRNA by processes of RNA silencing. One of the means by which cells are able to silence mRNA is by microRNAs. In plants, microRNAs are very target-specific and may thus cause the reduction or even the elimination of specific mRNA. As for the theme of the present chapter, it recently became evident that microRNAs have important roles in the regulation of expression of genes involved in flower patterning. A special section in this chapter will therefore be devoted to microRNAs.

No comprehensive and updated text on the various aspects of the angiosperm flower that contain information on the versatility of flowers, on the physiology of flowering, and on the molecular genetics of floral induction and flower patterning is available. Clearly, this short chapter will not replace such a nonexisting book. This chapter will thus focus on flower patterning; even within this theme, only a choice of subjects will be handled, with emphasis on novel knowledge that has accumulated in recent years.

A Note on the Versatility of Flowers

The term "perception" has two main meanings. In one meaning, it is the recording of phenomena, while the other meaning is to understand the phenomena and to accumulate knowledge on the processes that lead to a specific phenomenon. In this section, I shall briefly deal with the first meaning of perception of the angiosperm flower; while in all the subsequent sections of this chapter, I shall deal with the second meaning of perception of the flower.

The vast versatility of flowers is clearly recognized by every human being. A short visit in nature or in a garden will indicate this versatility. The latter will also become evident by looking at a catalog of a company that sells flower seeds. Basic variations in flowers are schematically presented in Fig. 65. The variations presented in this figure concern only forms of flowers and floral organs. It ignores color and scent, which can add variability. It should be noted that there are not only variations in form between plants of different species, genera and families — there are also variations between flowers of the same plants. Thus, within the same compound flower of composite plants (e.g. sunflower), there are different forms of flowers. Also, in some cucumber lines, there may be pistillate (female) and hermaphrodite ("perfect") flowers. These differ with respect to the location of the ovary: pistillate flowers have an inferior (epigynous) ovary, while the hermaphrodite flowers commonly have a superior ovary and are hypogynous (see Frankel and Galun, 1977 and Smyth, 2005).

Some indications on the variability of flowers were presented in the reviews of Soltis *et al.* (2002), a team that forms the Floral Genome Project Research Group, in which investigators from Florida, Oslo, Alabama, Pennsylvania, Michigan and Jena (Germany) participate. These investigators constructed a summary phylogenetic tree of the angiosperms (see Fig. 66) with some details in the respective "trees"

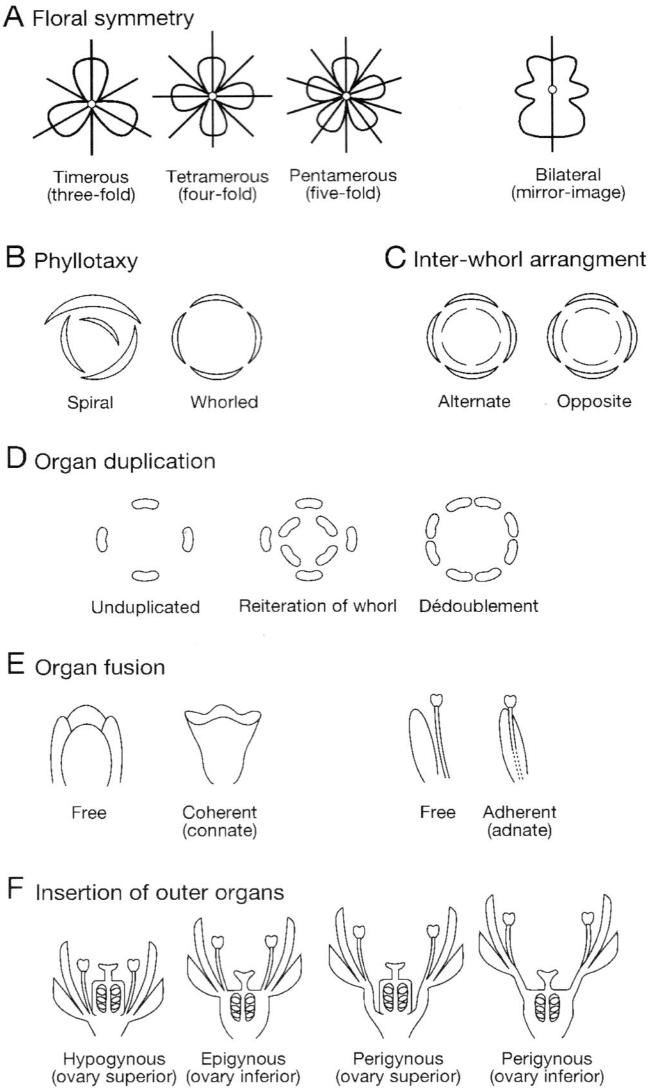

Fig. 65. Variations in floral structure. **(A)** The symmetry of flowers is defined by the number of similar axes that can be drawn through its plan. Radially symmetric flowers (also called regular or actinomorphic) have two or more axes and include many monocots (3), Arabidopsis (2, derived from 4), tomato, and Petunia (5). Bilaterally symmetric flowers (also called irregular, zygomorphic, or mirror-imaged) have only one and include *Antirrhinum* and the pea family (Papilionaceae).

of *Brassicales*, *Lamiales* and monocots. To indicate the variability in the photographs, eight representative angiosperm species are presented (see Fig. 67). We should note that there are several hundreds of thousands of known species of angiosperms. Even a detailed acquaintance with the floral structures of these species will not be helpful for the understanding of patterning of flowers. On the contrary, this plethora of flower structures may cause us confusion, and no common strategy of flower patterning will emerge.

The main contributors of information on flower patterning opted to focus on one or two model plants of angiosperms. Indeed, one of these main contributors, Eliot M. Meyerowitz of the California Institute of Technology (Pasadena, CA), had no previous botanical training and in the past studied *Drosophila* differentiation. Hence, if one wishes to understand the patterning of a nonmodel flower, one has to go "backwards", meaning to look for genes and processes involved in the patterning of a model flower (e.g. Arabidopsis) and then search for similar genes and processes in the nonmodel flower.

A Short History of Studies on Flower Morphogenesis

David R. Smyth of Monash University in Melbourne, Australia, provided a historical essay on the studies of flower morphogenesis

---

(B) Phyllotaxy of organs may be spiral (especially in the basal angiosperms) or whorled. **(C)** Organs may arise alternate with, or opposite to, those in the adjacent whorl. **(D)** Duplication of organs may occur by the reiteration of a whorl (as in poppies) or by two organs arising in place of one or vice versa (dédoublement, as in Lepidium [Brassicaceae]). **(E)** Organs may either be free or show different patterns of fusion with each other. Those within a whorl may be coherent in a tube (e.g. the corolla of *Antirrhinum*). Different types of organ may be adherent (e.g. the stamens and corolla of *Antirrhinum*). These forms are usually congenital. Post-genital fusions occur after organs have formed (e.g. anthers in the daisy family, Asteraceae). **(F)** The place of attachment of the perianth organs and stamens varies. They may be attached to the receptacle near the base of the ovary (hypogynous) or at the top of the ovary (epigynous). They are sometimes attached to the rim of a floral tube (perigynous), which itself may be basal or apical to the ovary. The ovary is either superior, if the other organs (or a floral tube) are attached to its base, or inferior if they are attached to its apex. The floral tube is sometimes called a hypanthium. (From Smyth, 2005.)

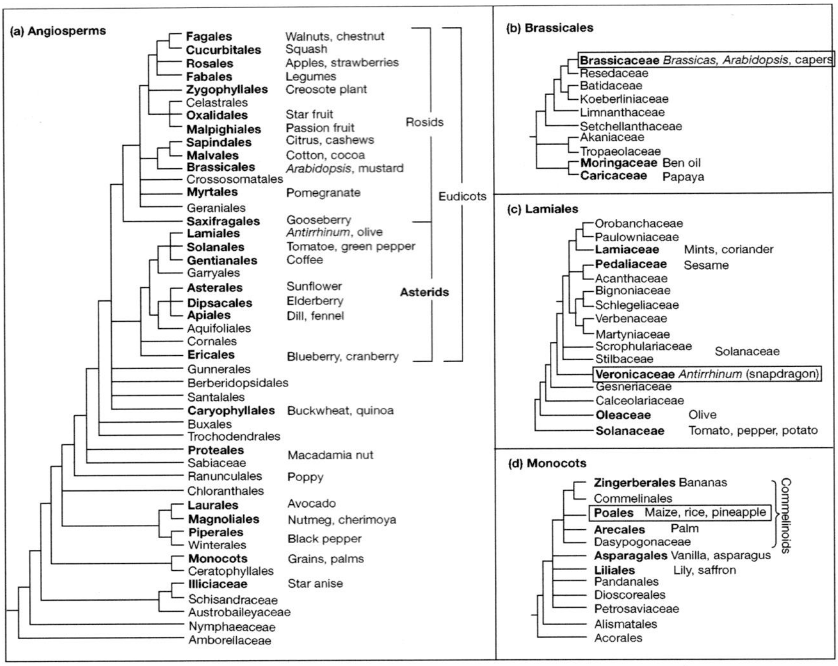

Fig. 66. The summary tree **(a)** is based on 567 taxa, each sequenced for three genes (~5000 bp per taxon), with relationships between basal angiosperms updated to reflect recent analyses based on 6000 > 15 000 bp of sequence data per taxon. Both *Arabidopsis* and *Antirrhinum* are members of the eudicot clade, and both appear in derived positions within well defined clades of core eudicots, the rosids and asterids, respectively. Brassicaceae (which contain *Arabidopsis*) are nested well within Brassicales; Brassicales are, in turn, deeply nested within the rosids. An expanded summary tree **(b)** is provided for the Brassicales. *Antirrhinum* belongs to the Scrophulariaceae (nested within the Lamiales), a family now known to be polyphyletic; the subclade containing *Antirrhinum* is nested well within the asterids. An expanded summary tree **(c)** is provided for the Lamiales. Maize and other grasses (Poaceae) are also in a highly derived phylogenetic position; Poaceae are a tip clade within Poales, which are in turn nested within the commelinoid clade of the monocots. A summary tree for the monocots is provided **(d)**. In all the figures, only well supported nodes (as measured by bootstrap or jack-knife values) are depicted. Major groups that contain plants for which some portion of the flower, fruit or inflorescence is of economic importance (but not including ornamentals) are listed in bold. Common names of one or a few such plant are provided following the clade name (e.g., Laurales – avocado). This is by no means a comprehensive list of economically important plants. Clades containing the model organisms *Antirrhinum*, *Arabidopsis* and maize are highlighted in boxes. (From Soltis *et al.*, 2002.)

(Smyth, 2005). Several steps in this history will be outlined below up to 1988–1990, when the great leap in understanding flower morphogenesis emerged, mainly due to focusing on model flowers (e.g. Arabidopsis and snapdragon) by Haughn and Sommerville (1988, then in East Lansing, Michigan) as well as by investigators of the laboratories of H. Saedler (Schwarz-Sommer *et al.*, 1990) of the MPI in Cologne and of Coen and Meyerowitz (1991) of the John Innes Institute and the California Institute of Technology, respectively. Starting from the studies of these research teams, investigations on flower patterning will be handled in detail in subsequent sections.

A description of plant components was provided by the "father" of botany, Theophrastus (~370–285 BC). Theophrastus was a colleague of Aristotle and succeeded the latter as the leader of the Lyceum. He considered plants to be made of *consistent* parts (roots, stems, branches and twigs) and *ephemeral* parts (leaves, flowers, fruits with seeds). The flowers included petals, stamens and the styles of the carpels. The sepals were excluded as being a kind of leaf and the ovary was excluded as the future fruit case. The role of flowers in sexual reproduction was not handled by Theophrastus, neither was the difference between sexual and vegetative reproduction. However, it is reasonable to assume that it was known since ancient times that flowers have such a role because for the production of certain crops, such as dates, transfer of pollen from one palm (that produces pollen but no fruits) to another palm (that is the potential carrier of dates) was known to be essential for the production of dates. Nevertheless, academic records on the role of pollen in sexual reproduction in plants did not appear until the 17th century, with the observations of Rudolf Camerarius (1665–1721), the director of the Botanic Garden of Tübingen, Germany.

A closer look at plants and their flowers started in the 16th century (e.g. Andrea Cesalpino, 1519–1603). Detailed observations of the plant components, including flowers, were performed by Italian scholars who initiated the Lincean Academy (founded by Prince Federico Cesi in 1603, of which Galileo Galilei was a member; see Freedberg, 2002). Only later, when simple microscopes became available, did scholars like Marcello Malpighi (1628–1694) and Nehemiah Grew (1641–1712) follow the initiation of mature plant organs, including the

Fig. 67. Photographs of basal angiosperms (a–g) illustrating floral diversity, with *Arabidopsis* (h) shown for comparison as a core eudicot. **(a)** *Amborella trichopoda* (Amorellaceae), the sister to all other angiosperms (Fig. 66), has unisexual flowers (this is a pistillate flower) with a spirally arranged, undifferentiated perianth of 6–15 tepals. Pisillate flowers have four to eight free carpels. Staminate flowers possess many (12–21) laminar stamens. (Photograph courtesy of Sandra Floyd.) **(b)** *Cabomba* sp. (Nymphaeaceae), which represents one of the earliest diverging lineages of extant angiosperms (Fig. 66)), has bisexual flowers and a perianth consisting of three sepals and three petals; there are six stamens and one to three carpels. (Photograph courtesy of Ed Schnieder.) **(c)** *Illicium* sp. (Illiciaceae), another early-diverging angiosperm lineage (Fig. 66), has bisexual flowers with an undifferentiated perianth with many tepals. The tepals are commonly arranged in several series (spiral within each series), with those of the outermost series small and sepal-like, and those of the inner series larger and more petaloid; the innermost tepals can also be reduced and transitional to the stamens. The stamens are numerous, distinct, spirally arranged and more or less laminar. Carpels are numerous and distinct. (Photograph courtesy of Douglas Soltis.) **(d)** *Austrobaileya scandens* (Austrobaileyaceae) has bisexual flowers with numerous, undifferentiated, spirally arranged perianth parts; stamens and carpels are also spirally arranged and numerous. (Photograph courtesy of Peter Endress.) **(e)** *Ceratophyllum demersum* (Ceratophyllaceae) has minute, unisexual flowers (two clusters of flowers are depicted). A series of scales (6–13), sometimes interpreted as a perianth, surrounds the andorecium or gynoecium. Male flowers have a variable number (3–46) of stamens that are arranged either spirally or in whorls; stamens are not differentiated into filament and anther. Female flowers each possess a single carpel. (Photograph courtesy of Peter Endress.) **(f)** *Eupomatia benettii* (Eupomatiaceae), a member of Magnoliales (Fig. 66), has distinctive bisexual flowers that do not have a perianth but are covered when young by a deciduous calyptra (interpreted as bract). There are many (~70) spirally arranged stamens that are attached at the rim of the receptacle, the inner stamens being sterile and more or less

initiation of flowers. They recognized that what was later termed parianth (sepals and petals) start as leaf-like initials. John Ray (1623–1705) termed the dust of the stamens *pollen* and also distinguished between the sepals and the internal corolla, terming the latter *petals*. Camerarius (of Tübingen) revealed the essential role of pollen through experimental work with unisexual flowers of dioecious and monoecious plant species (e.g. mulberry and castor, respectively). But, a full recognition of the roles of floral members (flower organs) in the process of fertilization, embryo development and seed production had to wait until the studies of Joseph Gottlieb Koelreuter (1733–1806) of Karlsruhe, Germany as well as Christian Konrad Sprengel (1750–1816), and Wilhlem Friedrich Hofmeister (1824–1877) of Berlin. The process of nuclear fusion between the sperm and the egg nucleus of plants was recorded by Eduard Strasburger (1844–1912) of Bonn in 1878 (more than 10 years after the genetic discoveries of Gregor Mendel). One may note that many of these masters of botany enjoyed a long life — far longer than the average life span of their contemporary humans. In a previous book (Galun, 2003), I noted that the philosopher Baruch Spinoza (1632–1677) and the microscopist Antony van Leuwenhoek (1632–1723), both of the Netherlands and born in the same year, lived 44.5 and 89 years, respectively. But the former made his living by polishing lenses, while the latter used lenses to observe botanical material.

In botanical texts and reviews on flower patterning, the German poet (and geologist, graduate of law school, etc.) Johann Wolfgang von Goethe is mentioned as the first to state in 1790 that the floral organs are leaves that underwent metamorphosis. The fact that

---

petaloid. The carpels are also numerous and spirally arranged, and connate (fused) by their margins to form a sort of compound pistil. (Photograph courtesy of Peter Endress.) **(g)** *Sarcandra chloranthoides* (Chloranthaceae) represents a distinct lineage of basal angiosperms (Fig. 66). This species has bisexual flowers that lack a perianth. Several flowers are shown here, each consisting of a single stamen and a single carpel. (Photograph courtesy of Peter Endress.) **(h)** *Arabidopsis* (Brassicaceae) is a core eudicot in the large rosid clade (Fig. 66) and is provided for comparison with the basal angiosperms shown above. Flowers have well differentiated sepals and petals (four of each), as well as six stamens and two fused carpels (Photograph courtesy of Hong Ma.) (From Soltis *et al.*, 2002.)

Caspar Friedrich Wolff submitted a doctoral thesis to the University of Halle (Germany) in 1759, in which he claimed that all plant organs except the shoot and the root are modified leaves (see Preface of this book), was usually ignored. While sepals and petals could be generally accepted as modified leaves, the claim that stamens and carpels are also modified leaves was substantiated by evidence provided by Robert Brown (mentioned above).

This evidence led to the present view that includes sepals, petals, stamens and carpels (including the styles) as flower organs. Further contributions on the modifications that led to the mature flower and on flower development were furnished by the Swiss scientist Augustin Pyram de Candolle (1788–1841) and the French scientist Jean-Baptist Payer (1818–1860). Two books (Wardlaw, 1965; Steeves and Sussex, 1989) provided details on the then existing knowledge on the shoot apical meristem and the flower meristem at the cellular level. Another book (Endress, 1994) suggested three aspects of flower constructure, that have different underlying rates of evolutionary change: *organization*, *construction*, and *mode*. The first is the floral "blueprint" that defines the number and position of floral organs, the second is the basic three-dimensional structure of the flower, and the third is the lateral elaboration of special characteristics (such as the style). The suggestion of Endress may be helpful in organizing the thoughts of botanists, but it was not based on molecular genetic data.

Looking Back at the Molecular Genetics of the Transition from Vegetative Growth to Floral Induction

An account of our present understanding of the molecular genetics of the transition from vegetative growth to floral induction was provided in Chapter 10 (see Figs. 51–53). In summary, the protein encoded by the gene *FLC* was found (mainly based on studies with Arabidopsis) to have a major role in floral induction. The expression of *FLC* was found to suppress the transition to flowering, and genes such as *FRI* and *FRL1* were found to cause an increase the expression of *FLC*. This latter expression was found to be repressed by the pathway of vernalization (cold treatments) and by the autonomous pathways (see

Boss *et al.*, 2004 for a detailed review). Moreover, it was revealed that the vernalization pathway exerts its impact through chromatin remodeling by causing histone deacetylation and methylation at a specific location of the chromatin and reducing the transcription of *FLC*. This epigenetic effect is long-lasting and thus can be considered as the reason for remembering the winter. The plant can therefore remember a cold treatment several weeks after the end of the cold treatment. Also, one of the autonomous pathways in which the genes *FLD* and *FVE* are involved is mediated through chromatin remodeling.

Another autonomous pathway in which genes, such as *FY*, *FCA*, *EPA* and *FL*, are involved is not mediated through chromatin remodeling. It seems that the photoperiod pathway, i.e. the pathway that permits the transition to flowering, under favorable day length regimes and in plants that have a requirement for such regimes, does not affect flowering through changes in the activity of the *FLC* gene. Rather, it activates the *CO* gene, and the latter imposes its effect downstream of *FLC* (see Fig. 53), directly activating the genes *SOC1* and *FT*; and consequently, the genes *LFY* and *AP1*, respectively, are then activated. The proteins encoded by the latter two genes, and probably mainly the interaction of these proteins, trigger the transition to flowering. From the latter two Arabidopsis genes and the interaction of the respective encoded proteins, it becomes clear that the transition to flowering and the floral organ identity genes are not independent components: the *AP1* gene is also a *bona fide* floral organ identity gene, as shall be detailed in subsequent sections.

It should be noted that while the process of transition to flowering was mainly studied in Arabidopsis, and this process is therefore clarified mainly in this plant, the general strategy of induction to flowering is similar among very distant plant species. For example, the wheat genes that act downstream of vernalization, *VRN2* and *VRN1*, encode a MADS-box protein that is not homologous to *AP1*, but have a similar role in the transition to flowering in wheat, such as the role of *AP1* in this transition in Arabidopsis.

Before leaving the subject of transition to flowering, it should be noted that the number of genes involved in this transition is by far greater than the 25 genes listed in Fig. 53A. As many as 2000 genes change their levels of expression while a vegetative plant is being shifted into flowering. How many of these genes are casually related to the shift into flowering is not yet clear, but this huge number of genes points to the possibility that more genes than those currently identified to be involved in the transition to flowering will be revealed in future studies. Recent studies revealed a novel group of plant hormones: short peptides (see review by Matsubayashi and Sakagami, 2006) that can move across several cell layers (e.g. the 12-amino-acid–long peptide encoded by *CLV3*; see Itoh *et al.*, 2006 and Kondo *et al.*, 2006). Incidentally, the *CLV* genes not only have a role in regulating cell division in SAMs, but also affect a floral organ: the pattern of the carpels. Short peptides may, therefore, be found in the future to be mobile signals involved in bridging the perception of environmental conditions in leaves and changes in the shoot apex that lead to floral induction.

Focusing on Model Plants for the Understanding of Floral Patterning

Up to about 20 years ago, geneticists reported on a great number of flower abnormalities that could be traced to specific genes. Such reports were scattered in the botanical, horticultural and genetic literature as either notable peculiarities, or for considerations that the respective genes may be useful for the breeding of ornamental flowers. It should be noted that, since the early 20th century, an abundance of mutations, either found in nature or produced experimentally by mutagenesis, has been accumulated. Among these are mutations that cause the formation of malformed flowers. Mutations that affect flower differentiation were mainly reported in three plants. Two of these are dicots: *Arabidopsis thaliana* (Arabidopsis) and *Antirrhinum majus* (snapdragon). The monocot that was searched for mutations was *Zea mays* (maize).

Studies on Arabidopsis have a rather long history. Already in 1907, the German botanist F. Laibach found this plant useful for different reasons, and then suggested it for cytogenetic and genetic studies. G. Redei was probably the one who transferred Arabidopsis from Hungary to the USA, when he immigrated to the US. He liked Arabidopsis particularly because of its short life cycle and the ease of obtaining interesting mutants of Arabidopsis. The latter was then adopted as a model plant by other investigators, such as C. Somerville (then in East Lansing, MI) and E. Meyerowitz at the California Institute of Technology (in Pasadena, CA). These two investigators previously studied another model organism: *Drosophila*. As for snapdragon, this plant was used during the early years of the 20th century by E. Bauer. His followers (in the laboratories of H. Stubbe and H. Saedler in Cologne, Germany) continued to use the snapdragon for various purposes (including the study of transposable elements), but later focused on flower development. Snapdragon also served the investigators at John Innes Institute in England (R. Carpenter and E. Coen). The collection of maize mutants was initiated in the laboratory of R.A. Emerson at Cornell University (Emerson *et al.*, 1935) without special focus on flower mutations.

By the early 1980s, an abundance of flower mutations was revealed, not only in snapdragon, Arabidopsis and maize, but also in several other plants (e.g. tomato and petunia). Marx (1983) reviewed these mutants and provided a great number of citations where these mutants were handled (without tabulated lists). But, these mutants did not contribute to the understanding of flower patterning. A leap in the understanding of the patterning of flowers happened only at the end of the 1980s. It appears that three factors contributed to this leap: (1) the focus on only two model plants (Arabidopsis and snapdragon); (2) talented investigators; and (3) the emergence of efficient molecular genetic methodologies.

The emergence of the ABC model of floral organ identity genes

Haughn and Somerville (1988) reviewed the genetic control of morphogenesis in Arabidopsis, and handled the respective mutants that are involved in morphogenesis under three headings: (a) embryo-lethal mutants; (b) mutants affecting trichomes; and (c) mutants altering floral morphogenesis. For our deliberations, the last group of mutants is of interest because these served the authors to suggest the first model for gene regulation of floral organ identity. For the mutants of group (c),

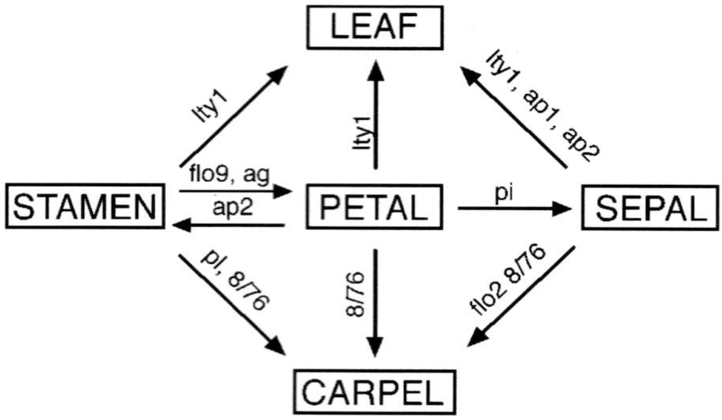

Fig. 68. A scheme indicating the homeotic transformation of flower organs caused by mutations of *Arabidopsis*. The arrow shows the direction of a homeotic transformation where an organ type observed in the wild-type flower (arrow points away) is replaced by a different organ type in the mutant flower (arrow points toward). Positioned above each arrow are the names of mutations that cause the homeotic transformation indicated by that arrow. (From Haughn and Somerville, 1988.)

the authors listed 12 mutations. Among these were several that retained their names, such as *Agamous1*, *Apetela1*, *Apetela2*, *Pistillata*, *Clavata1*, *Clavata2*, and *Leafy1*, as well as several flower development (*flo*) mutants that were subsequently given different names. None of these mutants were further characterized at that time, but were already known to have interesting features. For example, they were incompletely penetrant and, in many of them, one organ of the flower was replaced by another organ (i.e. homeotic mutations) such as carpels replacing sepals. The homeotic transformation of flower organs by the then known Arabidopsis mutants led Haughn and Somerville (1988) to suggest a preliminary scheme for flower differentiation (see Fig. 68).

The model thus suggested by these investigators was (with a few modifications) the following:

Organ	A	B	C
Sepals	On	Off	Off
Petals	On	On	Off
Stamens	Off	On	On
Carpels	Off	Off	On

When one replaces the ABC letters as follows: A=*AP1*; B=*AP3/PI*; C=*AG*, the ABC model later suggested by Schwarz-Sommer *et al.* (1990) and Coen and Meyerowitz emerges, as will be detailed below.

The MSU-DOE Plant Research Laboratory (where G.W. Haughn and C.R. Somerville performed their research) was not the only site where Arabidopsis served to investigate flower differentiation. One other site was the laboratory of Eliot M. Meyerowitz at the Division of Biology (established by T.H. Morgan and where G. Beadle performed important genetic studies) at the California Institute of Technology (Caltech) in Pasadena. Another site where flower development in Arabidopsis was studied was the Osborn Memorial Laboratories at Yale University in New Haven, where Y.F. Irish and I.M. Sussex were situated. Meyerowitz, who had advocated Arabidopsis as a model plant for molecular genetics since the mid-1980s, assembled a talented team of investigators (most of whom later established their own laboratories throughout the US).

An important contribution of the Meyerowitz team (Yanofsky *et al.*, 1990) was the characterization of the gene *AGAMOUS* (*AG*). They cloned and sequenced *AG*, and found that this homeotic gene encodes a protein with a high degree of similarity to the DNA-binding region of transcription factors (from yeast and man) and also to a previously found protein encoded by an *Antirrhinum* homeogene (*DEF*). The authors thus suggested that *AG* encodes a transcription factor that regulates genes determining stamen and carpel development in wild-type Arabidopsis flowers. In the eastern US, Irish and Sussex (1990) focused on the *APETALA-1* (*APT-1*) homeotic gene. They found that the homozygous mutant of *apt-2* lacked sepals and petals. Moreover, they concluded that *APT-1* (and *APT-2*) may act in concert with the product of *AG* to establish a determinate floral meristem, whereas other homeotic gene products are required for cells to differentiate correctly according to this position.

The team of H. Saedler at the Max-Planck Institute near Cologne was engaged for many years in studies on snapdragon. One of their interests was the development of the snapdragon flower. More specifically, they asked what the mechanisms were by which cells in a meristem sense and interpret their position, and then differentiate correctly into a specific floral organ. These investigators were well aware of the usefulness of homeotic genes in animals (e.g. *Drosophila*) for understanding the control of differentiation. Therefore, they started to use homeotic mutants affecting flower differentiation in snapdragon. They also assumed that the molecular analysis of homeotic mutants would be useful in elucidation of the molecular mechanisms that lead to the differentiation of specific organs of the snapdragon flower.

Based on the phenotype of numerous homeotic genes in snapdragon, and on information supplied by Arabidopsis, the investigators Schwarz-Sommer *et al.* (1990) suggested a comprehensive model for organ identity genes in the *Antirrhinum* flower. These authors considered three classes of genes: Class I includes genes involved in the transition to flowering; Class II includes genes that affect the symmetry of flowers (wild-type *A. majus* flowers are zygomorphic, while Arabidopsis flowers are actinomorphic with a radial symmetry); and Class III includes homeotic genes that specify floral organ identity. In our present discussion, we shall focus on Class III genes. The investigators of the Saedler laboratory used their own vocabulary. The meaning of their term "type" is the following:

- Type 1: plants in which the first (sepals) and the second (petals) whorls are affected;
- Type 2: plants in which the third (stamens) and the fourth (carpels) whorls are affected;
- Type 3: plants in which the second (petals) and the third (stamens) whorls are affected.

We shall see that by the nomenclature later used by the Meyerowitz laboratory, the translations of types to ABC genes is the following:

Type 1 = defective in A genes
Type 2 = defective in C genes
Type 3 = defective in B genes.

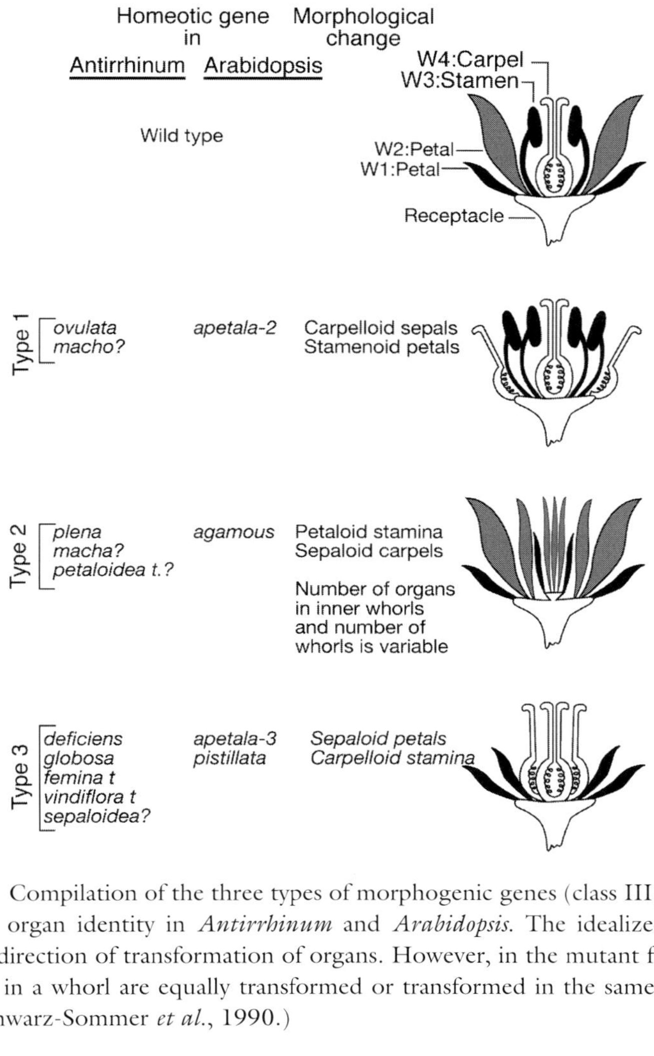

Fig. 69. Compilation of the three types of morphogenic genes (class III) that control floral organ identity in *Antirrhinum* and *Arabidopsis*. The idealized schemes show the direction of transformation of organs. However, in the mutant flowers not all organs in a whorl are equally transformed or transformed in the same direction. (From Schwarz-Sommer *et al.*, 1990.)

The consequences drawn by Schwarz-Sommer *et al.* (1990) are presented in Fig. 69. Briefly, the model of the latter authors indicates the following. In *Type 1*, when genes such as *OVULATA* (snapdragon) or *APETALA-2* are mutated, the sepals and the petals are not normal. In *Type 2*, when *PLENA* (snapdragon) or *AG* (Arabidopsis) are mutated, the stamens and the carpels are not normal. In *Type 3*, when *DEFICIENS* and *GLOBOSA* (snapdragon) or *AP-3* and *PISTILLATA* (Arabidopsis) are mutated, the petals and the stamens are not normal. What was novel in the model of Schwarz-Sommer *et al.* (1990) was that two adjacent whorls were simultaneously affected by a mutation of a single homeotic gene. Also, no homeotic flower identity genes in snapdragon were isolated that caused an alteration of identity only in one whorl.

These authors also noted the similarity (but not complete identity) of the roles of the homeotic class III genes in snapdragon and Arabidopsis. It should be noted that the flower phenotypes in a given mutant were not exactly the same even in the same plant. The flowers ranged in their phenotypes from (almost) normal to complete missing of flower organs. Also, the situation was sometimes confusing because external conditions (e.g. temperature) affect the mutated phenotypes. On the other hand, the basic similarity of the flower organ identity genes in snapdragon and Arabidopsis, which belong to genera of rather distant phylogeny, strongly suggested that the mechanism of floral member differentiation evolved in the early evolution of angiosperms. It should also be noted that the long experience of the Cologne investigators with transposable elements in snapdragon (see Galun, 2003) was very helpful in elucidating the mechanisms of flower patterning. For example, where a homeotic gene was mutated by a transposable element, the results of the excision of this element from the gene (and consequently a reversion to the wild-type gene) could be followed in specific cell lines.

There is one more important issue that emerged from the study of the Saedler laboratory: the recognition of the MADS-box genes. These investigators found that there are regions of *AG* (of Arabidopsis) and *DEF* (of snapdragon) that are very similar to the (animal) serum response element *SRE*, the DNA sequence motif to which the serum response factor (SRF) binds, and to the binding site of the yeast gene *MCM1* that encodes the minichromosome maintenance protein. Because a distinct family of genes was found in snapdragon and Arabidopsis that encodes proteins with homology to two known transcription factors, the authors termed this family MADS-box genes (*MCM1/AG/DEFA/SRF*). Furthermore, already at this early stage, the Cologne investigators suggested that the transcription factors affecting flower organs may have to interact with other factors in order to fulfill their role. The suggestion that MADS-box genes have to first establish heterodimers (or complexes with more than one additional MADS box) by Schwarz-Sommer *et al.* (1990) was prophetic, as shall be seen below. On the other hand, the authors' suggestion that floral meristem cells identify their position by eccentric gradients of hypothetical factors was not experimentally verified.

Less than 1 year after the publication of the article of Schwarz-Sommer *et al.* (1990) in *Science*, Coen and Meyerowitz (1991) published their ABC model of the control of flower development in *Nature*. Their review article was headed with an impressive title: "The War of the Whorls ...." It should be noted that Coen and Meyerowitz (especially the former) communicated during an extensive period with the Saedler laboratory and exchanged information on experimental results. In summary, Coen and Meyerowitz (1991) and Schwarz-Sommer *et al.* (1990) arrived at very similar conclusions with respect to the control of floral organ identity in Arabidopsis and snapdragon; and the former stated that in spite of being distantly related, these two plants use homologous mechanisms. However, the details of the inflorescence and floral meristem are not the same in the two plants. For example, in Arabidopsis, after the transition from vegetative growth to flowering, the main shoot will first produce inflorescence meristems from the axils of cauline leaves, and these inflorescences will then bear flowers. In snapdragon, the flowers are borne directly in the axils of bracts on the main shoot. Also, Arabidopsis flowers are radially symmetric, while snapdragon flowers are zygomorphic. On the other hand, both snapdragon and Arabidopsis have four whorls of flower organs: sepals, petals, stamens and carpels (there are four sepals and four petals in Arabidopsis; while in snapdragon there are five and five, respectively).

I shall not repeat the details of the ABC model of Coen and Meyerowitz (1991) because the principle is the same as the model of Schwarz-Sommer *et al.* (1990), and is also based on the principle that the same mutant (of the A, B, or C type of genes) affects two whorls

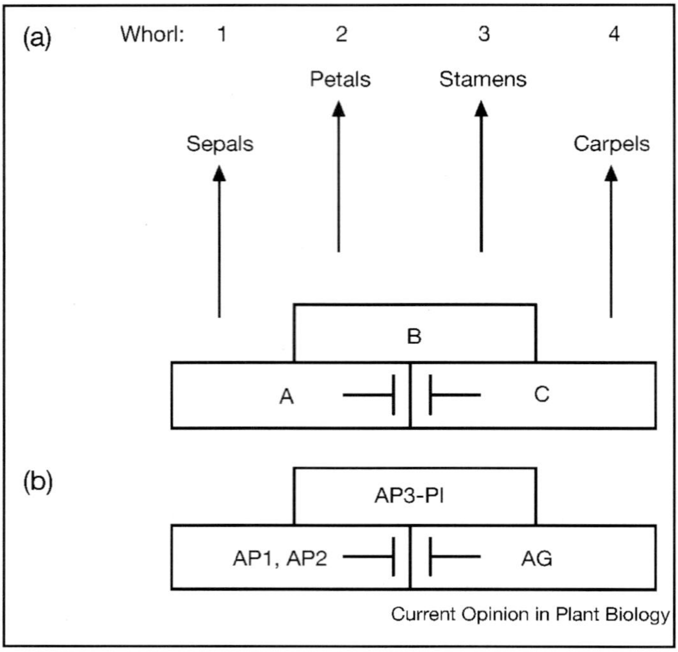

Fig. 70. The "classical ABC model" for flower organ identity in *Arabidopsis*. **(a)** Floral organ identity is specified by homeotic functions. A, B, and C, which are each active in two adjacent whorls. A alone specifies sepals in whorl 1; the combined activities of A and B specify petals in whorl 2; B and C specify stamens in whorl 3; and C alone specifies carpels in whorl 4. The activities A and C are mutually antagonistic: A prevents the activity of C in whorls 1 and 2, and C prevents the activity of A in whorls 3 and 4. **(b)** The proteins providing the floral homeotic functions in *Arabidopsis*. Except AP2, all are MADS-domain proteins, which are expressed in the regions in which they specify organ identity. Antagonistic interactions are indicated by barred lines. A hyphen indicates heterodimer formation; a comma symbolizes that the mode of interaction is unknown. (From Theissen, 2001.)

(e.g. *AP3* of Arabidopsis affects both the petals and the stamen whorls). The now classic ABC model of Coen and Meyerowitz was then also accepted by the MPI-Cologne investigators (see Fig. 70). It should be noted that in spite of the apparent homology between MADS-box genes in the two plants, there are also differences. For example, in the Arabidopsis *ag* mutants, there are sepals in the fourth whorl; while in the corresponding *ple* mutants of snapdragon, they are variable and often show carpelloid features. Coen and Meyerowitz (1991) already assumed that their model presented a simplification of the actual flora organ identity–controlling mechanism. For example, the gene *AG* may have several roles: it specifies carpel (in the absence of functional *AP3* and *P1*), counteracts *AP1/AP2,* and causes determinacy of the flower (preventing indeterminate flora meristem growth).

The "Golden Decade" of floral development

The basic ABC model elucidated by Haughn and Somerville (1988) and formulated by Schwarz-Sommer *et al.* (1990) and Coen and Meyerowitz (1991) initiated intensive studies during the next 10 years. These studies were reviewed by Theissen and Saedler (1999) under the title, "The Golden Decade of Molecular Floral Development (1990–1999): A Cheerful Obituary." Investigators of flower patterning have a remarkable talent for phrasing impressive titles to their reviews. Here is a sample of such titles: "The Unfolding Drama of Flower Development ..." (Ma, 1994); "Three Ways to Learn the ABCs" (Ng and Yanofsky, 2000); "Floral Quartets" (Theissen and Saedler, 2001); "... Stories from the MADS House" (Theissen, 2001); "Building Beauty: The Genetic Control of Floral Patterning" (Lohmann and Weigel, 2002); "Molecular Mechanisms of Flower Development: An Armchair Guide" (Krizek and Fletcher, 2005); and "To B or not to B a Flower: ... " (Zahn *et al.*, 2005a).

Those interested in the history of the accumulation of knowledge regarding floral organ identity genes should consult one or more of the above cited reviews, or other reviews such as that of Theissen *et al.* (2000). The latter review carries a hidden humor in its title: "A short history of MADS-box genes in plants." This "short" history spans 34 pages and devotes much of its discussion to a concept developed by the Saedler laboratory: evodevotics — evolutionary developmental genetics.

In the following years, several important pieces of information were added during this "Golden Decade" and shall be summarized.

It became evident that there is negative interaction between *AP1* and *AG*, hence there is a functional antagonism between the *A* function and the *C* function. Thus, in the presence of active *AG*, the *AP1* is exposed only in the outer two whorls; but in *ag* mutants, the expression of *AP1* is in all four whorls. Likewise, *AP2* (which is not a MADS-box gene) inhibits the expression of *AG* in the two outer whorls. The situation seemed to be complicated because there seemed to be a difference between the *function* of a gene (*AG*) and the *expression* of its transcript.

Several of the genes involved in floral organ identity were cloned during the 1990–1999 decade. The first of these was *DEF* of snapdragon (Sommer *et al.*, 1990). It was established that many of these genes (e.g. *AP1*, *AG*, *AP3* and *PI*) encode MADS-box proteins, while others (*AP2*) do not encode such proteins. It was also revealed (see Ng and Yanofsky, 2001) that *AG*, *AP1* and *AP3* are targets of *LFY*, although additional genes (e.g. *UFO*, *factorX*) may be required for the influence of *LFY* on these targets.

Important additional knowledge concerns the *AGAMOUS-LIKE* (*AGL*) genes. These genes were found to be mediators between the floral meristem and the floral organ identity genes. The three *AGL* genes, *AGL2*, *AGL4* and *AGL9* — later renamed *SEPALLATA1* (*SEP1*), *SEP2* and *SEP3*, respectively — were found by several investigators (e.g. Flanagan and Ma, 1994; Savidge *et al.*, 1995; Pelaz *et al.*, 2000) to be required for the expression of the B- and C-class floral organ identity genes. These *SEP* genes were consequently added to the ABC model (see Zahn *et al.*, 2005a, 2005b, 2006 for recent reviews), and the function of *SEP* genes was termed *E function*. Several genetic and molecular studies clearly indicated that the MADS-box proteins act in complexes to uniquely specify the organ identities in flowers. Physical interactions between the SEP and ABC MADS-box gene products have been demonstrated by the yeast two-hybrid and immunoprecipitation experiments (see reviews of Zik and Irish, 2003a, 2003b).

Another enigma was studied during the "Golden Decade": what restricts the effect of ABC genes to specific whorls? Several findings were made. For example, *LFY* interacts with *UNUSUAL FLORAL*

ORGANS (*UFO*) to limit the *AP3* activity to the second and third whorls. It should be noted that *UFO* encodes an F-box (SCF) ubiquitin ligase complex rather than a MADS-box protein. Another gene in Arabidopsis, *STERILE APETALA* (*SAP*), was identified (Byzova *et al.*, 1999) as limiting the *AG* expression to the third and fourth whorls. A special role was assigned to *HUA* and *HEN* genes in the correct processing of the mRNA of floral organ identity genes (see Cheng *et al.*, 2003). More recent studies clearly indicated that microRNAs also participate in restricting the effect of floral organ identity genes to specific sites of the developing flower. Another Arabidopsis gene, *SUPERMAN* (*SUP*) — which is not a MADS-box gene, but rather encodes a C2H2 zinc-finger transcription factor (Sakai *et al.*, 2000) — regulates the boundaries *between* the whorls and the genes *CUP-SHAPED COTYLEDON* (*CUC*) and *NO APICAL MERISTEM* of Arabidopsis and Petunia, respectively, and also control organ boundaries *within* whorls. It was more recently found that the function of *CUC* genes is regulated posttranscriptionally by microRNA (Laufs *et al.*, 2004; Mallory *et al.*, 2004).

Another class of flower organ identity genes was added during the Golden Decade: the D-class genes that specify placenta and ovule identity. These were first revealed in Petunia (FBP7 and FBP11). Placenta and ovules may represent fruit identity genes rather than flower organ identity genes. However, because of meagre molecular genetic information on the patterning of angiosperm fruits, this book shall not deal with the differentiation of fruits.

Recent findings and the updated ABC model

By 2001, the basic ABC model for the Arabidopsis floral organ specification was established, as shown in Figs. 71A and 71B (from Theissen and Saedler, 2001 and Jack, 2004, respectively). According to this model, the identities of the four floral organs — sepals, petals, stamens and carpels — are determined by combinations of MADS-box proteins that are bound to the promoters of the downstream targets. The expression of the target proteins is thus activated or repressed as appropriate for the differentiation of each type of organ.

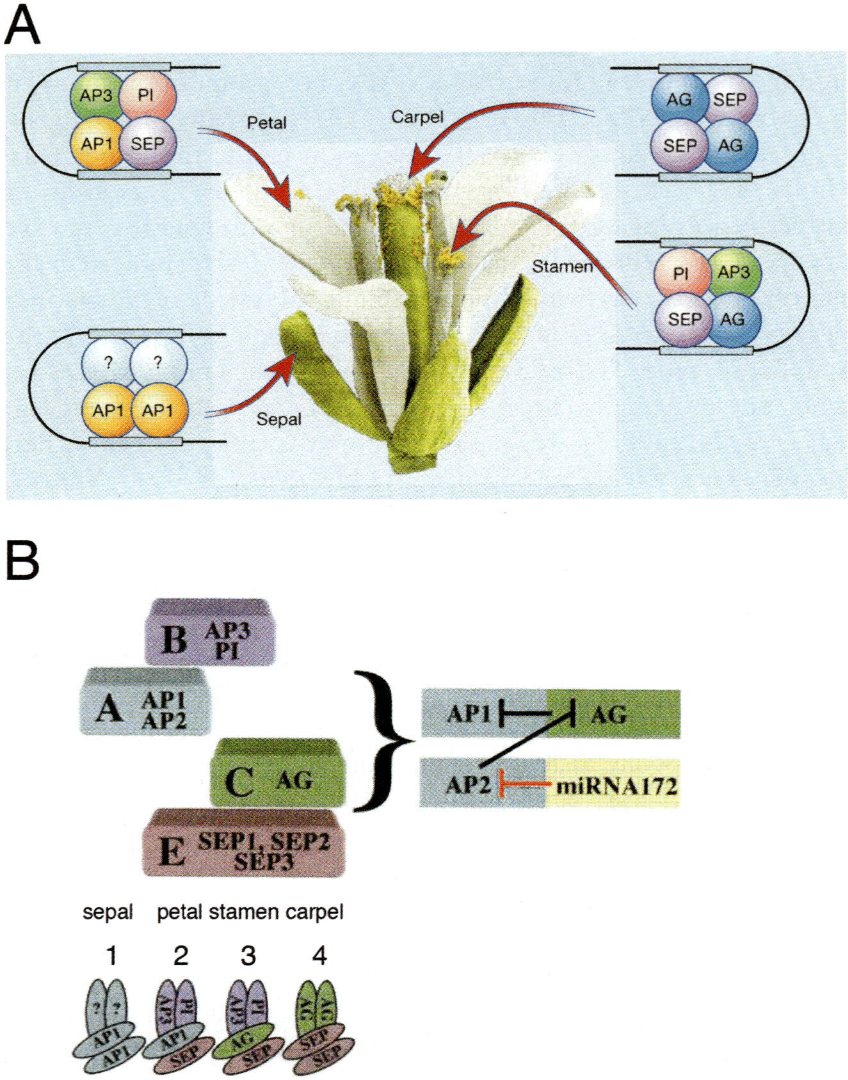

Fig. 71. (**A**) Flower structure and the "quartet model" of floral organ specification in Arabidopsis. According to this model, the identity of the different floral organs — sepals, petals, stamens and carpels — is determined by four combinations of floral homeotic proteins known as MADS-box proteins. The protein quartets, which are transcription factors, may operate by binding to the promoter regions of target genes, which they activate or repress as appropriate for the development of the different floral

Two dimers of each tetramer recognize two DNA sites (CArG boxes) on the same DNA that are brought into close proximity due to the binding of the DNA. The A (*AP1*) and C (*AG*) activities are mutually repressive (see Fig. 71B). *AP2*, which is not a MADS-box–encoding gene, is repressed by the microRNA172.

Another means to regulate gene expression in eukaryotes is by the regulation of splicing. The production of a specific protein, encoded by a specific transcript, requires that the transcript is spliced in the proper way. In 1998, it was reported that correct splicing of the transcript of *AP3* is essential for the formation of a functional *AP3* protein (Yi and Jack, 1998 and Sablowski and Meyerowitz, 1998). In Arabidopsis, an *AP3* mutant was found in which splicing is thermally conditioned. The respective *ap3-1* mutant, which was found to be temperature-sensitive, carries a missence mutation near a 5′ splice site. The domain of the protein that was affected notably interfered with the *AP3* protein's ability to participate in the functional protein–protein interaction (*AP3/PI/API/SEP* and *AP3/PI/AG/SEP*) that is essential for the role of *AP3* in the control of flower organ identity (Lamb and Irish, 2003). It thus appears that a single change in a

---

organs. According to the model, two dimers of each tetramer recognize two different DNA sites (termed CArG-boxes, shown here in grey) on the same strand of DNA, which are brought into close proximity by DNA bending. The exact structures of the multimeric complexes of the MADS-box proteins controlling the identity of flower organs are still hypothetical; question marks denote components whose identity is especially uncertain. Proteins: AG, *AGAMOUS*; AP1, *APETALA1*; AP3, *APETALA3*; PI, *PISTILLATA*; SEP, *SEPALLATA*. (From Theissen and Saedler, 2001.) **(B)** The revised ABC model of flower development. A, B, C, and E are four activities that are present in adjacent whorls of the flower. These four activities are postulated to function combinatorially to specify the identity of the four organs in the flower: sepals, petals, stamens, and carpels. A second major tenet of the ABC model is that A and C activities are mutually repressive. The specific molecular interactions between A and C class genes as well as their regulators are shown at the right. The majority of ABC genes encode MADS domain transcription factors. Recent evidence suggests that MADS proteins function together in complexes larger than a dimer. The "quartet" model postulates that tetramers of MADS proteins specify floral organ identity (shown as colored ovals). Interactions demonstrated to be direct are indicated in red. (From Jack, 2004.)

nucleotide of a transcript can lead to a great change in function, not by directly abolishing protein–protein interaction, but by changing the splicing, thus yielding nonfunctional proteins.

A further complication regarding *AP1*, *AP3* and *PI* was recently reported by the laboratory of V. Irish (of Yale University), who collaborated with investigators of the University of Agricultural Sciences in Uppsala, Sweden (Sundstrom *et al.*, 2006). It was previously found that *AP1* is initially expressed throughout the floral meristem, but later its expression is limited to sepal and petal primordia. The latter authors found that the protein encoded by *PI* has the ability to bind directly to the promoter region of *AP1.* These authors concluded that the proteins encoded by *AP1* and *AP3* regulate the transcription of *AP1.* This downregulation of the *AP1* transcript is apparently a very fast process: it happens within 2 hours of the *AP3/PI* activation, meaning that *AP1* may be a direct target of the *AP3/PI* heterodimer and that this heterodimer (possibly together with other factors) restricts the expression of *AP1* during stages of floral development. Another twist in the expression of floral organ identity genes was studied at the University of Maryland, College Park (Sridhar *et al.*, 2004, 2006). The investigators asked how it could be that two corepressors of *AG* (*LUG* and *SEU*), which encode proteins that do not recognize DNA-binding motifs on the AG protein, have the ability to repress its expression. It was found that the suppression is indeed not direct and rather requires interaction with AP1/SEP3 proteins, meaning that the AP1 and SEP proteins may serve as the DNA-binding partners of SEU/LUG and then together control *AG* expression.

To the previously mentioned *SEP* genes (*SEP1, SEP2, SEP3*), another *SEP* gene (*SEP4*, previously termed *AGL-3*) was added (Ditta *et al.*, 2004); and when all these *SEP* genes are mutated (*sep1*, *sep2*, *sep3*, *sep4*), the quadruple mutants develop vegetative leaves rather than floral organs. Ovules are initiated before the anthesis of flowers, thus ovules can also be considered floral organs. Two genes, *SHP* and *STK*, were found to be required for ovule differentiation; the triple mutants, *shp1*, *shp2*, *stk*, produce leaf-like structures rather than ovules (Favaro *et al.*, 2003). The determination of ovule differentiation was

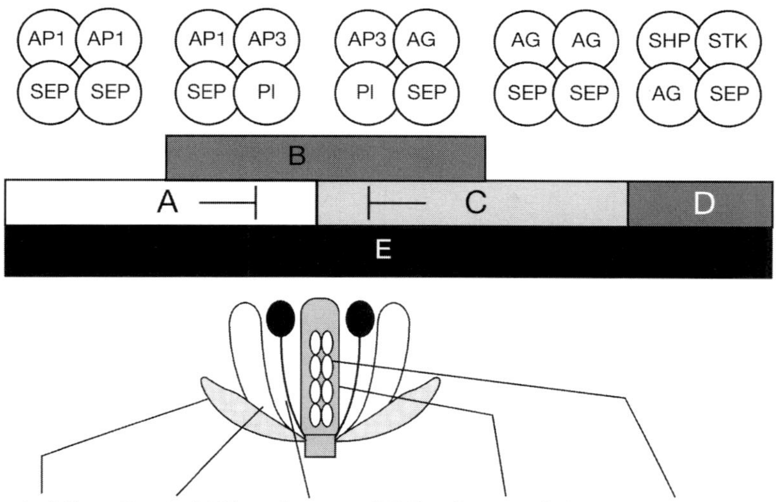

Fig. 72. The genetic ABC + DE model and the quartet model. The ABC model states that A-function genes, such as *AP1* in *Arabidopsis*, are necessary for the formation of the sepals, B-function genes, which include *AP3* and *PI* in *Arabidopsis*, along with A-function genes, are necessary for the formation of the petals and B-functions, along with C-function genes, which in Arabidopsis includes *AG*, are necessary for the formation of the stamens, and C-function genes alone are necessary for the formation of the carpels. This has been expanded to include class D- and E-function genes, which are necessary for the ovules and whorls of the flower, respectively. D-function genes in Arabidopsis include *SEEDSTICK* (*STK*) and *SHATTERPROOF1* and *SHATTERPROOF2* (*SHP1*, *SHP2*). E-function *sensu lato* requires at least one of the four *SEPALLATA* (*SEP1*, *SEP2*, *SEP3*, and *SEP4*) genes. The floral quartet model expands on this idea using data from protein interaction studies (Theissen, 2001; Theissen and Saedler, 2001.) In this figure, the hypothesized quartets, based on experimentally determined dimeric or multimeric protein interactions, necessary for each floral organ are presented. (From Zahn *et al.*, 2005a.)

termed D-function. The accumulated knowledge then led to a modified ABC + DE model (see Fig. 72).

Zahn *et al.* (2006) focused on the B-function genes of these *DEF* and *GLO* genes of snapdragon; and soon after, *AP3* and *PI* of Arabidopsis were sequenced. It was previously found that *DEF* and

GLO are orthologs of *AP3* and *PI*. The authors performed phylogenetic analysis of angiosperms with respect to the sequences of these genes. It was found that *DEF* and *GLO* are closely related paralogs that apparently originated from an early gene duplication — after the separation of the lineages that led to the extant gymnosperms and angiosperms (i.e. about 300 million years ago). Here, it should be noted that the present angiosperms represent only one branch of many that led to clades that disappeared. The other branches disappeared during evolution.

There is no firm answer for the origin of the angiosperm flower. Sequencing of orthologous genes indicated that there were many gene duplications that actually provided multiple potential opportunities for functional divergences. Indeed, such functional divergences were revealed even within the B-function genes that belong to the *DEF/GLO* MADS-box subfamily; for example, orthologs of *DEF/GLO* genes in plant families other than Brassicaceae and Leguminеae (e.g. *Medicago*), such as Solanaceae, have different sites of expression in the developing flower than *AP3/PI* of Arabidopsis. Specifically, in Petunia, the *pMADS1* (an ortholog of *GLO*) mutation (i.e. *pmads1*) converts petals to sepals (as expected by the ABC model). However, in the third whorl, the stamens of these Petunia flowers have petaloid cells on their filaments. In short, the analysis of data performed by Zahn *et al.* (2005a, 2005b) indicated that, although the primary role of the *DEF/GLO* MADS-box subfamily is generally to specify stamens and the inner whorl of the parianth (petals), there are several examples of neofunctionalization that can be related to orthologs of *DEF/GLO* that ramified from ancestral genes.

A recent review on the molecular mechanisms of flower development (Krizek and Fletcher, 2005) starts with a bewilderment: “An afternoon stroll through an English garden reveals the breathtaking beauty and enormous diversity of flowering plants.” On the other side of the globe, in Japanese gardens, one may also see beauty, but (almost) without flowers (although occasionally with cherry blossoms). Is it because flowers wilt and the Japanese refrain from observing wilting? Whatever the reason, this review provides a good update on flowering patterning and also provides updated information on links to databases and relevant websites.

There are several issues that were highlighted by this review and were not handled above in detail. First, the transition to flowering that is induced by the *LFY* and *AP1* genes was found to have partially overlapping roles, although *AP1* functions downstream of *LFY*. For example, there is *CAULIFLOWER* (*CAL*), whose function overlaps with the *AP1* function, as well as *UNUSUAL FLORAL ORGANS* (*UFO*) that functions in the same direction. All these are Arabidopsis genes. In snapdragon, the *FIMBRIATA* (*FIM*) gene has a similar function. It was also found that *LFY* and *AP1* repress *TEL1* and *AGL24* in floral meristems, whereas *TEL1* represses *LFY* and *AP1* in the inflorescence meristem. Moreover, in the *tfl1* mutants, *LFY* and *AP1* are ectopically expressed in the inflorescence, causing the normally indeterminate inflorescence to terminate with the production of a flower. The change of a single amino acid with the *TFL1*-encoded protein is sufficient to confer its activity as a repressor of flowering. Is it because this changes the binding of this protein to other proteins, to DNA or to RNA? If the latter is correct, one "smells" the involvements of microRNA.

Although Solanaceae are phylogenetically distant from Arabidopsis, orthologs of the Arabidopsis *SEP* genes were detected in Petunia and termed *FLORAL BINDING PROTEIN* (*FBP*); *FBP2* has a similar function in Petunia as *SEP3* in Arabidopsis.

It should be noted that all but two of the floral organ identity genes that were characterized are MADS-box genes. The two genes, which do not belong to the latter genes, are the genes *AP2* and *LIP1/2*. These are encoding transcription factors that belong to the ethylene-responsive element–binding factors. A comprehensive protein–protein interaction map of the Arabidopsis MADS-box transcription factors was constructed by a team of investigators from five different European laboratories (de Folter *et al.*, 2005).

How does a flower terminate the production of its whorls with the formation of the carpels? We should regard the development of floral organs as representing a short shoot that produces lateral outgrowths (the floral organs). This short shoot has a meristem from the flanks of which the organs develop. As long as this meristem maintains the division of its stem cells, new organs will be formed. The maintenance of this division is kept by the products of *LFY* and *WUS*. However, these products also activate *AG*. Then, there is a feedback: *AG* causes the repression of *WUS*, causing the cessation of stem cell division. The meristem of the short shoot is then arrested, and no further floral organs are produced (Lenhard *et al.*, 2001; Lohmann *et al.*, 2001).

On the other hand, it appears that the maintenance of the vegetative growth of the shoot requires genes that *prevent* the transition to flowering. Genes were identified as *EMBRYONIC FLOWER* (*EMF*) and *FERTILIZATION-INDEPENDENT ENDOSPERM* (*FIE*). When *EMF* and *FIE* are mutated, there is a premature transition to flowering, even soon after germination. It appears that the respective encoded proteins of these genes can interact to form a polycomb group complex that will modify the chromatin (by specific methylation), and thus repress the transcription of certain genes. Hence, the strategy of plants in patterning their organs occasionally resembles the strategy of the legendary emperor: to assure that he will not loose his pants, he used both a belt *and* suspenders.

An issue that has emerged in recent years is the boundary specifications in the developing flower. This is the issue of *cadastral genes*, meaning genes that define the boundaries within the flower. Some genes were revealed that specify such boundaries. For example, Sridhar *et al.* (2004) reported on transcriptional repression of target genes by *LEUNIG* (*LUG*) and *SEUSS* (*SEU*) that work together in Arabidopsis to repress *AG* expression. Also, the *STYLOSA* (*STY*) of snapdragon, an ortholog of *LUG*, has a similar function (Navaro *et al.*, 2004). Interestingly, the proteins encoded by *LUG* and *SEU* do not have DNA-binding activity, suggesting that additional factors are involved together in the *LUG–SEU* complex to regulate *AG* expression. There are candidate genes that may encode these additional factors. Among these may be the genes *AP2*, *AINTEGUMENTA* (*ANT*), *STERILE APETALA* (*SAP*), and *BELLRINGER* (*BLR*). Sridhar *et al.* (2006) suggested a model to explain how the inner whorl-specific activation of *AG* is achieved (see Fig. 73). The zinc-finger protein of Arabidopsis, encoded by *SUPERMAN* (*SUP*), may also be involved in establishing boundaries by restricting the *AP3* expression. When *SUP* is mutated, the *AP3* domain expression expands, and extra stamens are formed instead of carpels. However, the exact mechanism of

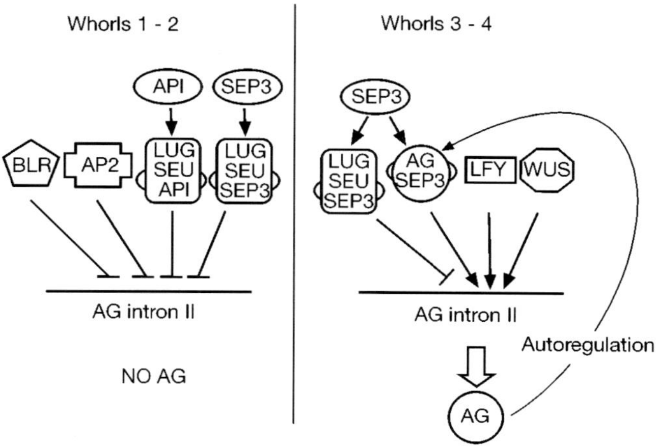

Fig. 73. A proposed model of how the inner whorl-specific activation of AG is achieved. In whorls 1 and 2, multiple negative regulatory activities impinge upon the *AG cis*-regulatory region (such as the *AG* intron II) to prevent *AG* transcription. These negative regulatory factors include *AP2*, *BLR* and *SEU/LUG/AP1* or *SEU/LUG/SEP3*. In whorls 3 and 4, multiple positive regulatory factors antagonize the negative effect of SEU/LUG/SEP3 to promote *AG* transcription. These positive regulatory factors include the combined activities of *LFY* and *WUS*, as well as positive autoregulation by AG/SEP3. Additionally, the AG/SEP3 complex inhibits *AP1* transcription and an interaction between AG and SEP3 may include the SEU/LUG co-repressors from interacting with SEP3. Arrows leading from AP1 or SEP3 to the respective LUG/SEU/AP1, LUG/SEU/SEP3 or AG/SEP3 complexes indicate the incorporation of these MADS box proteins into the respective protein complexes. (From Srighar *et al.*, 2006.)

boundary formation by *SUP* and other cadastral genes awaits further studies. There are also boundaries within whorls that await clarification. For example, why are some petals separated (e.g. Arabidopsis), while in other flowers the petals are fused (Petunia)?

Floral Organ Identity Genes in Dicots Other than Arabidopsis and Snapdragon

The bulk of information on floral organ identity genes in dicots was derived from studies on Arabidopsis and snapdragon. The trend to focus on one or two plants for gaining an understanding on flower patterning emerged in the late 1980s, and over the years it became obvious that this trend was justified. It became evident that, despite the differences in flower forms, the underlying mechanisms for the determination of the identity of floral organs are largely conserved in dicots (and actually in all angiosperms). However, the details of how functions of A, B, and C are specified vary among dicots. How much they vary we do not yet know because out of at least 250 000 angiosperm species detailed investigations on these genes were conducted on only a few specific species.

Before I provide summaries on the latter investigations, the readers are referred to Fig. 66 for acquaintance with the phylogenetic relationship of angiosperm taxa.

Stellari *et al.* (2004), of the laboratory of Elena Kramer at Harvard University, looked at *AP3* and *PI* homologs in the root of dicots' phylogeny (basal angiosperms). These investigators looked at the sequences of 29 representatives of the B-gene lineage from the basal angiosperms and the magnoliid dicots. This study indicated that gene duplications occurred at several phylogenetic levels, before and after the duplication that resulted in the separation of the *AP3* and *PI* lineages. As for *PI* homologs, a 12-nucleotide deletion, symptomatic for the whole *PI* lineage, arose very early: after the split of the *Nymphaeales* from other basal angiosperms, and before the separation of the *Austrobaileyales* from the branch that then ramified into monocots and dicots. These authors also found evidence for alternative splicing in the *Nymphaea AP3* homolog.

The eudicots can be divided into two main branches: the Rosids and the Asterids. Arabidopsis and other Brassicaceae (Cruciferae) belong to the Rosid branch (together with the Rosaceae and the legumes); while the snapdragon (*Antirrhinum*) belong to the Asterid branch, together with the Solanaceae and the Asteraceae (previously termed Compositae, and these include *Gerbera* and sunflower).

Let us first look at the Asterid branches that include one of the two main model plants for flower patterning, the snapdragon (*Antirrhinum mayus*). The family that is closely related to the snapdragon is the Solanaceae. Several plants of this family were studied with respect to their MADS-box genes, such as Petunia and tomato (now termed *Solanum lycopersicum*). The tomato MADS-box genes and their homology with MADS-box genes of the main model plant were intensively studied since the early 1990s by E. Lifschitz and associates of the Technion in Haifa, Israel (Pnueli *et al.*, 1991, 1994a, 1994b). A recent collaborative study of V. Irish of Yale University in New Haven (CT), A. Levy of the Weizmann Institute of Science in Rehovot (Israel) and their associates (de Martino *et al.*, 2006) focused on two *AP3* genes of tomato: the tomato *AP3* (*TAP3*) and the tomato MADS-box gene 6 (*TM6*). These genes have functionally diversified from the ancestral *AP3*. A loss of function of *TAP3* results in homeotic transformation of both petals and stamens. RNA interference induced the reduction of the *TM6* function, resulting primarily in defects in the stamens. The functional difference between *TAP3* and *TM6* could be partially ascribed to different expression domains. When overexpressed, both genes could partially rescue the *tap3* mutant. The proteins encoded by these two genes have different biochemical capabilities, indicating that the two genes play qualitatively different roles in floral patterning.

The authors suggested that the original role of ancestral *AP3* lineage genes was to specify stamen development, and that the duplication and subsequent divergence led to the acquisition of role in petal specification in plants of the core eudicots. Additional information on the results of de Martino *et al.* (2006) will be provided in a subsequent section, where I shall deal with gene duplication. Another recent publication from the laboratory of V. Irish at Yale reported on a wide-range molecular and phylogenetic analysis of MIKC^c-type MADS-box genes in tomato (Hileman *et al.*, 2006), and identified 24 such genes (17 of which were novel). The results suggested that there have been a number of gene duplications and loss events in tomatoes relative to Arabidopsis.

Successful merchants tend to praise their merchandise. In their review, Gerats and Vandenbussche (2005) advocated Petunia as a proper model plant for a variety of purposes. They provided 11 reasons why Petunia is an excellent model system. Beyond their praise, these authors also summarized the contributions of Petunia research for understanding floral development. One example is related to the presumed duplication event and the ancestral preframeshift function of the *DEF/AP3* MADS-box lineage. This issue cannot be studied in either Arabidopsis or snapdragon because both probably lost the *TM6* representative, which does exist in Solanaceae species (see above). In Petunia, both an *enAP3* (*PhDEF*) and the homolog of *TM6* exist (as *PhTM6*). It was found that the *PhTM6* function is significantly different from the classical *DEF/AP3* genes, leading to different expression patterns and genetic interactions (Vandenbussche *et al.*, 2004).

Another example of an issue that can be studied in Petunia, but not in Arabidopsis, is petal fusion; there are no fused petals in Arabidopsis flowers. The *MAEWEST* (*MAW*) gene of Petunia was found to be required for fused petals; in the *maw* mutant, this fusion is distorted. Vandenbussche, Gerats and associates also used transposon insertion mutagenesis in Petunia to analyze the E-function genes' (*FBP2*, *FBP5*) role in flower organ identities, and found that the E-genes in Petunia have a somewhat different function than the E-genes (*SEP*) of Arabidopsis, and that the manifestation of the D-function (ovules) in Petunia also requires the presence of the E-function genes (Vandebussche *et al.*, 2003a, 2003b). Another, more recent, study by these investigators (Rijpkema *et al.*, 2006) focused on the Petunia homolog (*GP*) of *DEF*, and revealed differences in function between *GP* and *DEF*. Moreover, the Petunia *TM6* was found to be regulated like a C-class gene rather than a B-class gene: it is expressed mainly on whorls 3 and 4, but is repressed in the perianth by the *BLIND* (*BL*) gene.

Another plant family that belongs to the Asterids, but is less related to snapdragon than the Solanaceae, is the Asteraceae that includes several ornamental genera (such as Gerbera) and crop plants (such as sunflower). Plants of the Asteraceae have a composite inflorescence that is not found in the main model plants for flower patterning. In the same flower head (capitulum) there are different kinds of flowers that differ in the differentiation of the floral organs (e.g. stamens may be aborted in marginal flowers, and petals and anthers may be fused). To learn more about flower patterning of the Asteraceae, Laitinen *et al.* (2005) from Finland and Norway analyzed the floral transcriptome and revealed novel regulators of organ determination in flowers of *Gerbera*. These authors also reviewed previous studies by their team and others on *Gerbera*. This review will lead the interested readers to literature on patterning in Asteraceae. A study on the identification of *Chrysanthemum* MADS-box genes that have similarities to *AP1* and *SEP3* was performed by Shchennikova *et al.* (2004). It was found that, as in Arabidopsis, the respective proteins form heterodimers.

The families Cucurbitaceae, Fabaceae (legumes) and Rosaceae contain major crop plants and ornamentals (e.g. roses, apples, legumes, cucurbits), and as such also served in studies on MADS-box genes that regulate flower patterning. These three families, as well as the family Myrtaceae, belong to Rosids and are distantly related to Arabidopsis (but less related to snapdragon, which belongs to the Asterids). Studies on the MADS boxes of plants from these families contributed to the reproductive biology of the respective plants, but their contribution to the understanding of flower patterning in angiosperm is meagre. The following are some examples of such studies: MADS-box genes that affect flower organ identity in pea (*Pisum sativum*, Fabaceae). These studies were conducted by Berbel *et al.* (2005) with emphasis on the pea homolog of *PI* (*PsPI*). In apple (*Malus domestica*, Rosaceae), Sung *et al.* (1999) studied the apple homolog of *SQUAMOSA* (*SQUA*), termed it *MdMADS2*, and suggested that *MdMADS2* plays an important role during the early stages of flower development.

The cucumber (*Cucumis sativus*, Cucurbitacea) MADS boxes were also studied. In a relatively early study (Kater *et al.*, 1998), it was found that certain MADS-box genes of cucumber play a similar role in flower organ identity as specific Arabidopsis MADS-box genes.

The species *Eucalyptus grandis* belongs to the Myrtaceae (and thus to the Rosid branch of dicots). *Eucalyptus* is native in Australia, and its MADS-box genes were investigated by a team from Queensland in northern Australia (Southerton *et al.*, 1998). Three MADS-box genes were thus revealed in *Eucalyptus*: *EGM1*, *EGM2* and *EGM3*. Of these, *EGM2* was expressed in petals and stamens, and was most homologous to the *GLO* genes that regulate (in snapdragon and Arabidopsis) the organogenesis of the second and third floral whorls.

As noted above, this information is of value for the understanding of Eucalypt floral differentiation, but does not contribute new basic knowledge on the patterning of the dicot flower.

Flower Patterning in Monocots

The monocots (monocotyledons, i.e. plants having a single seed leaf during embryogenesis) were separated from other angiosperms, such as core eudicots (plants that form two seed leaves during embryogenesis) and magnoliales, about 140 million years ago. The former group consists of plant families that have impressive and colorful flowers such as Iridaceae (e.g. *Iris* species), Orchidaceae and Liliaceae (e.g. *Lilium* and *Tulipa* species), many of which have commercial value as ornamentals. One monocot family, the Poaceae (previously termed Gramineae), includes the most important crop plants (e.g. wheat, rice, and maize), but their flowers lack aesthetic value; these plants are commonly wind-pollinated (for pollination mechanisms, see Frankel and Galun, 1977). The flowers of monocots vary vastly in their structure, hence the patterning of their flowers cannot be handled under one heading. Moreover, only in very few monocot genera was flower patterning studied by molecular genetic investigation. Such investigations were performed mainly in plants that belong to the Poaceae (e.g. maize, rice) and in plants of a few other families (e.g. Lilaceae, Orchidaceae). In spite of the vast variability of monocot flowers, the basic strategy of flower patterning in all studied monocots has great similarities to the flower patterning of the model plants of dicots (Arabidopsis and snapdragon). Hence, the patterning of monocot flowers will be handled below rather briefly, on the basis of a few examples.

As detailed above, the typical dicot model for flower development describes flower organ identity genes for sepals, petals, stamens and carpels. Although monocot flowers also contain stamens and carpels, they differ from dicots with respect to the outer whorls.

Flowers of the Liliaceae often have two outer whorls of conspicuous petal-like *tepals*; whereas flowers of the grass family (Poaceae) have *paleas*, *lemmas* and *ludicules* in place of sepals and petals (see Fig. 54

in Chapter 10). We shall see that flower patterning of both Liliaceae and Gramineae can be understood on the basis of modified ABC models.

Flower patterning in Liliaceae

The MADS-box genes of Liliaceae were already discussed in "A short history of MADS-box genes in plants." This "short history" (of 34 pages) was presented by the team of Saedler and Theissen (Theissen *et al.*, 2000) of the Max Planck Institute near Cologne, and a modified ABC model for *Lilium* flowers was suggested by these authors. More detailed studies on MADS boxes of *Lilium logniflorum*, *Tulipa gesneriana* and *Eustoma gradiflorum* (lisianthus) were thereafter conducted by several research teams (e.g Tzeng *et al.*, 2002, 2003; Kanno *et al.*, 2003; Benedito *et al.*, 2004). These studies discovered homologies between MADs-box genes in Liliaceae and those of Arabidopsis. For example, it could be concluded that the *LlAG1* gene is the functional *AG* ortholog of lily (Benedito *et al.*, 2004).

The latter Dutch team, being obviously aware of the commercial value of double lily flowers, had remarked that knowledge about *AG*-like genes could lead to novel lily cultivars with double-flower phenotypes (see my remark on the poem of Goethe, on the wild rose and its conversion to the typical double-rose cultivar, in the Epilogue). Also, the sequences of MADs-box genes of lily *LMADS2* and of lisianthus *EgMADS1* indicated (Tzeng *et al.*, 2002) that these genes which are specifically expressed in the carpels fulfill the function of D-genes in Arabidopsis. The functional similarity between specific MADs-box genes of Liliaceae could be shown when the coding sequences of the Liliaceae genes were fused to the respective Arabidopsis promoters. Arabidopsis plants with specific nonfunctional MADS-box genes could then be compensated by the transgenic hybrid genes. Figure 74 shows the proposed modified ABC model of floral patterning in *Tulipa gesneriana*.

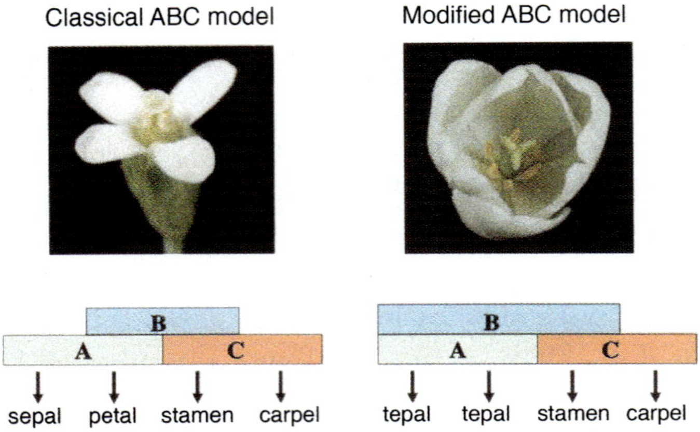

Fig. 74. Classical and modified ABC model. In higher eudicotyledonous flowering plants (left, Classical ABC model, *Arabidopsis thaliana*), expression of an A-function gene specifies sepal formation in whorl 1. The combination of A- and B-function genes specifies the formation of petals in whorl 2; B- and C-function genes specify stamen formation in whorl 3; and expression of a C-function alone determines the formation of carpels in whorl 4. In contrast, Liliaceae plants (right, Modified ABC model, *Tulipa gesneriana*) have two whorls of almost identical petaloid organs, called tepals. A modified ABC model has been proposed to explain the flower morphology of tulip. In this model, class B genes are expressed in whorl 1 as well as whorls 2 and 3; thus the organs of whorl 1 and whorl 2 have the same petaloid structure. (From Kanno *et al.*, 2003.)

Flower patterning in Orchidaceae

Including more than 20 000 species, the Orchidaceae is the largest angiosperm family. Species of this family developed elaborate mechanisms that promote cross-pollination by insects. A vast variety of flower structures that are probably beautiful not only to the human eye, but also to the eyes of insects, exist in rain forests such as in Amazonia. Possibly the competition for appealing to the pollinators rendered the orchid flowers very attractive.

The first report on the sequence of a MADs-box gene from an orchid was published by a team from the National University of Singapore (Lu *et al.*, 1993). A cDNA with similarity (95% homology) to the Arabidopsis gene AG was thus isolated from the *Aranda* deborah orchid. A later publication from the same laboratory reported on three orchid MADS-box genes involved in the transition to flowering in the orchid *Dendrobium* cv Madam Thong-In (Yu and Goh, 2000). Progress in the molecular genetic studies on the orchid's reproductive system followed advances in methodological methods such as vastly shortening the life cycle and genetic transformation of orchids (Yu and Goh, 2001).

A team from Taiwan studied another orchid, *Oncidium* cv Gower Ramsey (Hsu and Yung, 2002), and isolated a cDNA that had sequence homologies to the B-class MADS-box genes of dicots: *AP3* and *TM6*. This orchid gene, termed *OMADS3*, in spite of its overall sequence similarity to B-class MADS-box genes of dicots, had clearly different expression profiles and a different function than the B-class genes of dicots. The authors suggested that *OMADS3* represents an ancestral form of *TM6* which was conserved in monocots (including orchids) with a function similar to A-class functional genes, and regulates flower differentiation and floral initiation. The same team from Taiwan (Hsu *et al.*, 2003) subsequently reported on another gene from *Oncidium*: *OMADS1* that seemed to be involved in the transition to flowering.

Another team from Taiwan, Tsai *et al.* (2004) studied MADS-box genes from the orchid *Phalaenopsis equestris*. They characterized four MADS-box genes from this orchid that are *DEF*-like genes: *PeMADS2*, *PeMADS3*, *PeMADS4* and *PeMADS5*. Each of these seemed to have a distinct role in flower patterning and their expression in the flower organs varied. For example, *PeMADS4* was expressed in the "lips" (petals) and columns, while *PeMADS5* was mainly expressed in the petals and less in the columns. Note that in *Phalaenopsis* there are sepals and petals; while in the monocot families mentioned above, there are tepals, without the division into sepals and petals.

In a recent study by a team from Singapore (Xu *et al.*, 2006b), the investigators attempted to relate the ABCDE model of dicots to the flower differentiation in orchids. As before, their orchid model plant was *Dendrobium crumenatum*. The investigators thus isolated from their orchid genes that had similarities to *AP2*, *PI/GLO*, *AP3/DEF*, *AG* and

SEP of dicot model plants. The investigators first revealed that the locations of gene expression of the orchid genes differed considerably from the locations of expression of the orthologous Arabidopsis genes. On the other hand, the orchid genes *DcOPI* and *DcOAG1* could replace the *PI* and *AG* genes in Arabidopsis. It was also found that the orchid gene *DcOAP3A* was a putative B-function gene. Furthermore, DcOAP3A/B and DcOPI proteins could form heterodimers that could further interact with the DcOSEP proteins to form higher protein complexes. In short, in spite of the great phylogenetic distance between orchids and eudicots, there is a partial conservation in the B- and the C-class function genes between orchids and Arabidopsis. Clearly, the basic strategy of using flower organ identity genes for the differentiation of the flower is the same in orchids as in Arabidopsis.

Flower patterning in the Poaceae

The Poaceae contains the most important crop plants (cereals) such as wheat, rice and maize. The flowers of Poaceae, although being botanically true flowers, are inconspicuous because grasses are wind-pollinated or apomictic and the attraction of animals (such as insects) does not play a role in fertilization. The two grasses in which most molecular genetic studies on flower patterning were performed are rice (*Oryza sativa*) and maize (*Zea mays*). Rice and maize have very different inflorescences. While rice has a branched inflorescence, termed panicle; maize, which is a man-made plant, has two types of inflorescences. One usually bears the female flower, the ear, and the other is panicle-like and usually bears the male flowers (see Fig. 75). In spite of the vast morphological difference between the two maize inflorescences, a single gene can have a decisive impact so that the tassel will be able to bear grains. Fig. 75C schematically describes the florets of rice and maize. It should be noted that rice and maize belong to two different subfamilies of the Poaceae (Kellog, 2001): while rice belongs to the Ehrhartoideae (the BEP clade), the maize belongs to the Panicoideae (the PACC clade), and these clades probably separated about 40–50 million years ago. The genetics and evolution of the inflorescences and the flowers of rice, maize and other grasses

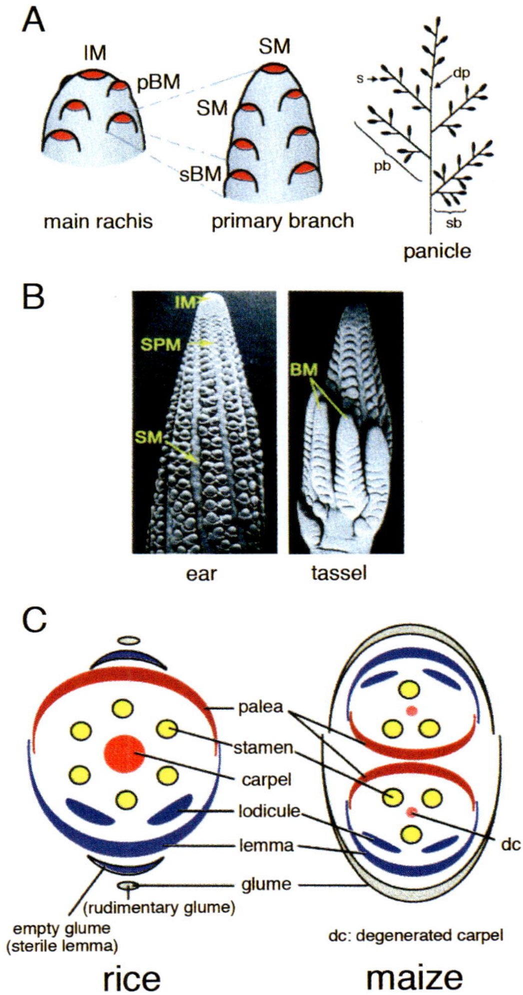

Fig. 75. **(A)** Rice inflorescence development. The inflorescence meristem (IM) initiates primary branch meristem (BMs) in a spiral phyllotaxy. Primary BMs initiate secondary BMs (sBM) and spikelet meristems (SMs) in a biased distichous phyllotaxy and terminate as an SM. Each meristem is colored red. dp, degenerate point; s, spikelet; pb, primary branch; sb, secondary branch. **(B)** Maize inflorescence development. The IM initiates files of spikelet pair meristems (SPMs). Each SPM then produces two SMs. In the tassel, the IM initiates several BMs first and then produces SPMs and SMs in the same way as in the ear. **(C)** Schematic models of rice and maizema spikelets. (From Bommert *et al.*, 2005.)

were thoroughly reviewed by Kellogg (2001), Bommert *et al.* (2005) and Preston and Kellogg (2006); and there is a Grass Phylogeny Working Group (GPWG; www.virtualherbarium.org/GPWG/) that updates the phylogeny of Poaceae.

The rice spikelet contains a single functional floret that is bisexual. It includes a pistil, six stamens, two *lodicules*, a *palea* and a *lemma* (the lodicules are considered homologs of the petals of dicots; the lemma is probably homologous to the sepal of dicots or to a bract of dicots). Outside of these, there are two empty glumes, and then two rudimentary glumes substand three florets: one fertile and two strongly reduced and sterile florets.

In maize spikelets, two florets are initiated, but only the upper one is functional in the ear spikelet. Both male and female florets initiate one pistil, three stamens, two lodicules, a palea and a lemma. However, carpel primordia in male florets and stamen primordia in female florets abort during the early stages of flower development, producing unisexual florets. While the homology of the first whorl is not clear, there is an agreement that lodicules, stamens and the pistils of rice and maize are considered homologous to whorls 2, 3 and 4 of dicots. Studies on homology between floral organ identity genes in maize and rice and those of Arabidopsis and snapdragon were performed mainly on B-class and C-class genes (see Bommert *et al.*, 2005; Preston and Kellogg, 2006 and Yamaguchi *et al.*, 2006 for detailed literature reviews). With respect to B-class homeotic genes, there is a conservatism between maize and rice. For example, in the *silky1* (*si1*) of maize, stamens are replaced by carpels and lodicules are replaced by bracts (that resemble palea/lemma). Similarly, in the mutant *superwoman1* (*spw1*) of rice, there is a homeotic transformation of stamens and lodicules into carpels and palea-like organs, respectively. Furthermore, molecular studies indicated that *SI1* and *SPW1* encode AP3-like proteins. More generally, experimental results suggested a conservation between the B-class gene function, at the protein level, of Arabidopsis and maize. The rice genome (which is now fully sequenced) contains two *PI* orthologs: *OsMADS2* and *OsMADS4*. *In vitro* interaction of rice B-class proteins (*OsMADS2*, *OsMADS4* and *OsMADS16*) can form homodimers and heterotrimers, but there is no exact homology in the interactions of the rice proteins to those of the proteins encoded by the Arabidopsis *AP3* and *PI* genes. It was thus suggested that the *OsMADS* genes underwent diversification during evolution.

The diversification of C-class genes, between the grasses and the dicots, seems to be greater than the diversification of B-class genes. However, the same general trend could be revealed: an ancestral MADs-box gene underwent duplication followed by either diversification in sequence and function, or one of the duplicated genes became redundant and was gradually lost. Such cycles of gene duplication and diversification were repeated several times during the evolution of angiosperms, and some MADS-box genes had their roots in the gymnosperms.

An interesting example of the functional diversification of two C-class MADS-box genes in rice was provided by a team composed of Korean and Japanese investigators (Yamaguchi *et al.*, 2006). Such a collaboration has significance for those who have historically followed the interactions between the two countries, since the transfer of Chinese characters to Japan was probably mediated some 1400 years ago by Koreans. This team focused its attention on the *AG*-like MADs-box genes in rice. As detailed in previous sections of this chapter, the C-class MADS-box gene *AG* plays a major role in flower differentiation of Arabidopsis. It regulates the organ identity of stamens and carpels, represses A-class genes and causes floral meristem determinacy. Yamaguchi *et al.* (2006) isolated two C-class genes from rice: *OSMADS3* and *OSMADS58*. The respective sequences suggested that these two genes have arisen by gene duplication that occurred before the divergence between rice and maize.

The activity of those two genes was disrupted by the investigators. They established a knockout line of *OSMADS3* by T-DNA insertion, and reduced the production of protein encoded in *OSMADS52* by RNA silencing. The knockout of *OSMADS3* caused homeotic transformation of stamens into lodicules and ectopic development of lodicules in the second whorl of rice, near the palea, where no lodicules develop in the wild-type rice. The silencing of *OSMADS58* led to the formation of very unusual flowers: there was a reiteration of floral organs, including lodicules, stamens and carpel-like organs. The flowers seemed to have lost their unusual determinacy. The results suggested to the authors that the two C-class genes were partially subfunctionalized during the evolution of rice. In other words, the function, originally regulated by the gene *AG*, has been partially partitioned into the two paralogous genes, *OSMADS3* and *OSMADS58*, which was caused by a recent gene duplication. For a general background on gene duplication in plants, the readers are referred to Chapter 4 of this book.

Maize was the main subject of plant genetics and cytogenetics in the US during the early years of the 20th century. Being a major crop plant in the US, the development of efficient cytogenetic methodology attracted many investigators to study the genetics and cytogenetics of maize. These investigators received a boost after the discovery of the transposable elements in maize by Barbara McClintock (see Galun, 2003). A huge number of mutations was identified and the respective genes were characterized. Among these were genes that affect flowering and flower patterning. By 1993, Veit *et al.* (1993), a team of investigators from the Plant Gene Expression Center in Albany (CA) and the University of California at San Diego, listed over 40 mutations with inflorescence phenotypes. Three years later, a team of investigators from the University of California at San Diego and Pioneer Hi-Bred International (in Johnson, IA) isolated the first genes that are homologus to MADS-box genes of Arabidopsis (Mena *et al.*, 1996). When the maize homology of *AG* (*ZAG1*) was mutated, the maize inflorescence lost the indeterminacy, but the organ identity (as a C-class gene) was largely unaffected; while a mutation in another maize gene, *ZMM2*, did disrupt the C-function. The authors thus concluded that in maize the task of *AG* was orchestrated by two closely related genes, *ZAG1* and *ZMM2*, which have overlapping but nonidentical activities.

Another US team (Chuck *et al.*, 1998) focused on a maize homolog of *AP2*. The latter is not a MADS-box gene, but has an A-class function in Arabidopsis. The maize inflorescence meristem branches several times to produce lateral meristems, each of which produces two flower meristems. The authors identified a gene termed *Indeterminate spikelet1* (*ids1*) that specifies a determinate spikelet meristem fate and thereby limits the number of floral meristems. When *ids1* does not function, the spikelet meristem becomes indeterminate and produces additional florets. Thus, while by sequence *IDS1* has similarity to *AP2*, its function differs considerably from the function of *AP2*.

The San Diego team (Ambrose *et al.*, 2000) handled a MADS-box gene of maize that has a B-class function: *SILKY1* (*Si1*). They cloned this gene. The mutant *si1* is male-sterile and homeotic: stamens are converted into carpels, and lodicules are converted into palea/lemma-like structures. In the double mutant *si1,zag1*, the spikelets have a peculiar phenotype in which glumes enclose reiterated palea/lemma-like organs. Hence, *Si1* seems to function as a B-class gene of dicots. This homeotic conversion also supports the assumption that the lodicules of maize are modified petals, and that the basic strategy of flower patterning in maize is the same as in dicots. By sequence, *Si1* appears to be an ortholog of the Arabidopsis B-class gene *AP3*. Another study (Whipple *et al.*, 2004) identified an additional B-class gene in maize, *ZMM16*, which seems to be an ortholog of *PI* in Arabidopsis.

A collaborative team of investigators from two universities (in Madison, WI and in San Diego, CA) with investigators from two seed companies (Syngenta and Pioneer) revealed two maize genes that function upstream of the ABC system: *ZFL1* and *ZFL2* (Bomblies *et al.*, 2003). The latter genes shared conserved roles with *FLO* of snapdragon and *LFY* of Arabidopsis. This indicated that even functions upstream of ABC are conserved between maize and dicots.

Whippe and Schmidt (2006) thoroughly reviewed the genetics of grass flower development; and the A-, B-, C-, D- and E-class genes in maize and rice were summarized in a table (see Table 9). This table indicates that for all these five classes of floral organ function genes of dicots, there are parallel genes in maize and rice.

Targets of MADS-Box Genes

After a group of cells at the flank of a floral meristem is assigned to become a specific floral organ (e.g. sepal, petal, stamen, carpel) by

Table 9. Summary of Grass A, B, C, D, and E Class MADS-box Genes Classified According to their Phylogenetic Relationships (a)

	Eudicot Lineage	Maize	Rice	Other
A class	*AP1/CAL/FUL*	*Zap1a, Zap1b (ZmMADS3)**	*OsMADS15*	*BM8, LtMADS2*
			OsMADS14	*BM5, LtMADS1*
		Zmm4, Zmm15	*OsMADS18*	*Vrn1**
		Zmm28		*BM3*
B class	*AP3/DEF*	***Si1***	***SPW***	
	PI/GLO	*Zmm16*	*OsMADS2**	
		Zmm18, Zmm29	*OsMADS4**	
C class	*AG*	***Zag1***	*OsMADS58**	
		Zmm2, Zmm23	***OsMADS3****	
D class	*FBP11/STK*	*Zag2, Zmm1*	*OsMADS13*	
		Zmm25	*P0408G07.14*	
E class	*SEP3*	*Zmm6*	*OsMADS7*	
		Zmm27	*OsMADS8*	
	SEP1, 2, 4	*Zmm24, Zmm31*	*OsMADS34*	
		ZmLHS1a, ZmLHS1b	***LHS1****	
		Zmm3	*OsMADS5*	

(a) The first column indicates the closest eudicot lineage to each of the grass genes in that row. Bold genes have characterized mutants, and genes with a * indicate co-suppression or other studies of gene function. (From Whipple and Schmidt, 2006.)

the respective organ identity MADS-box gene(s), it should proceed in its development to become a mature organ. Hence, target genes have to enter into the process of organ formation. What are these target genes? Moreover, there probably is a cascade of such target genes expressed downstream of the organ identity genes. Presently, very little is known about these target genes and, as already pointed out in the introduction of this book, we know almost nothing about the mechanisms that lead from the linear sequence of nucleotides in genes to the three-dimensional structures. However, there does exist some information on target genes involved in the differentiation of petals, stamens and carpels. These will briefly be mentioned below.

To reveal target genes that are regulated by *AP3* and *PI*, Zik and Irish (2003b) of Yale University used a microarray analysis. A total of

9216 ESTs of Arabidopsis were screened with probes corresponding to mRNAs from different mutants and transgenic plants that misexpressed *AP3* and/or *PI*. The investigators found that *AP3* and *PI* regulate a small number of genes. A very low percentage of these genes encoded transcription factors, and the investigators suggested that *AP3* and *PI* act relatively directly to regulate genes that are required for the basic cellular processes responsible for shaping petals and stamens.

While the molecular genetic processes of patterning the shape of petals are still enigmatic, investigators of the E. Coen laboratory at the John Innes Centre in Norwich (UK) suggested a description for the dynamics of petal shape asymmetry in snapdragon (Rolland-Lagan *et al.*, 2003). While the incipial petals of a snapdragon flower are similar to the petals of other flowers (e.g. Arabidopsis), the final shape of the snapdragon petals are highly asymmetric and unique to the petal lobes of the genus *Antirrhinum*. How does this happen? The authors analyzed cell divisions and cell growth in the developing petal lobes, and claimed that the key process is the control of the direction of growth of cells rather than regional differences in growth rates. The direction of growth is maintained parallel to the proximodistal axis of the flower, irrespective of changes in shape, implying that the long-range signals orient growth along the petal as a whole. Still, what the signal is and how it is spatiotemporally regulated is yet unknown. A further study by the John Innes Centre team (Costa *et al.*, 2005) revealed the controlling of the floral asymmetry in snapdragon, and identified two genes that are involved. One is the *CYCLOIDEA* (*CYC*) that encodes a DNA-binding protein. This protein seems to recognize a downstream gene, *RADIALIS* (*RAD*). This interaction could be reconstituted in transgenic Arabidopsis, causing petal asymmetry, although the Arabidopsis *RAD* is not compatible with the *CYC* of snapdragon.

Another team of investigators from the John Innes Centre and the University of Cambridge (Perez-Rodriguez *et al.*, 2005) also looked at target genes of the B-class genes of snapdragon, *DEF* and *GLO*. The *MIXTA* gene of snapdragon directs the formation of conical epidermis cells in the petals, but this gene can also promote trichome development, depending on when it is expressed. A very early expression of a *MIXTA*-like gene (*AmMYBML1*) probably fulfills three functions: trichome formation; conical cell development in the petal hinge epidermis; and reinforcement of hinge through differential mesophyll cell expansion. The *AmMYBML1* gene seemed to be regulated by an interaction between the product of the *DIV* gene and the two B-class genes *DEF* and *GLO*.

A step towards the detection of target genes that are controlled by the protein encoded in *AG* and by the interaction of *AG* with other gene products was initiated by Gomez-Mena *et al.* (2005). One of the experimental approaches of these investigators was to conduct a global analysis of gene expression during *AG*-induced organogenesis. Such analyses are complicated, require a lot of controls, and then furnish a huge amount of data that has to be further analyzed in order to come up with reasonable models.

Flower Patterning and Sex Expression

The sex expression of plants has attracted the interest of botanists since at least 100 years ago. Correns (1906) already reviewed the inheritance and physiology of sex in angiosperms. Numerous later reviews were published on this subject (e.g. Allen, 1932; Heslop-Harrison, 1957; Westergaard, 1958), and a detailed discussion on sex expression in plants was presented in my book (Frankel and Galun, 1977). Up to the early 1990s, studies on sex expression in plants handled the effects of genetics, environmental effectors and plant hormones on sex determination. Plants can have a plethora of sex expressions. I shall mention only some of them. All the plants of a given species may produce bisexual (hermaphroditic) flowers with stamens and pistil(s). Plants may also be monoecious, bearing male (staminate) and female (pistillate) flowers. In some cases, the same plant bears male, female and hermaphrodite flowers. Some plant species are dioecious; among these, there are plants that bear only male flowers and other plants that bear only female flowers. However, there are additional sex types, such as plants with male flowers and hermaphrodite flowers (andromonoecious plants).

The study of a team of investigators from the University of California in San Diego and the Plant Gene Expression Center in

Albany, CA (Schmidt *et al.*, 1993) heralded the era of the molecular approach to sex expression in plants. The latter study was conducted with maize and handled the then newly discovered MADS-box genes. This study was not yet aimed at revealing sex determination, but two genes with sequence similarity to the Arabidopsis gene *AG* were isolated: *ZAG1* and *ZAG2*. The *ZAG1* transcript was found to accumulate early in stamen and carpel primordia (as *AG* in Arabidopsis). The accumulation of the *ZAG2* transcript starts later and is restricted to developing carpels. This was the first report that indicated a differential expression of a MADS-box gene in male versus female floral organs.

Another early study using molecular methods on sex expression was conducted at the Max Planck Institute by Hardenack *et al.* (1994). This study was on the dioecious species (*Silene latifolia* syn *Melandrium album*, or White Campion). The investigators isolated five genes from *S. latifolia* plants and termed them accordingly *SLM1*, *SLM2*, *SLM3*, *SLM4* and *SLM5*. These were MADS-box genes, and their similarity to snapdragon MADS-box genes was determined: *SLM1* was similar to *PLENA* (a C-class gene); *SLM2* was similar to *GLOBOSA* (a B-class gene); *SLM3* was similar to *DEFICIENS* (a C-class gene); and *SLM4* and *SLM5* were similar to *SQUAMOSA* (an A-class gene). The locations of the expression of the *S. latifolia* genes were the same whorls as those of the snapdragon genes. The transcripts of *SLM2* and *SLM3* were found in the petal and stamen whorls of the *S. latifolia* floral bud, and the male flowers had a smaller fourth whorl than the female flowers, indicating a repressed gynoecium in male flowers. However, it was not concluded whether the change in expression was the result or the cause of the *SLM* genes. Furthermore, the pattern of expression of *SLM4* and *SLM5* differed from the pattern of *SQUAMOSA* expression in snapdragon.

A group of investigators from the United Kingdom (Ainsworth *et al.*, 1995) studied the expression of MADS-box genes in the whorls of the dioecious plant *Rumex acetosa*, and also could not decide whether the different expressions of such genes in the developing flowers was the cause or the consequence of the different sex of the flowers.

Another team of investigators (Perl-Treves *et al.*, 1998; and see Perl-Treves, 1999) isolated three *AG* (a C-class gene) homologs from cucumber floral buds and sequenced these genes. The patterns of expression of these genes in male (androecious) and female (gynoecious) flowers did not indicate that these MADS-box genes were the cause of the different sex expressions. Also, the transcript levels of these three genes were not changed by hormonal treatments that change the sex of cucumber flowers (such as ethephone, an ethylene-releasing compound causing femalization, and gibberellin which causes maleness). The authors concluded that the pathway leading to reproductive organ arrest in the staminate (androecious) or in the pistillate (gynoecios) flowers of cucumber is independent of the MADS-box gene expression.

Tanurdzic and Banks (2004) of Purdue University, ID, reviewed the mechanisms of sex determination in land plants. In their section on sex determination in monoecious angiosperms, they summarized the respective mechanisms in maize and mentioned numerous genes that affect sex expression in this plant. It appears that several genes regulating the level of gibberellins strongly affect sex expression, and the authors concluded that increased endogenous gibberellin has a femalization effect in maize. The *TASSELSEED1* (*TS1*) and *TS2* genes encode a putative short-chain alcohol dehydrogenase that is correlated with the suppression of the pistil in wild-type tassels. In the respective mutants (*ts1* and *ts2*), there is no abortion of the pistil in the florets of the tassel. These results supported the notion that endogenous gibberellins play a role in sex determination in maize florets. However, direct evidence on the exact distribution and degradation of gibberellins at the cellular and subcellular levels in the developing maize flower is still missing.

Cucumber was probably the first plant in which it was reported that the application of gibberellins (e.g. *GA4*, *GA7* and *GA9*) shifted sex expression towards maleness (see Frankel and Galun, 1977). But, the detailed examination of gibberellin-treated gynoecious plants revealed an interesting picture. In cucumber, there are very short inflorescences in the axils of leaves. In each inflorescence, there is a gradient from female to male flowers. If the lower female flower (or two of them) develops into pistillate flower, it will suppress the further (male) flowers. However, if the first female flowers are suppressed (by gibberellin), the later (male) flowers will be able to develop. On the other hand, when gynoecious plants are treated by an antiethylene compound (e.g. aminoethoxyvinyl glycine), the pistils of the female flowers are suppressed, permitting the differentiation of the stamens in the same flower (Atsmon and Tablak, 1979). T. Trebitsh and her associates (now at Ben Gurion University of the Negev, Israel) studied the role of ethylene synthesis in the sex expression of cucumber (Trebitsh *et al.*, 1990; Knopf and Trebitsh, 2006), and found a female-specific gene (*Cs-ACS2*) that is expressed in female plants and is involved in sex expression. Additional information on sex determination in cucumber was provided by Perl-Treves (1999). As for the question whether the abortion of sex organs in cucumber is based on their position or on their identity, the findings of Kater *et al.* (2001) suggested that it is based on their position.

Will certain dioecious plant species lead to an understanding of the molecular genetic mechanisms that control sex expression in plants? Di Stilio and Kramer of Harvard University and Braun of the University of Wisconsin (Di Stilio *et al.*, 2005) believe that this is possible. Their model plant is *Thalictrum dioicum*, a dioecious species of the lower dicot family Ranunculaceae. In this plant, there are no initials of stamens in female plants and no initials of pistils in male plants, meaning that sex-determining mechanisms operate at or above the level of organ identity by a homeotic-like mechanism. The homologs of the Arabidopsis MADS-box genes *AP3*, *PI* and *AG* were isolated and characterized from *T. dioicum* and the hermaphrodite *T. thalictroides* in order to obtain information on the origin of dioecy in *Thalictrum* in further studies.

Duplication in Genes Controlling Flower Patterning

The general role of gene duplication in plant patterning was handled in Chapter 4 of this book. The availability of the full DNA sequences of several model plants, as well as the existence of many partial genomic sequences in other plants, served investigators to trace gene duplication during angiosperm evolution (de Bodt *et al.*, 2005). This led to evidence suggesting that angiosperms underwent two whole-genome duplication events early during their evolutionary history. In recent years, attention was focused on gene duplications during the evolution of floral homeotic genes (such as MADS-box genes). For example, investigators of the MPI in Cologne (Winter *et al.*, 2002) found evidence for duplications of B-class genes. There are two types of these genes: *DEF*-like and *GLO*-like, which specify the identity of petals and stamens (in Arabidopsis and snapdragon). The proteins encoded by these two types form only heterodimeric complexes, and such heterodimers are obligatory for the binding to target gene DNA. The authors found evidence that these two types originated from one ancestral gene that duplicated, and then the duplicates underwent diversification. The time of the duplication was *after* the gymnosperm/angiosperm split (about 300 million years ago) but *before* the monocot/eudicot split (about 140–200 million years ago). The heterodimerization may have become obligate in the eudicots *after* the monocot/eudicot split.

During recent years, it became evident that due to the accumulation of knowledge on the molecular biology of flower patterning, especially on floral organ identity genes, the patterning of flowers furnishes an important subject for the study of evolution of morphology in angiosperms (Kramer and Hull, 2005). Special attention was thus paid to the duplication of floral homeotic genes (Causier *et al.*, 2005). A recent review by a team of 13 scientists from Pennsylvania State University and various other institutes in the US and Norway summarized the widespread genome duplications that happened throughout the evolutionary history of angiosperms (Cui *et al.*, 2006). They surveyed cases in which pairs of duplicated genes underwent diversification. After gene duplication, some paralogs may be silenced and ultimately eliminated. In other cases, the two paralogs are preserved and each can be subject to changes in DNA sequence or gene expression, leading to subfunctionalization or neofunctionalization. Interestingly, Lynch and Conery (2000) estimated that in Arabidopsis it will take about 3.2 million years for a duplicated gene (that was not converted into a useful gene) to be eliminated. This is very close to the estimate for *Drosophila* of 2.9 million years. A summary of paleopolyploidy events, according to Cui *et al.* (2006), is shown in Fig. 76.

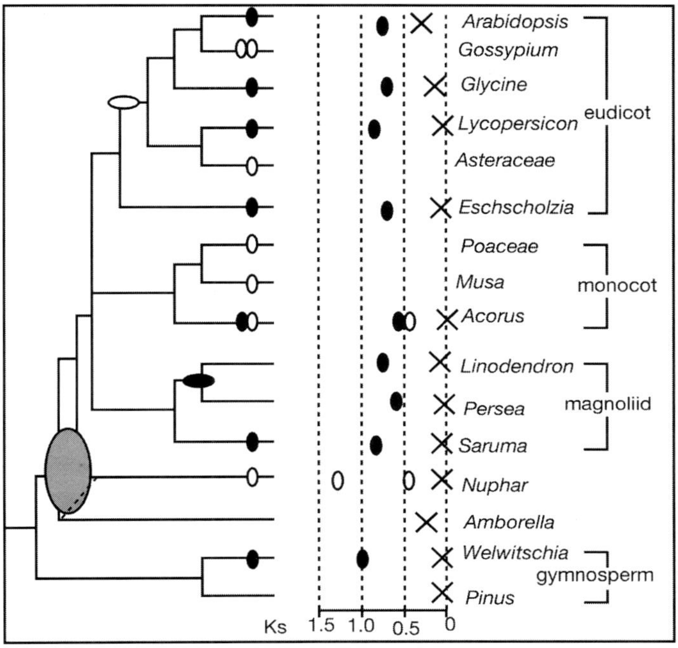

Fig. 76. Phylogenetic summary of paleopolyploidy events estimated by the mixture model approach and their distribution among angiosperm and gymnosperm lineages. Scaled graph in center with Xs corresponding to median K_S of pairs from background gene duplications, while small ovals indicate the median K_S of possible concentrated duplication in the history of particular lineages. The phylogenetic tree at left shows the likely placement of detected genome-scale duplications. Uncertainty in phylogenetic timing of what may be a single duplication event at the base of the angiosperms is indicated with a wide oval that covers possible branching points compatible with the K_S evidence. Hollow ovals indicate duplications identified in previous studies using paralogous genes or genomic data from those lineages. (From Cui *et al.*, 2006.)

Investigators of the laboratory of V. Irish of Yale University (Kramer *et al.*, 1998) studied the B-class genes *AP3* and *PI* of Arabidopsis in comparison with their orthologs in a variety of plants including higher eudicots, lower eudicots and magnoliid dicots. They found in these plants characteristic *PI*- and *AP3*-specific motifs. While the *PI*-specific motif was found in all the analyzed *PI* genes, the case of *AP3*-specific motifs was different in the lower eudicots and the magnoliid dicots. *AP3* homologs contained distinctly different motifs from those revealed in the higher eudicots. Moreover, the *AP3* and *PI* sequences uncovered multiple duplication events within each of the two gene lineages. A major duplication event in the *paleoAP3* lineage coincided with the base of the higher eudicot radiation and this, according to these authors, reflects the evolution of the petal-specific *AP3* function in the higher eudicot lineage. This duplication led to two major *AP3* lineages: *euAP3* and *TOMATO MADS-BOX GENE6* (*TM6*).

As noted above, 8 years later, the subject was re-evaluated and the results were published in two consecutive articles in the same *Plant Cell* issue: Rijpkema *et al.* (2006) and de Martino *et al.* (2006). Nancy A. Eckard, the News and Review editor of *Plant Cell*, nicely summarized these two publications in the same issue. The two gene lineages, *euAP3* and *TM6*, are distinguished mainly by a difference in the C-terminal regions. The C-terminal motif of *TM6* is more similar to that of the ancestral *paleoAP3*, while the C-terminal motif of *euAP3* lineage genes appears to have occurred by a frameshift mutation.

Not all core eudicots contain both *euAP3* and *TM6* lineages. The *TM6* homologs are missing from Arabidopsis. However, plants of the Solanaceae family, such as petunia, tomato and potato, have both *euAP3* and *TM6* homologs. The two mentioned publications focused on Solanaceae species. Rijpkema *et al.* (2006) examined *TM6* and *DEF* (the *AP3* homolog in petunia) in *Petunia hybrida*, while de Martino *et al.* (2006) investigated the homologous genes in *Solanum lycopersicum* (tomato). Rijpkema *et al.* (2006) of Nijmegen University looked at an insertional mutant, *tm6-1* (probably a null mutation of *TM6*), and downregulated *TM6* by producing a transgenic petunia in which RNA interference silenced the *TM6*. Neither of these two types of petunia plants had abnormal floral phenotypes. The *DEF* gene probably contributed to the normal phenotypes. Genetic manipulations, such as the establishment of double (homozygous) mutants (e.g. *def-1*, *tm6-1*), resulted in phenotypes that strongly suggested that petunia *DEF* and *TM6* act together to determine stamen identity in the third whorl and fulfill B-class function (petal and stamen identity in the second and third whorls, respectively).

The authors also sequenced the putative 5′ regulatory sequences of these two genes from petunia, and also sequenced the ortholog of the regulatory sequence from tomato (*TAP3* and *TM6*). The sequences were compared to the sequences of other *euAP3* and *paleoAP3* regulatory (5′) sequences. They found the presence of a distinct and highly conserved domain between the *euAP3* and *TM6* lineages. Together with previously described differences in protein–protein interactions between petunia *DEF* and *TM6*, this led to the conclusion that the two lineages diverted in function during evolution. The difference in promoter sequences (of *DEF* and *TM6*) indicated that the change in the regulatory elements happened during or after the gene duplication that resulted in *euAP3*- and *TM6*-type genes. Whatever the case, it became evident that the regulatory control of petunia *TM6* and *DEF* expression is fundamentally different; and that in petunia *TM6* is involved in the determination of stamen, *but not in petal identity*!

De Martino *et al.* (2006), a team from the laboratory of V. Irish and collaborators (A. Levy and E. Emmanuel) from the Weizmann Institute of Science in Rehovot, Israel, analyzed the orthologous tomato genes *TAP3* and *TM6*. They produced a loss of function of the *TAP3* gene and a tomato plant with a silenced *TM6*: *TM6i*. The phenotype of the loss of function *tap3* mutant of tomato was very different from the loss of function *def* mutant of petunia. The former tomato mutant exhibited complete transformational petals to sepaloid structures (in the second whorl) and transformation of stamens into carpelloid structures (in the third whorl). The gene expression also differed between the two tomato genes, *TAP3* and *TM6*. *TAP3* was transcribed mainly in petals and stamens, while the *TM6* transcript was found mainly in stamens and carpels. This clearly showed that the 5′ controlling element of these two genes differs. Also, the encoded proteins of these two genes were found to be different by analyses of protein–protein interactions. It was found that there was a difference in the interaction capability between *TAP3* and *TM6*. Both TAP3 and TM6 interacted strongly with TP1. The TAP3/TP1 dimers and the TM6/TP1 dimers could bind to TM5. However, while TAP3 alone could bind TM5, the protein TM6 alone could not bind TM5. Also, TAP3 could form a quaternary complex (in a yeast assay) with the other three proteins, but TM6 could not form this complex. A plausible hypothesis was that gene duplication in an ancestral gene was followed by subfunctionalization in the *AP3* lineage. The diversification is evident by different functions and different expression patterns, clearly indicating that diversification can affect the promoter as well as the coding sequence.

RNA Silencing and Flower Patterning

The impact of RNA silencing on flower patterning was already reported by Chuang and Meyerowitz (2000), soon after the discovery of the interference of double-stranded RNA, with the synthesis of specific proteins in nematodes (see Galun, 2005, for detailed information). These authors from the California Institute of Technology investigated the effects of specific double-stranded RNAs on the expression of Arabidopsis genes that are involved in the patterning of flowers and meristems: *AG*, *CLV3*, *AP3* and *PERIANTHIA*. Silencing such genes by the respective double-stranded RNAs led to severe phenotypes and opened the way to interfere in a specific way with the expression of genes controlling organ differentiation. Later, it became evident that the "wisdom" gained by millions of years of evolution had led organisms to use RNA silencing without outside intervention. It was found that eukaryotic organisms (and among them angiosperms) contain genes that lead to the production of specific double-stranded RNA (microRNA genes). The activation of these genes at regulated sites and times constitutes a regular way for plants to control their differentiation. Numerous reports on the roles of endogenous microRNAs (miRNAs or miRs) in angiosperms have appeared in the literature in recent years. Here, I shall mention only a few of these reports.

Xuemei Chen, then at the Waksman Institute of Rutgers University in New Jersey, studied the effect of miRNA172 on the expression of *AP2* in Arabidopsis (Chen, 2004). She found that miRNA172 guides a cleavage mechanism for the mRNA of *AP2*, and thus reduces the production of the AP2 protein. Such a reduction at certain locations, during flower differentiation, is essential for the normal patterning of the Arabidopsis flower. In a mutant with a coding region that leads to the normal AP2, but has nucleotide changes (which do not change the coding) that prevent recognition by *miRNA172*, there was no cleavage of the mRNA and the AP2 protein accumulated to high levels. This resulted in abnormal flower phenotypes, because the controlled repression of *AP2* is crucial for the proper development of floral organs and for the timely termination of floral stem cells.

Alvarez *et al.* (2006) at the laboratory of Y. Eshed at the Weizmann Institute of Science launched an extensive study on the role of plant microRNAs in regulating differentiation. I shall not detail this study that intended to explore the potency, autonomy, target range and functional conservation of plant miRNA genes, because most of the targets chosen in this study are not directly involved in flower patterning. The overall strategy was to manipulate the sequences of natural microRNAs, and to bind them to specific promoters of regulatory genes. Generally, one of the conclusions of this study was that the effect of microRNAs can be quantitative silencing. Another conclusion was that, at least in several tested cases, the effect of the silencing was autonomous, meaning it did not move to other plant organs nor across graft connections. It also became evident that Arabidopsis-based endogenous and synthetic pre-miRNAs were processed effectively in tomato and tobacco. As for flower patterning, these investigators used the promoter of the flower meristem gene *AP1* (an A-class floral identity gene) and attached to it *miR164b* or *miR165b*. The hybrid genes were transformed into Arabidopsis plants and resulted in gross malformations of the flowers.

The Arabidopsis genome contains over 120 R2R3-MYB genes. These are transcription factors that affect a variety of metabolic processes and differentiation phenomena, such as GA-induced α-amylase expression in rice (by *GAMYB* genes) and anther sterility in barley. Millar and Gubler (2005) of the CSIRO in Canberra,

Australia, explored the *GAMYB*-like genes of Arabidopsis. The regulation of these genes by microRNA and their impact on anther development were explored. The Arabidopsis genes *MYB33* and *MYB65* have great similarity to the barley *GAMYB*. Mutating the *MYB33* and *MYB65* genes (by T-DNA insertion) showed that the Arabidopsis genes are essential for normal anther differentiation; but that if only one of them is mutated, there will be no sterile phenotype. The sterility was expressed in the double-homozygous mutants *myb33*, *myb65*. Moreover, even in the double-homozygous mutants, there was occasional fertility of the male organ and selfed seeds were sporadically obtained, especially under high light intensity and low temperature.

The fertility of the double-homozygous mutants could also be restored when a genomic (wild-type) clone of *MYB33* was transformed into the double mutant. The anthers of the double mutant developed normally up to the stage of the division of the archesporial cells that gives rise to the pollen mother cells, tapetum, middle layer and endothecium. Then, in the mutant the tapetum began to enlarge and the products of the meiosis — the tetrads — were not visible in the mutant. Introduction of the *MYB33:GUS* construct into Arabidopsis plants showed GUS expression only in the anthers, especially after the stage of pollen mother cells. On the other hand, fusion of GUS to the promoter of *MYB33* showed a very different expression of GUS. Further experiments strongly suggested that *miR159* was involved in the localization of the *MYB33* and *MYB65* expressions. The wild-types *MYB33* and *MYB65* have *miR159* target motifs. When this motif in the *MYB33* was mutated (and thus not recognized by miR159), dramatic pleotropic defects were observed. In summary, this study furnished evidence indicating that microRNA has a vital role in localizing, spatially and temporally, the production of a protein that is essential for flower differentiation. It should be noted that plant microRNAs are generally more target-specific in silencing genes than animal microRNAs. About 120 microRNAs were identified in Arabidopsis, and similar or greater numbers of microRNAs were identified in other plants (for an update, see miRbase, http://microrna.sanger.ac.uk). It is thus expected that a great number of regulatory genes is controlled by plant microRNAs.

EPILOGUE

Sah ein Knab' ein Roeslein stehn,
Roeslein auf der Heiden,
War so jung und morgenschoen,
Lief er schnell es nah zu sehn,
Sah's mit vielen Freuden.
Roeslein, Roeslin, Roeslein rot, Roeslin auf der Heiden.

This relatively early poem of Johann W. von Goethe (written in 1771) features a lad who encounters a red flower of a wild rose plant in the meadow. Wild rose flowers are "single", having only one whorl of five petals; while most cultivated roses have "double" flowers with several whorls of petals. In the Preface of this book, I noted that J.W. von Goethe already expressed the idea that all the floral members are leaves that underwent metamorphosis. Goethe wrote the Roeslein poem when he was 22 years old and studying law at the University of Strasbourg; he had then a sweetheart in a neighboring village, Sesenheim. According to a Strasbourg legend, it was on a stormy night, when Goethe was riding to meet Friederike, that he was inspired to write his ballad, *Der Erlkoenig*. Goethe expressed the idea of leaf metamorphosis in 1790, after he returned from an extensive visit in Italy. It took about 170 more years to get from Goethe's idea to the awareness that a mutation in a homeotic gene, such as *AGAMOUS1*, can change a flower from having a single whorl of petals to a flower with several whorls of petals.

Sir Arthur Conan Doyle (1850–1930), in one of his detective stories about Sherlock Holmes, narrated a discussion between Holmes and his assistant, Dr. Watson, in which the latter informs Holmes about a scientific fact. Holmes is angered because he claimed that filling his memory with irrelevant information would cause the exit of relevant information from his memory. This is obviously a false claim because — according to the Law of Memory of Amiel Ben David Halevy (my pseudonym) — if one adds more information in a man's memory (especially at a young age), he makes room for much more memory storage. In a far-fetched analogy, this happens with information and knowledge on plant patterning: The more knowledge we gain on this subject, the more additional questions come to mind. The ex-Secretary of Defense of the US, Donald Rumsfeld, once stated that there are "known unknowns" and "unknown unknowns" (referring to the war in Iraq). In the case of plant patterning research, some "unknowns" are changed to "knowns" and some "unknown unknowns" are converted to "known unknowns." Such conversions should lead to additional knowledge about nature and, in the spirit of Baruch Spinoza (1632–1677), pure human pleasure.

References

Abe, M., Y. Kobayashi, *et al.* (2005). "FD, a bZIP protein mediating signals from the floral pathway integrator *FT* at the shoot apex." *Science* **309**(5737): 1052–6.

Achard, P., W. H. Vriezen, *et al.* (2003). "Ethylene regulates Arabidopsis development via the modulation of DELLA protein growth repressor function." *Plant Cell* **15**(12): 2816–25.

Adams, K. L. and J. F. Wendel (2005). "Polyploidy and genome evolution in plants." *Curr Opin Plant Biol* **8**(2): 135–41.

Aderem, A. (2005). "Systems biology: its practice and challenges." *Cell* **121**: 511–3.

Aebersold, R. and M. Mann (2003). "Mass spectrometry-based proteomics." *Nature* **422**(6928): 198–207.

Aida, M. and M. Tasaka (2006). "Morphogenesis and patterning at the organ boundaries in the higher plant shoot apex." *Plant Mol Biol* **60**(6): 915–28.

Aida, M., T. Vernoux, *et al.* (2002). "Roles of *PIN-FORMED1* and *MONOPTEROS* in pattern formation of the apical region of the Arabidopsis embryo." *Development* **129**(17): 3965–74.

Ainsworth, C., S. Crossley, *et al.* (1995). "Male and female flowers of the dioecious plant sorrel show different patterns of MADS box gene expression." *Plant Cell* 7(10): 1583–98.

Alexander, D. L., E. A. Mellor, *et al.* (2005). "*CORKSCREW1* defines a novel mechanism of domain specification in the maize shoot." *Plant Physiol* **138**(3): 1396–408.

Al-Ghazi, Y., B. Muller, *et al.* (2003). "Temporal responses of Arabidopsis root architecture to phosphate starvation: evidence for the involvement of auxin signalling." *Plant Cell Environ* **26**(7): 1053–66.

Allen, C. E. (1932). "Sex inheritance and sex determination." *Am Nat* **66**: 97–107.

Alon, U. (2003). "Biological networks: the tinkerer as an engineer." *Science* **301**(5641): 1866–7.

Alon, U., M. G. Surette, *et al.* (1999). "Robustness in bacterial chemotaxis." *Nature* **397**(6715): 168–71.

Aloni, R., K. Schwalm, *et al.* (2003). "Gradual shifts in sites of free-auxin production during leaf-primordium development and their role in vascular differentiation and leaf morphogenesis in Arabidopsis." *Planta* **216**(5): 841–53.

Alvarez, J. P., I. Pekker, *et al.* (2006). "Endogenous and synthetic microRNAs stimulate simultaneous, efficient, and localized regulation of multiple targets in diverse species." *Plant Cell* **18**(5): 1134–51.

Amasino, R. (2004). "Vernalization, competence, and the epigenetic memory of winter." *Plant Cell* **16**(10): 2553–9.

Amasino, R. (2005). "1955: kinetin arrives. The 50th anniversary of a new plant hormone." *Plant Physiol* **138**: 1177–84.

Ambrose, B. A., D. R. Lerner, *et al.* (2000). "Molecular and genetic analyses of the *silky1* gene reveal conservation in floral organ specification between eudicots and monocots." *Mol Cell* **5**(3): 569–79.

Angenent, G. C., J. Stuurman, *et al.* (2005). "Use of *Petunia* to unravel plant meristem functioning." *Trends Plant Sci* **10**: 243–59.

Aoki, K., N. Suzui, *et al.* (2005). "Destination-selective long distance movement of phloem proteins." *Plant Cell* **17**: 1801–14.

Araki, T. (2001). "Transition from vegetative to reproductive phase." *Curr Opin Plant Biol* **4**(1): 63–8.

Atsmon, D. and C. Tabbak (1979). "Comparative effects of gibberellin, silver nitrate and aminoethoxyvinyl glicine on sexual tendency and ethylene evolution in the cucumber plant (*Cucumis sativus*)." *Plant Cell Physiol* **20**: 1547–55.

Aviv, D., P. Arzee-Gonen, *et al.* (1984a). "Novel alloplasmic *Nicotiana* plants by 'donor recipient' protoplast fusion." *Mol Gen Genet* **196**: 244–53.

Aviv, D., S. Bleichman, *et al.* (1984b). "Intersectional cytoplasmic hybrids in *Nicotiana*: identification of plastomes and chondriomes in *N.sylvestris* + *N.rustica* cybrids having *N.sylvestris* nuclear genomes." *Theor Appl Genet* **67**: 499–504.

Babb, S. and G. J. Muehlbauer (2003). "Genetic and morphological characterization of the *barley uniculm2* (*cul2*) mutant." *Theor Appl Genet* **106**(5): 846–57.

Baluska, F., P. W. Barlow, *et al.* (2005b). "What is apical and what is basal in plant root development?" *Trends Plant Sci* **10**(9): 409–11.

Baluska, F., D. Volkmann, *et al.* (2005a). "Plant synapses: actin-based domains for cell-to-cell communication." *Trends Plant Sci* **10**: 106–11.

Bannigan, A. and T. I. Baskin (2005). "Directional cell expansion — turning toward actin." *Curr Opin Plant Biol* **8**(6): 619–24.

Bao, F., J. J. Shen, *et al.* (2004b). "Brassinosteroids interact with auxin to promote lateral root development in Arabidopsis." *Plant Physiol* **134**(4): 1624–31.

Bao, N., K. W. Lye, *et al.* (2004a). "MicroRNA binding sites in Arabidopsis class IIIHD-ZIP mRNAs are required for methylation of the template chromosome." *Dev Cell* 7(5): 653–62.

Bao, Y. Q., B. Kost, *et al.* (2001). "Reduced expression of α-tubulin genes in *Arabidopsis thaliana* specifically affects root growth and morphology, root hair development and root gravitropism." *Plant J* **28**(2): 145–57.

Barkai, N. and S. Leibler (1997). "Robustness in simple biochemical networks." *Nature* **387**(6636): 913–7.

Barrier, M., R. H. Robichaux, *et al.* (2001). "Accelerated regulatory gene evolution in an adaptive radiation." *Proc Natl Acad Sci USA* **98**(18): 10208–13.

Baskin, T. I. (2000). Cytoskeleton. In: *Biochemistry and Molecular Biology of Plants.* Eds. B. B. Buchanan, W. Gruissem and R. L. Jones. Rockville, MD, American Society of Plant Biologists: pp. 202–58.

Bass, B. L., S. Hellwig, *et al.* (2005). "A nuclear RNA is cut out for translation." *Cell* **123**(2): 181–3.

Bassani, M., P. M. Neumann, *et al.* (2004). "Differential expression profiles of growth-related genes in the elongation zone of maize primary roots." *Plant Mol Biol* **56**: 367–80.

Baurle, I. and T. Laux (2003). "Apical meristems: the plant's fountain of youth." *Bioessays* **25**(10): 961–70.

Baurle, I. and T. Laux (2005). "Regulation of *WUSCHEL* transcription in the stem cell niche of the Arabidopsis shoot meristem." *Plant Cell* **17**(8): 2271–80.

Beilstein, M. A., I. A. Al-Shehbaz, *et al.* (2006). "Brassicaceae phylogeny and trichome evolution." *Am J Bot* **93**: 607–19.

Belostotsky, D. A. and A. B. Rose (2005). "Plant gene expression in the age of systems biology: integrating transcriptional and post-transcriptional events." *Trends Plant Sci* **10**(7): 347–53.

Benedito, V. A., P. B. Visser, *et al.* (2004). "Ectopic expression of *LLAG1*, an *AGAMOUS* homologue from lily (*Lilium longiflorum Thunb.*) causes floral homeotic modifications in Arabidopsis." *J Exp Bot* **55**(401): 1391–9.

Benjamins, R., A. Quint, *et al.* (2001). "The PINOID protein kinase regulates organ development in Arabidopsis by enhancing polar auxin transport." *Development* **128**(20): 4057–67.

Bennett, T. and O. Leyser (2006). "Something on the side: axillary meristems and plant development." *Plant Mol Biol* **60**(6): 843–54.

Bennetzen, J. L. (2005). "Transposable elements, gene creation and genome rearrangement in flowering plants." *Curr Opin Genet Dev* **15**(6): 621–7.

Bentwich, I., A. Avniel, *et al.* (2005). "Identification of hundreds of conserved and nonconserved human microRNAs." *Nat Genet* **37**(7): 766–70.

Berbel, A., C. Navarro, *et al.* (2005). "Functional conservation of *PISTILLATA* activity in a pea homolog lacking the PI motif." *Plant Physiol* **139**: 174–85.

Berg, P. and M. Singer (2003). *George Beadle — An Uncommon Farmer.* Cold Spring Harbor, NY, Cold Spring Harbor Laboratory Press.

Berger, D. and T. Altmann (2000). "A subtilisin-like serine protease involved in the regulation of stomatal density and distribution in *Arabidopsis thaliana*." *Genes Dev* **14**(9): 1119–31.

Bergmann, D. C., W. Lukowitz, *et al.* (2004). "Stomatal development and pattern controlled by a MAPKK kinase." *Science* **304**(5676): 1494–7.

Berleth, T., J. Mattsson, *et al.* (2000). "Vascular continuity and auxin signals." *Trends Plant Sci* **5**(9): 387–93.

Bernhardt, C., M. Z. Zhao, *et al.* (2005). "The bHLH genes *GL3* and *EGL3* participate in an intercellular regulatory circuit that controls cell patterning in the Arabidopsis root epidermis." *Development* **132**(2): 291–8.

Bewley, J. D., F. D. Hampel, *et al.* (2000). Reproductive development. In: *Biochemistry and Molecular Biology of Plants.* Eds. B. B. Buchanan, W. Gruissem and R. L. Jones. Rockville, MD, American Society Plant of Physiologists: pp. 988–1043.

Bharathan, G., T. E. Goliber, *et al.* (2002). "Homologies in leaf form inferred from *KNOXI* gene expression during development." *Science* **296**(5574): 1858–60.

Binns, A. N. (1994). "Cytokinin accumulation and action: biochemical, genetic and molecular approaches." *Annu Rev Plant Physiol* **45**: 173–96.

Birnbaum, K., D. E. Shasha, *et al.* (2003). "A gene expression map of the Arabidopsis root." *Science* **302**(5652): 1956–60.

Blanc, G. and K. H. Wolfe (2004a). "Widespread paleopolyploidy in model plant species inferred from age distributions of duplicate genes." *Plant Cell* **16**(7): 1667–78.

Blanc, G. and K. H. Wolfe (2004b). "Functional divergence of duplicated genes formed by polyploidy during Arabidopsis evolution." *Plant Cell* **16**(7): 1679–91.

Blazquez, M. A. (2005). "Plant science. The right time and place for making flowers." *Science* **309**(5737): 1024–5.

Blazquez, M. A., C. Ferrandiz, *et al.* (2006). "How Floral Meristems are Built." *Plant Mol Biol* **60**(6): 855–70.

Blilou, I., J. Xu, *et al.* (2005). "The PIN auxin efflux facilitator network controls growth and patterning in Arabidopsis roots." *Nature* **433**(7021): 39–44.

Bomblies, K., R. L. Wang, *et al.* (2003). "Duplicate *FLORICAULA/LEAFY* homologs *zfl1* and *zfl2* control inflorescence architecture and flower patterning in maize." *Development* **130**(11): 2385–95.

Bommert, P., N. Satoh-Nagasawa, *et al.* (2005). "Genetics and evolution of inflorescence and flower development in grasses." *Plant Cell Physiol* **46**(1): 69–78.

Bonner, J., R. C. Huang, *et al.* (1961). "The physical state of newly synthesized RNA." *Proc Natl Acad Sci USA* **47**: 1548–54.

Bortiri, E., G. Chuck, *et al.* (2006). "*ramosa2* Encodes a LATERAL ORGAN BOUNDARY Domain protein that determines the fate of stem cells in branch meristems of maize." *Plant Cell* **18**(3): 574–85.

Boss, P. K., R. M. Bastow, *et al.* (2004). "Multiple pathways in the decision to flower: enabling, promoting, and resetting." *Plant Cell* **16**: S18–S31.

Boudolf, W., R. Barroco, *et al.* (2004). "B1-type cyclin-dependent kinases are essential for the formation of stomatal complexes in *Arabidopsis thaliana*." *Plant Cell* **16**(4): 945–55.

Bouquin, T., C. Meier, *et al.* (2001). "Control of specific gene expression by gibberellin and brassinosteroid." *Plant Physiol* **127**(2): 450–8.

Bowers, J. E., B. A. Chapman, *et al.* (2003). "Unravelling angiosperm genome evolution by phylogenetic analysis of chromosomal duplication events." *Nature* **422**(6930): 433–8.

Bowman, J. L. and Y. Eshed (2000). "Formation and maintenance of the shoot apical meristem." *Trends Plant Sci* **5**(3): 110–5.

Brazma, A., M. Krestyaninova, *et al.* (2006). "Standards for systems biology." *Nat Rev Genet* 7(8): 593–605.

Breiman, A. and E. Galun (1990). "Nuclear — mitochondrial interrelation in angiosperms." *Plant Sci* **71**(1): 3–19.

Brembu, T., P. Winge, *et al.* (2004). "*NAPP* and *PIRP* encode subunits of a putative wave regulatory protein complex involved in plant cell morphogenesis. *Plant Cell* **16**(11): 2335–49.

Brown, R. C., B. E. Lemmon, *et al.* (1999). "Development of endosperm in *Arabidopsis thaliana*." *Sex Plant Reprod* **12**(1): 32–42.

Brukhin, V., J. Gheyselinck, *et al.* (2005). "The RPN1 subunit of the 26S proteasome in Arabidopsis is essential for embryogenesis." *Plant Cell* **17**(10): 2723–37.

Buchanan, B. B., W. Gruissem, *et al.* (2000). *Biochemistry and Molecular Biology of Plants.* Rockville, MD, American Society of Plant Physiologists.

Bunning, E. (1951). "Uber die differenzierungsvorgange in der cruciferenwurzel." *Planta* **39**: 126–53.

Bunning, E. (1956). General processes of differentiation. In: *The Growth of Leaves.* Ed. F. Milthorpe. London, Butterworths: 18–30.

Byrne, M. E., A. T. Groover, *et al.* (2003). "Phyllotactic pattern and stem cell fate are determined by the Arabidopsis homeobox gene *BELL-RINGER*." *Development* **130**(17): 3941–50.

Byzova, M. V., J. Franken, *et al.* (1999). "Arabidopsis *STERILE APETALA*, a multifunctional gene regulating inflorescence, flower, and ovule development." *Genes Dev* **13**(8): 1002–14.

Callis, J. (2005). "Auxin action." *Nature* **435**(7041): 436–7.

Callos, J. D., M. Dirado, *et al.* (1994). "The *forever young* gene encodes an oxidoreductase required for proper development of the Arabidopsis vegetative shoot apex." *Plant J* **6**(6): 835–47.

Cann, R. L., M. Stoneking, *et al.* (1987). "Mitochondrial DNA and human evolution." *Nature* **325**(6099): 31–6.

Carland, F. M., B. L. Berg, *et al.* (1999). "Genetic regulation of vascular tissue patterning in Arabidopsis." *Plant Cell* **11**(11): 2123–37.

Carland, F. M. and N. A. McHale (1996). "*LOP1*: a gene involved in auxin transport and vascular patterning in Arabidopsis." *Development* **122**(6): 1811–9.

Carles, C. C. and J. C. Fletcher (2003). "Shoot apical meristem maintenance: the art of a dynamic balance." *Trends Plant Sci* **8**(8): 394–401.

Carraro, N., A. Peaucelle, *et al.* (2006). "Cell differentiation and organ initiation at the shoot apical meristem." *Plant Mol Biol* **60**(6): 811–26.

Casimiro, I., T. Beeckman, *et al.* (2003). "Dissecting Arabidopsis lateral root development." *Trends Plant Sci* **8**(4): 165–71.

Casimiro, I., A. Marchant, *et al.* (2001). "Auxin transport promotes Arabidopsis lateral root initiation." *Plant Cell* **13**(4): 843–52.

Castellano, M. M. and R. Sablowski (2005). "Intracellular signalling in the transition from stem cells to organogenesis in meristems." *Curr Opin Plant Biol* **8**: 26–31.

Causier, B., R. Castillo, *et al.* (2005). "Evolution in action: following function in duplicated floral homeotic genes." *Curr Biol* **15**(16): 1508–12.

Chan, S. W.-L., I. R. Henderson, *et al.* (2005). "Gardening the genome: DNA methylation in *Arabidopsis thaliana*." *Nat Rev Genet* **6**: 351–60.

Chen, X. M. (2004). "A microRNA as a translational repressor of *APETALA2* in Arabidopsis flower development." *Science* **303**(5666): 2022–5.

Chen, Y. F., M. D. Randlett, *et al.* (2002). "Localization of the ethylene receptor ETR1 to the endoplasmic reticulum of Arabidopsis." *J Biol Chem* **277**(22): 19861–6.

Cheng, Y. L., N. Kato, *et al.* (2003). "Two RNA binding proteins, HEN4 and HUM, act in the processing of *AGAMOUS* pre-mRNA in *Arabidopsis thaliana*." *Dev Cell* **4**(1): 53–66.

Chory, J., P. Nagpal, *et al.* (1991). "Phenotypic and Genetic Analysis of *Det2*, a New Mutant That Affects Light-Regulated Seedling Development in Arabidopsis." *Plant Cell* **3**(5): 445–59.

Chory, J., C. Peto, *et al.* (1989b). "*Arabidopsis thaliana* mutant that develops as a light-grown plant in the absence of light." *Cell* **58**(5): 991–9.

Chory, J. and C. A. Peto (1990). "Mutations in the *Det1* gene affect cell-type-specific expression of light-regulated genes and chloroplast development in Arabidopsis." *Proc Natl Acad Sci USA* **87**(22): 8776–80.

Chory, J., C. A. Peto, *et al.* (1989a). "Different roles for phytochrome in etiolated and green plants deduced from characterization of *Arabidopsis thaliana* mutants." *Plant Cell* **1**(9): 867–80.

Chuang, C. F. and E. M. Meyerowitz (2000). "Specific and heritable genetic interference by double-stranded RNA in *Arabidopsis thaliana*." *Proc Natl Acad Sci USA* **97**(9): 4985–90.

Chuck, G., R. B. Meeley, *et al.* (1998). "The control of maize spikelet meristem fate by the *APETALA2*-like gene *indeterminate spikelet1*." *Genes Dev* **12**(8): 1145–54.

Clark, S. E. (2001). "Cell signalling at the shoot meristem." *Nat Rev Mol Cell Biol* **2**(4): 276–84.

Clark, S. E., S. E. Jacobsen, *et al.* (1996). "The *CLAVATA* and *SHOOT MERISTEMLESS* loci competitively regulate meristem activity in Arabidopsis." *Development* **122**(5): 1567–75.

Clark, S. E., M. P. Running, *et al.* (1993). "*Clavata1*, a Regulator of Meristem and Flower Development in Arabidopsis." *Development* **119**(2): 397–418.

Clarke, J. D. and T. Zhu (2006). "Microarray analysis of the transcriptome as a stepping stone towards understanding biological systems: practical considerations and perspectives." *Plant J* **45**(4): 630–50.

Clouse, S. D., M. Langford, *et al.* (1996). "A brassinosteroid-insensitive mutant in *Arabidopsis thaliana* exhibits multiple defects in growth and development." *Plant Physiol* **111**(3): 671–8.

Clouse, S. D. and J. M. Sasse (1998). "BRASSINOSTEROIDS: Essential Regulators of Plant Growth and Development." *Annu Rev Plant Physiol Plant Mol Biol* **49**: 427–51.

Clyton, W. D. and S. A. Renvoire (1986). "Triticeae." Kew Bull Add Ser XIII: 146–7.

Coen, E. (1999). *The Art of Genes.* Oxford, Oxford University Press.

Coen, E. S. and E. M. Meyerowitz (1991). "The war of the whorls: genetic interactions controlling flower development." *Nature* **353**(6339): 31–7.

Coen, E. S. and J. M. Nugent (1994). "Evolution of flowers and inflorescences." *Development Suppl*: 107–16.

Comai, L. (2005). "The advantages and disadvantages of being polyploid." *Nat Rev Genet* **6**: 836–46.

Cormack, D. J. (1962). "Development of root hairs in angiosperms." *Bot Rev* **28**: 446–64.

Correa-Aragunde, N., M. Graziano, *et al.* (2004). "Nitric oxide plays a central role in determining lateral root development in tomato." *Planta* **218**(6): 900–5.

Correa-Aragunde, N., M. Graziano, *et al.* (2006). "Nitric oxide modulates the expression of cell cycle regulatory genes during lateral root formation in tomato." *J Exp Bot* **57**(3): 581–8.

Correns, C. (1906). "Die vererbung der Gecshlechtsformen bei den gynodiozischen Pflanzen." *Ber Dtsch.Bot Ges* **24**: 459–74.

Costa, M. M., S. Fox, *et al.* (2005). "Evolution of regulatory interactions controlling floral asymmetry." *Development* **132**(22): 5093–101.

Costa, S. and P. Shaw (2006). "Chromatin organization and cell fate switch respond to positional information in Arabidopsis." *Nature* **439**(7075): 493–6.

Croxdale, J. L. (2000). "Stomatal patterning in angiosperms." *Am J Bot* **87**(8): 1069–80.

Crozier, A., Y. Kamiya, *et al.* (2000). Biosynthesis of hormones and elicitor molecules. In: *Biochemistry and Molecular Biology of Plants.* Eds. B. B. Buchanan, W. Gruissem and R. L. Jones. Rockville, MD, Americsn Society of Plant Physiologists: pp. 850–929.

Cui, L., P. K. Wall, *et al.* (2006). "Widespread genome duplications throughout the history of flowering plants." *Genome Res* **16**(6): 738–49.

Dai, Y., H. Wang, *et al.* (2006). "Increased expression of MAP KINASE KINASE7 causes deficiency in polar auxin transport and leads to plant architectural abnormality in Arabidopsis." *Plant Cell* **18**(2): 308–20.

De Bodt, S., S. Maere, *et al.* (2005). "Genome duplication and the origin of angiosperms." *Trends Ecol Evol* **20**(11): 591–7.

de Folter, S., R. G. H. Immink, *et al.* (2005). "Comprehensive interaction map of the Arabidopsis MADS box transcription factors." *Plant Cell* **17**: 1424–33.

de Martino, G., I. Pan, *et al.* (2006). "Functional analyses of two tomato *APETALA3* genes demonstrate diversification in their roles in regulating floral development." *Plant Cell* **18**(8): 1833–45.

Deak, K. I. and J. Malamy (2005). "Osmotic regulation of root system architecture." *Plant J* **43**(1): 17–28.

Deeks, M. J. and P. J. Hussey (2003). "Arp2/3 and 'the shape of things to come'." *Curr Opin Plant Biol* **6**(6): 561–7.

Deeks, M. J., P. J. Hussey, *et al.* (2002). "Formins: intermediates in signal-transduction cascades that affect cytoskeletal reorganization." *Trends in Plant Sci* 7(11): 492–8.

Deeks, M. J., D. Kaloriti, *et al.* (2004). "Arabidopsis NAP1 is essential for Arp2/3-dependent trichome morphogenesis." *Curr Biol* **14**(15): 1410–14.

Dharmasiri, N., S. Dharmasiri, *et al.* (2005). "The F-box protein TIR1 is an auxin receptor." *Nature* **435**(7041): 441–5.

Di Stilio, V. S., E. M. Kramer, *et al.* (2005). "Floral MADS box genes and homeotic gender dimorphism in *Thalictrum dioicum* (Ranunculaceae) — a new model for the study of dioecy." *Plant J* **41**: 755–66.

Ditta, G., A. Pinyopich, *et al.* (2004). "The *SEP4* gene of *Arabidopsis thaliana* functions in floral organ and meristem identity." *Curr Biol* **14**(21): 1935–40.

Dixit, R. and R. Cyr (2004). "The cortical microtubule array: from dynamics to organization." *Plant Cell* **16**(10): 2546–52.

Djakovic, S., J. Dyachok, *et al.* (2006). "BRICK1/HSPC300 functions with SCAR and the ARP2/3 complex to regulate epidermal cell shape in Arabidopsis." *Development* **133**(6): 1091–100.

Doebley, J., A. Stec, *et al.* (1997). "The evolution of apical dominance in maize." *Nature* **386**(6624): 485–8.

Doerner, P. (2000). Cell division regulation. In: *Biochemistry and Molecular Biology of Plants*. Eds. B. B. Buchanan, W. Gruissem and R. L. Jones. Rockville, MD, American Society of Plant Physiology: pp. 528–65.

Doerner, P., J. E. Jorgensen, *et al.* (1996). "Control of root growth and development by cyclin expression." *Nature* **380**(6574): 520–3.

Dogterom, M., J. W. J. Kerssemakers, *et al.* (2005). "Force generation by dynamic microtubules." *Curr Opin Cell Biol* **17**: 67–74.

Dolan, L., C. M. Duckett, *et al.* (1994). "Clonal Relationships and Cell patterning in the Root Epidermis of Arabidopsis." *Development* **120**(9): 2465–74.

Dolan, L., K. Janmaat, *et al.* (1993). "Cellular organization of the *Arabidopsis thaliana* root." *Development* **119**(1): 71–84.

Doust, A. N., K. M. Devos, *et al.* (2004). "Genetic control of branching in foxtail millet." *Proc Natl Acad Sci USA* **101**(24): 9045–50.

Dunlap, J. C. (1999). "Molecular bases for circadian clocks." *Cell* **96**(2): 271–90.

Ecker, J. R. (1995). "The ethylene signal-transduction pathway in plants." *Science* **268**(5211): 667–75.

Eckhardt, N. A. (2001). "A sense of self: the role of DNA sequence elimination in allopolyploidization." *Plant Cell* **13**: 1699–704.

El-Assal, S. E.-D., J. Le, *et al.* (2004). "Arabidopsis *GNARLED* encodes a *NAP125* homolog that positively regulates *ARP2/3*." *Curr Biol* **14**(15): 1405–9.

Emerson, R. A., G. W. Beadle, *et al.* (1935). "A summary of linkage studies in maize." *Memoir Cornell Univ Agric Exp Station* **180**: 1–83.

Emery, J. F., S. K. Floyd, *et al.* (2003). "Radial patterning of Arabidopsis shoots by class III HD-ZIP and KANADI genes." *Curr Biol* **13**(20): 1768–74.

Endress, P. K. (1994). *Diversity and Evolutionary Biology of Tropical Flowers*. Cambridge, UK, Cambridge University Press.

Ermolaeva, M. D., M. Wu, *et al.* (2003). "The age of the *Arabidopsis thaliana* genome duplication." *Plant Mol Biol* **51**(6): 859–66.

Esau, K. (1965). *Plant Anatomy*. New York, John Wiley and Sons.

Eshed, Y., S. F. Baum, *et al.* (2001). "Establishment of polarity in lateral organs of plants." *Curr Biol* **11**(16): 1251–60.

Eshed, Y. and J. L. Bowman (2004). "MicroRNAs guide asymmetric DNA modifications guiding asymmetric organs." *Dev Cell* 7(5): 629–30.

Eshed, Y., A. Izhaki, *et al.* (2004). "Asymmetric leaf development and blade expansion in Arabidopsis are mediated by KANADI and YABBY activities." *Development* **131**(12): 2997–3006.

Evans, M. M. S. and M. K. Barton (1997). "Genetics of angiosperm shoot apical meristem development." *Ann Rev Plant Physiol Plant Mol Biol* **48**: 673–701.

Favaro, R., A. Pinyopich, *et al.* (2003). "MADS-box protein complexes control carpel and ovule development in Arabidopsis." *Plant Cell* **15**(11): 2603–11.

Feldman, M., B. Liu, *et al.* (1997a). "Rapid elimination of low-copy DNA sequences in polyploid wheat: a possible mechanism for differentiation of homoeologous chromosomes." *Genetics* **147**(3): 1381–7.

Feldman, R. M. R., C. C. Correll, *et al.* (1997b). "A complex of Cdc4p, Skp1p, and Cdc53p/cullin catalyzes ubiquitination of the phosphorylated CDK inhibitor Sic1p." *Cell* **91**(2): 221–30.

Feldman, K. A. and D. A. Coury (1977). Exceptional segregation of a selectable marker (Kan R) in Arabidopsis identifies genes important for gametophytic growth and development. *Genetics* **147**(3): 1411–22.

Ferl, R. and A. L. Paul (2000). Genome organization and expression. In: *Biochemistry and Molecular Biology of Plants*. Eds. B. B. Buchanan, W. Gruissem and R. L. Jones. Rockville, MD, American Society of Plant Physiologists: pp. 312–57.

Ferreira, F. J. and J. J. Kieber (2005). "Cytokinin signaling." *Curr Opin Plant Biol* **8**(5): 518–25.

Fields, S. and O. K. Song (1989). "A Novel Genetic System to Detect Protein–Protein Interactions." *Nature* **340**(6230): 245–6.

Flanagan, C. A. and H. Ma (1994). "Spatially and temporally regulated expression of the MADS-box gene *Agl2* in wild-type and mutant Arabidopsis flowers." *Plant Mol Biol* **26**(2): 581–95.

Fleet, C. M. and T.-P. Sun (2005). "A DELLAcate balance: the role of gibberellin in plant morphogenesis." *Curr Opin Plant Biol* **8**: 77–85.

Fleming, A. J. (2005). "Formation of primordia and phyllotaxy." *Curr Opin Plant Biol* **8**(1): 53–8.

Fletcher, J. C., U. Brand, *et al.* (1999). "Signaling of cell fate decisions by *CLAVATA3* in Arabidopsis shoot meristems." *Science* **283**(5409): 1911–4.

Frankel, R. and E. Galun (1977). *Pollination Mechanisms, Reproduction and Plant Breeding*. Berlin, Germany, Springer Verlag.

Freedberg, D. (2002). *The Eye of the Lynx*. Chicago, IL, University of Chicago Press.

Freeling, M. (1992). "A conceptual framework for maize leaf development." *Dev Biol* **153**(1): 44–58.

Freeling, M. and S. Hake (1985). "Developmental genetics of mutants that specify Knotted leaves in maize." *Genetics* **111**(3): 617–34.

Friedrichsen, D. M., C. A. P. Joazeiro, *et al.* (2000). "Brassinosteroid-insensitive-1 is a ubiquitously expressed leucine-rich repeat receptor serine/threonine kinase." *Plant Physiol* **123**(4): 1247–55.

Friis, E. M., K. R. Pedersen, *et al.* (2005). "When earth started blooming: insights from fossil record." *Curr Opin Plant Biol* **8**: 5–12.

Friml, J. (2003). "Auxin transport — shaping the plant." *Curr Opin Plant Biol* **6**(1): 7–12.

Friml, J., P. Benfey, *et al.* (2006). "Apical–basal polarity: why plant cells don't stand on their heads." *Trends Plant Sci* **11**(1): 12–4.

Fu, Y., Y. Gu, *et al.* (2005). "Arabidopsis interdigitating cell growth requires two antagonistic pathways with opposing action on cell morphogenesis." *Cell* **120**(5): 687–700.

Fu, Y., G. Wu, *et al.* (2001). "Rop GTPase-dependent dynamics of tip-localized F-actin controls tip growth in pollen tubes." *J Cell Biol* **152**(5): 1019–32.

Fu, Y. and Z. B. Yang (2001). "Rop GTPase: a master switch of cell polarity development in plants." *Trends Plant Sci* **6**(12): 545–7.

Fu, Y., M. Yuan, *et al.* (2000). "Changes in actin organization in the living egg apparatus of *Torenia fournieri* during fertilization." *Sex Plant Reprod* **12**(6): 315–22.

Fukuda, H. (2004). "Signals that control plant vascular cell differentiation." *Nat Rev Mol Cell Biol* **5**(5): 379–91.

Gallagher, K. L., A. J. Paquette, *et al.* (2004). "Mechanisms regulating SHORT-ROOT intercellular movement." *Curr Biol* **14**(20): 1847–51.

Galun, E. (1959a). "Effects of gibberellic acid and naphthaleneacetic acid on sex expression and some morphological characters in the cucumber plant." *Phyton* **13**: 1–8.

Galun, E. (1959b). "The cucumber tendril — a new test organ for gibberellic acid." *Experientia* **15**: 184–5.

Galun, E. (1981). "Plant protoplasts as physiological tools." *Annu Rev Plant Physiol Plant Mol Biol* **32**: 237–66.

Galun, E. (1988). Application of molecular methods to modern Citrus taxonomy. In: *Proceedings of the Sixth International Citrus Congress.* Eds. R. Goren and K. Mendel. Philadelphia, PA, Balaban Publishers: pp. 295–301.

Galun, E. (2003). *Transposable Elements.*Dordrecht, The Netherlands, Kluwer Academic Publishers.

Galun, E. (2005a). *RNA Silencing.* Singapore, World Scientific Publishing.

Galun, E. (2005b). "RNA silencing in plants." *In Vitro Cell Dev Biol Plant* **41**: 113–23.

Galun, E., P. Arzee-Gonen, *et al.* (1982). "Cytoplasmic hybridization in *Nicotiana*: mitochondrial analysis in progenies resulting from fusion between protoplasts having different organelle constitutions." *Mol Gen Genet* **186**: 50–6.

Galun, E. and D. Aviv (1986). "Organelle transfer." *Methods Enzymol* **118**: 113–23.

Galun, E. and A. Breiman (1997). *Transgenic Plants.* London, Imperial College Press.

Galun, E. and E. Galun (2001). *Manufacture of Medical and Health Products by Transgenic Plants.* London, Imperial College Press.

Galun, E., J. Gressel, *et al.* (1964). "Suppression of floral induction by actinomycin D — an inhibitor of messenger RNA synthesis." *Life Sci* **3**: 911–5.

Galun, E., Y. Jung, *et al.* (1962). "Culture and sex modification of male cucumber buds *in vitro*." *Nature* **194**: 595–8.

Galun, E., Y. Jung, *et al.* (1963). "Morphogenesis of floral buds of cucumber cultured *in vitro*." *Develop Biol* **6**: 370–87.

Galun, E. and J. G. Torrey (1969). "Initiation and suppression of apical hairs of *Fucus* embryos." *Dev Biol* **19**: 447–59.

Gamble, R. L., M. L. Coonfield, *et al.* (1998). "Histidine kinase activity of the ETR1 ethylene receptor from Arabidopsis." *Proc Natl Acad Sci USA* **95**(13): 7825–9.

Gao, Z., Y. F. Chen, *et al.* (2003). "Localization of the Raf-like kinase CTR1 to the endoplasmic reticulum of Arabidopsis through participation in ethylene receptor signaling complexes." *J Biol Chem* **278**(36): 34725–32.

Gazzani, S., A. R. Gendall, *et al.* (2003). "Analysis of the molecular basis of flowering time variation in Arabidopsis accessions." *Plant Physiol* **132**(2): 1107–14.

Ge, H., A. J. Walhout, *et al.* (2003). "Integrating 'omic' information: a bridge between genomics and systems biology." *Trends Genet* **19**(10): 551–60.

Geisler, M., J. Nadeau, *et al.* (2000). "Oriented asymmetric divisions that generate the stomatal spacing pattern in Arabidopsis are disrupted by the *too many mouths* mutation." *Plant Cell* **12**(11): 2075–86.

Geldner, N., N. Anders, *et al.* (2003). "The Arabidopsis GNOM ARF-GEF mediates endosomal recycling, auxin transport, and auxin-dependent plant growth." *Cell* **112**(2): 219–30.

Geldner, N., J. Friml, *et al.* (2001). "Auxin transport inhibitors block PIN1 cycling and vesicle trafficking." *Nature* **413**(6854): 425–8.

Gendrel, A. V., Z. Lippman, *et al.* (2002). "Dependence of heterochromatic histone H3 methylation patterns on the Arabidopsis gene *DDM1*." *Science* **297**(5588): 1871–3.

George, M. J. and O. A. Ryder (1986). "Mitochondrial DNA evolution in the genus *Equus*." *Mol Biol Evol* **3**: 535–46.

Gerats, T. and M. Van den Bussche (2005). "A model system for comparative research: *Petunia*." *Trends Plant Sci* **10**: 251–6.

Gernert, W. (1912). "A new subspecies of *Zea mays*." *Am Nat* **46**: 616–22.

Gifford, E. M. (1954). "The shoot apex in angiosperms." *Bot Rev* **20**: 477–529.

Giulini, A., J. Wang, *et al.* (2004). "Control of phyllotaxy by the cytokinin-inducible response regulator homologue *ABPHYL1*." *Nature* **430**(7003): 1031–4.

Golz, J. F. (2006). "Signalling between the shoot apical meristem and developing lateral organs." *Plant Mol Biol* **60**(6): 889–903.

Golz, J. F. and A. Hudson (2002). "Signalling in plant lateral organ development." *Plant Cell* **14** (Suppl): S277–88.

Gomez-Mena, C., S. de Folter, *et al.* (2005). "Transcriptional program controlled by the floral homeotic gene *AGAMOUS* during early organogenesis." *Development* **132**: 429–38.

Gomi, K. and M. Matsuoka (2003). "Gibberellin signalling pathway." *Curr Opin Plant Biol* **6**(5): 489–93.

Goodrich, J., P. Puangsomlee, *et al.* (1997). "A Polycomb-group gene regulates homeotic gene expression in Arabidopsis." *Nature* **386**(6620): 44–51.

Grandjean, O., T. Vernoux, *et al.* (2004). "*In vivo* analysis of cell division, cell growth, and differentiation at the shoot apical meristem in Arabidopsis." *Plant Cell* **16**(1): 74–87.

Grant, D., P. Cregan, *et al.* (2000). "Genome organization in dicots: genome duplication in Arabidopsis and synteny between soybean and Arabidopsis." *Proc Natl Acad Sci USA* **97**(8): 4168–73.

Grbic, V. and A. B. Bleecker (2000). "Axillary meristem development in *Arabidopsis thaliana*." *Plant J* **21**(2): 215–23.

Greb, T., O. Clarenz, *et al.* (2003). "Molecular analysis of the *LATERAL SUPPRESSOR* gene in Arabidopsis reveals a conserved control mechanism for axillary meristem formation." *Genes Dev* **17**(9): 1175–87.

Green, P. B. (1999). "Expression of pattern in plants: combining molecular and calculus-based biophysical paradigms." *Am J Bot* **86**(8): 1059–76.

Grennan, A. K. (2006). "Variations on a theme. Regulation of flowering time in Arabidopsis." *Plant Physiol* **140**: 399–400.

Grewal, S. I. S. and S. C. R. Elgin (2002). "Heterochromatin: new possibilities for the inheritance of structure." *Curr Opin Genet Dev* **12**: 178–87.

Grewal, S. I. S. and J. C. Rice (2004). "Regulation of heterochromatin by histone methylation and small RNAs." *Curr Opin Cell Biol* **16**(3): 230–8.

Grossniklaus, U., C. Spillane, *et al.* (2001). "Genomic imprinting and seed development: endosperm formation with and without sex." *Curr Opin Plant Biol* **4**(1): 21–7.

Guitton, A. E., D. R. Page, *et al.* (2004). "Identification of new members of Fertilisation Independent Seed Polycomb Group pathway involved in the control of seed development in *Arabidopsis thaliana*." *Development* **131**(12): 2971–81.

Guo, H. and J. R. Ecker (2004). "The ethylene signaling pathway: new insights." *Curr Opin Plant Biol* 7(1): 40–9.

Guo, H.-S., Q. Xie, *et al.* (2005). "MicroRNA directs mRNA cleavage of the transcription factor *NAC1* to downregulate auxin signals for Arabidopsis lateral root development." *Plant Cell* **17**: 1376–86.

Guyomarc'h, S., C. Bertrand, *et al.* (2005). "Regulation of meristem activity by chromatin remodelling." *Trends Plant Sci* **10**(7): 332–8.

Haecker, A., R. Gross-Hardt, *et al.* (2004). "Expression dynamics of *WOX* genes mark cell fate decisions during early embryonic patterning in *Arabidopsis thaliana*." *Development* **131**(3): 657–68.

Haecker, A. and T. Laux (2001). "Cell–cell signaling in the shoot meristem." *Curr Opin Plant Biol* **4**(5): 441–6.

Hake, S., H. M. S. Smith, *et al.* (2004). "The role of *knox* genes in plant development." *Annu Rev Cell Dev Biol* **20**: 125–51.

Hao, D., M. Ohme-Takagi, *et al.* (1998). "Unique mode of GCC box recognition by the DNA-binding domain of ethylene-responsive element-binding factor (ERF domain) in plant." *J Biol Chem* **273**(41): 26857–61.

Harberd, N. P., K. E. King, *et al.* (1998). "Gibberellin: inhibitor of an inhibitor of ... ?" *Bioessays* **20**(12): 1001–8.

Hardenack, S., D. Ye, *et al.* (1994). "Comparison of MADS box gene expression in developing male and female flowers of the dioecious plant White Campion." *Plant Cell* **6**(12): 1775–87.

Hardtke, C. S. (2006). "Root development — branching into novel spheres." *Curr Opin Plant Biol* **9**: 66–71.

Harrison, C. J., S. B. Corley, *et al.* (2005a). "Independent recruitment of a conserved developmental mechanism during leaf evolution." *Nature* **434**(7032): 509–14.

Harrison, J., M. Moller, *et al.* (2005b). "The role of *KNOX* genes in the evolution of morphological novelty in Streptocarpus." *Plant Cell* **17**(2): 430–43.

Haughn, G. W. and C. R. Somerville (1988). "Genetic control of morphogenesis in Arabidopsis." *Dev Genet* **9**(2): 73–89.

He, J. X., J. M. Gendron, *et al.* (2005). "BZR1 is a transcriptional repressor with dual roles in brassinosteroid homeostasis and growth responses." *Science* **307**(5715): 1634–8.

He, X. and J. Zhang (2005). "Gene complexity and gene duplicability." *Curr Biol* **15**: 1016–21.

He, Y. and R. M. Amasino (2005). "Role of chromatin modification in flowering-time control." *Trends Plant Sci* **10**(1): 30–5.

Hedden, P. and N. Kamiya (1997). "Gibberellin biosynthesis: enzymes, genes and their regulation." *Annu Rev Plant Physiol Plant Mol Biol* **48**: 431–60.

Heisler, M. G., C. Ohno, *et al.* (2005). "Patterns of auxin transport and gene expression during primordium development revealed by live imaging of the Arabidopsis inflorescence meristem." *Curr Biol* **15**(21): 1899–911.

Hepler, P. K., L. Vidali, *et al.* (2001). "Polarized cell growth in higher plants." *Annu Rev Cell Dev Biol* **17**: 159–87.

Herbert, R. J., B. Vilhar, *et al.* (2001). "Ethylene induces cell death at particular phases of the cell cycle in the tobacco TBY-2 cell line." *J Exp Bot* **52**(361): 1615–23.

Hershko, A. and A. Ciechanover (1998). "The ubiquitin system." *Annu Rev Biochem* **67**: 425–79.

Hertel, R., K. Thomson, *et al.* (1972). "*In vitro* auxin binding to particulate cell fractions from corn coleoptiles." *Planta* **107**: 325–40.

Heslop-Harrison, J. (1957). "The experimental modification of sex expression in flowering plants." *Biol Rev* **32**: 38–90.

Heywood, V. (1993). *Flowering Plants of the World.* New York, NY, Oxford University Press.

Higashiyama, T., H. Kuroiwa, *et al.* (1998). "Guidance *in vitro* of the pollen tube to the naked embryo sac of *Torenia fournieri.*" *Plant Cell* **10**(12): 2019–31.

Higashiyama, T., H. Kuroiwa, *et al.* (2000). "Explosive discharge of pollen tube contents in *Torenia fournieri.*" *Plant Physiol* **122**(1): 11–4.

Higashiyama, T., H. Kuroiwa, *et al.* (2003). "Pollen-tube guidance: beacons from the female gametophyte." *Curr Opin Plant Biol* **6**(1): 36–41.

Hileman, L. C., J. F. Sundstrom, *et al.* (2006). "Molecular and Phylogenetic Analyses of the MADS-box Gene Family in Tomato." *Mol Biol Evol* **23**(11): 2245–58.

Himanen, K., M. Vuylsteke, *et al.* (2004). "Transcript profiling of early lateral root initiation." *Proc Nat Acad Sci USA* **101**(14): 5146–51.

Himmelbach, A., Y. Yang, *et al.* (2003). "Relay and control of abscisic acid signaling." *Curr Opin Plant Biol* **6**(5): 470–9.

Hochholdinger, F., W. J. Park, *et al.* (2004a). "From weeds to crops: genetic analysis of root development in cereals." *Trends Plant Sci* **9**(1): 42–8.

Hochholdinger, F., K. Woll, *et al.* (2004b). "Genetic dissection of root formation in maize (*Zea mays*) reveals root-type specific developmental programmes." *Ann Bot* **93**(4): 359–68.

Hong, Z., M. Ueguchi-Tanaka, *et al.* (2005). "The rice *brassinosteroid-deficient dwarf2* mutant, defective in the rice homolog of Arabidopsis DIMINUTO/DWARF1, is rescued by the endogenously accumulated alternative bioactive brassinosteroid, dolichosterone." *Plant Cell* **17**(8): 2243–54.

Hsieh, T. F., O. Hakim, *et al.* (2003). "From flour to flower: how Polycomb group proteins influence multiple aspects of plant development." *Trends Plant Sci* **8**(9): 439–45.

Hsu, H. F., C. H. Huang, *et al.* (2003). "Ectopic expression of an orchid (*Oncidium* Gower Ramsey) *AGL6*-like gene promotes flowering by activating flowering time genes in *Arabidopsis thaliana*." *Plant Cell Physiol* **44**(8): 783–94.

Hsu, H. F. and C. H. Yang (2002). "An orchid (*Oncidium* Gower Ramsey) *AP3*-like MADS gene regulates floral formation and initiation." *Plant Cell Physiol* **43**(10): 1198–209.

Hua, J. and E. M. Meyerowitz (1998). "Ethylene responses are negatively regulated by a receptor gene family in *Arabidopsis thaliana*." *Cell* **94**(2): 261–71.

Huang, T., H. Bohlenius, *et al.* (2005). "The mRNA of the Arabidopsis gene *FT* moves from leaf to shoot apex and induces flowering." *Science* **309**(5741): 1694–6.

Hubbard, L., P. McSteen, *et al.* (2002). "Expression patterns and mutant phenotype of *teosinte branched1* correlate with growth suppression in maize and teosinte." *Genetics* **162**(4): 1927–35.

Hudson, A. (2005). "Plant meristems: mobile mediators of cell fate." *Curr Biol* **15**(19): R803–5.

Ideker, T., T. Galitski, *et al.* (2001). "A new approach to decoding life: systems biology." *Annu Rev Genomics Hum Genet* **2**: 343–72.

Ingouff, M., J. N. Fitz Gerald, *et al.* (2005). "Plant formin AtFH5 is an evolutionarily conserved actin nucleator involved in cytokinesis." *Nat Cell Biol* 7: 374–80.

Ingram, G. C. (2005). "Plant development: spacing out stomatal pores." *Curr Biol* **15**(17): R663–5.

Inukai, Y., T. Sakamoto, *et al.* (2005). "*Crown rootless1*, which is essential for crown root formation in rice, is a target for an AUXIN RESPONSE FACTOR in auxin signaling." *Plant Cell* **17**: 1387–96.

Irish, E. E. and T. M. Nelson (1991). "Identification of Multiple Stages in the Conversion of Maize Meristems from Vegetative to Floral Development." *Development* **112**(3): 891–8.

Irish, V. F. and I. M. Sussex (1990). "Function of the *Apetala-1* gene during Arabidopsis floral development." *Plant Cell* **2**(8): 741–53.

Ito, Y., I. Nakanomyo, *et al.* (2006). "Dodeca-CLE peptides as suppressors of plant stem cell differentiation." *Science* **313**(5788): 842–5.

Itoh, J., K. Nonomura, *et al.* (2005). "Rice plant development: from zygote to spikelet." *Plant Cell Physiol* **46**(1): 23–47.

Jack, T. (2004). "Molecular and genetic mechanisms of floral control." *Plant Cell* **16**: S1–S17.

Jackson, D., B. Veit, *et al.* (1994). "Expression of maize *Knotted1* related homeobox genes in the shoot apical meristem predicts patterns of morphogenesis in the vegetative shoot." *Development* **120**(2): 405–13.

Janssen, B. J., L. Lund, *et al.* (1998). "Overexpression of a homeobox gene, *LeT6*, reveals indeterminate features in the tomato compound leaf." *Plant Physiol* **117**(3): 771–86.

Jasinski, S., P. Piazza, *et al.* (2005). "KNOX action in Arabidopsis is mediated by coordinate regulation of cytokinin and gibberellin activities." *Curr Biol* **15**(17): 1560–5.

Jenik, P. D. and M. K. Barton (2005). "Surge and destroy: the role of auxin in plant embryogenesis." *Development* **132**(16): 3577–85.

Johnson, C. S., B. Kolevski, *et al.* (2002). "*TRANSPARENT TESTA GLABRA2*, a trichome and seed coat development gene of Arabidopsis, encodes a WRKY transcription factor." *Plant Cell* **14**(6): 1359–75.

Johnson, K. L., K. A. Degnan, *et al.* (2005). "*AtDEK1* is essential for specification of embryonic epidermal cell fate." *Plant J* **44**: 114–27.

Jonsson, H., M. Heisler, *et al.* (2005). "Modeling the organization of the *WUSCHEL* expression domain in the shoot apical meristem." *Bioinformatics* **21**: I232–40.

Juarez, M. T., J. S. Kui, *et al.* (2004b). "microRNA-mediated repression of *rolled leaf1* specifies maize leaf polarity." *Nature* **428**(6978): 84–8.

Juarez, M. T., R. W. Twigg, *et al.* (2004a). "Specification of adaxial cell fate during maize leaf development." *Development* **131**(18): 4533–44.

Jurgens, G. (2001). "Apical–basal pattern formation in Arabidopsis embryogenesis." *EMBO J* **20**(14): 3609–16.

Kakimoto, T. (2003). "Perception and signal transduction of cytokinins." *Ann Rev Plant Biol* **54**: 605–27.

Kanno, A., H. Saeki, *et al.* (2003). "Heterotopic expression of class B floral homeotic genes supports a modified ABC model for tulip (*Tulipa gesneriana*)." *Plant Mol Biol* **52**(4): 831–41.

Kasahara, R. D., M. F. Portereiko, *et al.* (2005). "*MYB98* is required for pollen tube guidance and synergid cell differentiation in Arabidopsis." *Plant Cell* **17**(11): 2981–92.

Kater, M. M., L. Colombo, *et al.* (1998). "Multiple *AGAMOUS* homologs from cucumber and petunia differ in their ability to induce reproductive organ fate." *Plant Cell* **10**(2): 171–82.

Kater, M. M., J. Franken, *et al.* (2001). "Sex determination in the monoecious species cucumber is confined to specific floral whorls." *Plant Cell* **13**(3): 481–93.

Kawakatsu, T., J. Itoh, *et al.* (2006). "*PLASTOCHRON2* regulates leaf initiation and maturation in rice." *Plant Cell* **18**(3): 612–25.

Kaya, H., K. Shibahara, *et al.* (2001). "*FASCIATA* genes for chromatin assembly factor-1 in Arabidopsis maintain the cellular organization of apical meristems." *Cell* **104**(1): 131–42.

Keck, E., P. McSteen, *et al.* (2003). "Separation of genetic functions controlling organ identity in flowers." *EMBO J* **22**(5): 1058–66.

Kellogg, E. a. (2001). "Evolutionary history of the grasses." *Plant Physiology* **125**(3): 1198–205.

Kellogg, E. a., R. Appels, *et al.* (1996). "When genes tell different stories: the diploid genera of Triticeae (Gramineae)." *Syst Bot* **21**(3): 321–47.

Kepinski, S. (2006). "Integrating hormone signaling and patterning mechanisms in plant development." *Curr Opin Plant Biol* **9**: 28–34.

Kepinski, S. and O. Leyser (2005a). "Plant development: auxin in loops." *Curr Biol* **15**: R208–10.

Kepinski, S. and O. Leyser (2005b). "The Arabidopsis F-box protein TIR1 is an auxin receptor." *Nature* **435**(7041): 446–51.

Kerstetter, R. A. and S. Hake (1997). "Shoot meristem formation in vegetative development." *Plant Cell* **9**(7): 1001–10.

Kessler, S., S. Seiki, *et al.* (2002). "*Xcl1* causes delayed oblique periclinal cell divisions in developing maize leaves, leading to cellular differentiation by lineage instead of position." *Development* **129**(8): 1859–69.

Kieber, J. J., M. Rothenberg, *et al.* (1993). "*CTR1*, a negative regulator of the ethylene response pathway in Arabidopsis, encodes a member of the Raf family of protein kinases." *Cell* **72**(3): 427–41.

Kim, I. and P. C. Zambryski (2005). "Cell-to-cell communication via plasmodesmata during Arabidopsis embryogenesis." *Curr Opin Plant Biol* **8**(6): 593–9.

Kim, M., T. Pham, *et al.* (2003). "Reduced leaf complexity in tomato *wiry* mutants suggests a role for *PHAN* and *KNOX* genes in generating compound leaves." *Development* **130**(18): 4405–15.

Kim, T. W., J. Y. Hwang, *et al.* (2005). "Arabidopsis CYP85A2, a cytochrome P450, mediates the Baeyer–Villiger oxidation of castasterone to brassinolide in brassinosteroid biosynthesis." *Plant Cell* **17**(8): 2397–412.

Kim, V. N. (2005). "MicroRNA biogenesis: coordinated cropping and dicing." *Nat Rev Mol Cell Biol* **6**: 376–85.

Kinder, C. A. and R. A. Martienssen (2005). "The developmental role of microRNA in plants." *Curr Opin Plant Biol* **8**: 38–44.

Kinoshita, T., A. C. Cano-Delgado, *et al.* (2005). "Binding of brassinosteroids to the extracellular domain of plant receptor kinase BRI1." *Nature* **433**(7022): 167–71.

Kirch, T., R. Simon, *et al.* (2003). "The *DORNROSCHEN/ENHANCER OF SHOOT REGENERATION1* gene of Arabidopsis acts in the control of meristem cell fate and lateral organ development." *Plant Cell* **15**(3): 694–705.

Kirschner, M. W. (2005). "The meaning of systems biology." *Cell* **121**: 503–4.

Kitano, H. (2002a). "Looking beyond the details: a rise in system-oriented approaches in genetics and molecular biology." *Curr Genet* **41**(1): 1–10.

Kitano, H. (2002b). "Systems biology: a brief overview." *Science* **295**(5560): 1662–4.

Kitano, H. (2002c). "Computational systems biology." *Nature* **420**(6912): 206–10.

Kitano, H. (2004). "Biological robustness." *Nat Rev Genet* **5**(11): 826–37.

Knopf, R. R. and T. Trebitsh (2006). "The female-specific *Cs-ACS1G* gene of cucumber. A case of gene duplication and recombination between the non-sex-specific 1-aminocyclopropane-1-carboxylate synthase gene and a branched-chain amino acid transaminase gene." *Plant Cell Physiol* **47**(9): 1217–28.

Kochian, L. V. (2000). Molecular physiology of mineral nutrient acquisition, transport and utilization. In: *Biochemistry and Molecular Biology of Plants.* Eds. B. B. Buchanan, W. Gruissem and R. L. Jones. Rockville, MD, American Society of Plant Physiologists: pp. 1204–49.

Koka, C. V., R. E. Cerny, *et al.* (2000). "A putative role for the tomato genes *DUMPY* and *CURL-3* in brassinosteroid biosynthesis and response." *Plant Physiology* **122**(1): 85–98.

Komatsu, M., A. Chujo, *et al.* (2003). "*FRIZZY PANICLE* is required to prevent the formation of axillary meristems and to establish floral meristem identity in rice spikelets." *Development* **130**(16): 3841–50.

Kondo, T., S. Sawa, *et al.* (2006). "A plant peptide encoded by *CLV3* identified by *in situ* MALDI-TOF MS analysis." *Science* **313**(5788): 845–8.

Koornneef, M., J. Van Eden, *et al.* (1983). "A linkage map of *Arabidopsis thaliana*." *J Hered* **74**: 265–72.

Koshino-Kimura, Y., T. Wada, *et al.* (2005). "Regulation of *CAPRICE* transcription by MYB proteins for root epidermis differentiation in Arabidopsis." *Plant Cell Physiol* **46**(6): 817–26.

Kramer, E. M., R. L. Dorit, *et al.* (1998). "Molecular evolution of genes controlling petal and stamen development: duplication and divergence within the *APETALA3* and *PISTILLATA* MADS-box gene lineages." *Genetics* **149**(2): 765–83.

Kramer, E. M. and J. C. Hall (2005). "Evolutionary dynamics of genes controlling floral development." *Curr Opin Plant Biol* **8**: 13–8.

Krizek, B. A. and J. C. Fletcher (2005). "Molecular mechanisms of flower development: an armchair guide." *Nature Rev Genet* **6**(9): 688–98.

Ku, H. M., T. Vision, *et al.* (2000). "Comparing sequenced segments of the tomato and Arabidopsis genomes: large-scale duplication followed by selective gene loss creates a network of synteny." *Proc Natl Acad Sci U S A* **97**(16): 9121–6.

Kuriyama, H. and H. Fukuda (2001). "Regulation of tracheary element differentiation." *J Plant Growth Regul* **20**(1): 35–51.

Kutz, A., A. Muller, *et al.* (2002). "A role for nitrilase 3 in the regulation of root morphology in sulphur-starving *Arabidopsis thaliana*." *Plant J* **30**(1): 95–106.

Kwon, C. S., C. B. Chen, *et al.* (2005). "*WUSCHEL* is a primary target for transcriptional regulation by SPLAYED in dynamic control of stem cell fate in Arabidopsis." *Genes Dev* **19**(8): 992–1003.

Lai, L. B., J. A. Nadeau, *et al.* (2005). "The Arabidopsis R2R3 MYB proteins FOUR LIPS and MYB88 restrict divisions late in the stomatal cell lineage." *Plant Cell* **17**(10): 2754–67.

Laitinen, R. A. E., J. Immanen, *et al.* (2005). "Analysis of the floral transcriptome uncovers new regulators of organ determination and gene families related to flower organ differentiation in *Gerbera hybrida* (Asteraceae)." *Genome Res* **15**(4): 475–86.

Lamb, R. S. and V. F. Irish (2003). "Functional divergence within the *APETALA3/PISTILLATA* floral homeotic gene lineages." *Proc Natl Acad Sci U S A* **100**(11): 6558–63.

Lange, T., J. Kappler, *et al.* (2005). "Gibberellin biosynthesis in developing pumpkin seedlings." *Plant Physiol* **139**: 213–23.

Laufs, P., A. Peaucelle, *et al.* (2004). "MicroRNA regulation of the *CUC* genes is required for boundary size control in Arabidopsis meristems." *Development* **131**(17): 4311–22.

Laux, T., K. F. X. Mayer, *et al.* (1996). "The *WUSCHEL* gene is required for shoot and floral meristem integrity in Arabidopsis." *Development* **122**(1): 87–96.

Laux, T., T. Wurschum, *et al.* (2004). "Genetic regulation of embryonic pattern formation." *Plant Cell* **16**: S190–S202.

Lee, D., E. Ezhkova, *et al.* (2005). "The Proteasome Regulatory Particle Alters the SAGA Coactivator to Enhance Its Interactions with Transcriptional Activators." *Cell* **123**(3): 423–36.

Lee, M. M. and J. Schiefelbein (2002). "Cell pattern in the Arabidopsis root epidermis determined by lateral inhibition with feedback." *Plant Cell* **14**(3): 611–8.

Lee, Y.-R. J. and B. Liu (2004). "Cytoskeletal motors in Arabidopsis. Sixty-one kinesins and seventeen myosins." *Plant Physiol* **136**: 3877–83.

Lenhard, M., A. Bohnert, *et al.* (2001). "Termination of stem cell maintenance in Arabidopsis floral meristems by interactions between *WUSCHEL* and *AGAMOUS*." *Cell* **105**(6): 805–14.

Lenhard, M. and T. Laux (1999). "Shoot meristem formation and maintenance." *Curr Opin Plant Biol* **2**(1): 44–50.

Lenhard, M. and T. Laux (2003). "Stem cell homeostasis in the Arabidopsis shoot meristem is regulated by intercellular movement of *CLAVATA3* and its sequestration by *CLAVATA1*." *Development* **130**(14): 3163–73.

Letham, D. S. and C. O. Miller (1965). "Identity of kinetin-like factors from *Zea mays*." *Plant Cell Physiol* **6**: 355–9.

Levy, Y. Y. and C. Dean (1998). "The transition to flowering." *Plant Cell* **10**(12): 1973–89.

Leyser, O. (2002). "Molecular genetics of auxin signaling." *Annu Rev Plant Biol* **53**: 377–98.

Li, J. (2005). "Brassinosteroid signalling: from receptor kinases to transcription factors." *Curr Opin Plant Biol* **8**: 526–31.

Li, J. M., M. G. Biswas, *et al.* (1997). "Conservation of function between mammalian and plant steroid 5 alpha-reductases." *Proc Natl Acad Sci USA* **94**(8): 3554–9.

Li, J. M. and J. Chory (1997). "A putative leucine rich repeat receptor kinase involved in brassinosteroid signal transduction." *Cell* **90**: 929–38.

Li, J. M. and J. Chory (1999). "Brassinosteroid actions in plants." *J Exp Bot* **50**: 275–82.

Li, J. M., P. Nagpal, *et al.* (1996). "A role for brassinosteroids in light-dependent development of Arabidopsis." *Science* **272**(5260): 398–401.

Li, J. M., K. H. Nam, *et al.* (2001). "*BIN2*, a new brassinosteroid-insensitive locus in Arabidopsis." *Plant Physiol* **127**(1): 14–22.

Li, M., C. Qin, *et al.* (2006). "Double Knockouts of Phospholipases *D{zeta}1* and *D{zeta}2* in Arabidopsis Affect Root Elongation during phosphate-limited Growth But Do not Affect Root Hair Patterning." *Plant Physiol* **140**(2): 761–70.

Li, S. M., C. M. Armstrong, *et al.* (2004). "A map of the interactome network of the metazoan *C. elegans*." *Science* **303**(5657): 540–3.

Li, X., Q. Qian, *et al.* (2003). "Control of tillering in rice." *Nature* **422**(6932): 618–21.

Lifschitz, E., T. Eviatar, *et al.* (2006). "The tomato *FT* ortholog triggers systemic signals that regulate growth and flowering and substitute for diverse environmental stimuli." *Proc Natl Acad Sci USA* **103**(16): 6398–403.

Lincoln, C., J. Long, *et al.* (1994). "*A knotted1-like* homeobox gene in Arabidopsis is expressed in the vegetative meristem and dramatically alters leaf morphology when overexpressed in transgenic plants." *Plant Cell* **6**(12): 1859–76.

Little, D. Y., H. Rao, *et al.* (2005). "The putative high-affinity nitrate transporter NRT2.1 represses lateral root initiation in response to nutritional cues." *Proc Natl Acad Sci U S A* **102**(38): 13693–8.

Ljung, K., A. K. Hull, *et al.* (2005). "Sites and regulation of auxin biosynthesis in Arabidopsis roots." *Plant Cell* **17**: 1090–104.

Lohmann, J. U., R. L. Hong, *et al.* (2001). "A molecular link between stem cell regulation and floral patterning in Arabidopsis." *Cell* **105**(6): 793–803.

Lohmann, J. U. and D. Weigel (2002). "Building beauty: the genetic control of floral patterning." *Dev Cell* **2**(2): 135–42.

Long, J. and M. K. Barton (2000). "Initiation of axillary and floral meristems in Arabidopsis." *Dev Biol* **218**(2): 341–53.

Lopez-Bucio, J., A. Cruz-Ramirez, *et al.* (2003). "The role of nutrient availability in regulating root architecture." *Curr Opin Plant Biol* **6**(3): 280–7.

Lord, E. M. and S. D. Russell (2002). "The mechanisms of pollination and fertilization in plants." *Annu Rev Cell Dev Biol* **18**: 81–105.

Lorenzo, O. and R. Solano (2005). "Molecular players regulating the jasmonate signalling network." *Curr Opin Plant Biol* **8**(5): 532–40.

Lu, Z. X., M. Wu, *et al.* (1993). "Nucleotide sequence of a flower-specific MADS box cDNA clone from orchid." *Plant Mol Biol* **23**(4): 901–4.

Lukowitz, W., A. Roeder, *et al.* (2004). "A MAPKK kinase gene regulates extra-embryonic cell fate in Arabidopsis." *Cell* **116**(1): 109–19.

Luo, M., P. Bilodeau, *et al.* (2000). "Expression and parent-of-origin effects for *FIS2*, *MEA*, and *FIE* in the endosperm and embryo of developing Arabidopsis seeds." *Proc Natl Acad Sci USA* **97**(19): 10637–42.

Lynch, M. and J. S. Conery (2000). "The evolutionary fate and consequences of duplicate genes." *Science* **290**(5494): 1151–5.

Ma, H. (1994). "The unfolding drama of flower development — recent results from genetic and molecular analyses." *Genes Dev* **8**(7): 745–56.

Magallon, S. and M. J. Sanderson (2001). "Absolute diversification rates in angiosperm clades." *Evol Int J Org Evol* **55**(9): 1762–80.

Maheshwari, P. (1950). *An Introduction to the Embryology of Angiosperms.* New York, NY, McGraw-Hill.

Mahonen, A. P., A. Bishopp, *et al.* (2006a). "Cytokinin signaling and its inhibitor AHP6 regulate cell fate during vascular development." *Science* **311**(5757): 94–8.

Mahonen, A. P., M. Bonke, *et al.* (2000). "A novel two-component hybrid molecule regulates vascular morphogenesis of the Arabidopsis root." *Genes Dev* **14**(23): 2938–43.

Mahonen, A. P., M. Higuchi, *et al.* (2006b). "Cytokinins regulate a bidirectional phosphorelay network in Arabidopsis." *Curr Biol* **16**(11): 1116–22.

Malamy, J. E. (2005). "Intrinsic and environmental response pathways that regulate root system architecture." *Plant Cell Environ* **28**(1): 67–77.

Malamy, J. E. and P. N. Benfey (1997a). "Organization and cell differentiation in lateral roots of *Arabidopsis thaliana*." *Development* **124**(1): 33–44.

Malamy, J. E. and P. N. Benfey (1997b). "Down and out in Arabidopsis: the formation of lateral roots." *Trends Plant Sci* **2**(10): 390–6.

Mallory, A. C., D. V. Dugas, *et al.* (2004). "MicroRNA regulation of NAC-domain targets is required for proper formation and separation of adjacent embryonic, vegetative, and floral organs." *Curr Biol* **14**(12): 1035–46.

Mandava, N. B. (1988). "Plant growth–promoting brassinosteroids." *Annu Rev Plant Physiol Plant Mol Biol* **39**: 23–52.

Marchant, A., R. Bhalerao, *et al.* (2002). "AUX1 promotes lateral root formation by facilitating indole-3-acetic acid distribution between sink and source tissues in the Arabidopsis seedling." *Plant Cell* **14**(3): 589–97.

Marks, M. D. (1997). "Molecular genetic analysis of trichome development in Arabidopsis." *Annu Rev Plant Physiol Plant Mol Biol* **48**: 137–63.

Marx, G. A. (1983). "Developmental mutants in some annual seed plants." *Annu Rev Plant Physiol* **34**: 389–417.

Mathur, J. (2004). "Cell shape development in plants." *Trends Plant Sci* **9**: 583–90.

Mathur, J. (2006). "Local interactions shape plant cells." *Curr Opin Cell Biol* **18**(1): 40–6.

Mathur, J. and N. H. Chua (2000). "Microtubule stabilization leads to growth reorientation in Arabidopsis trichomes." *Plant Cell* **12**(4): 465–77.

Mathur, J. and M. Hulskamp (2002). "Microtubules and microfilaments in cell morphogenesis in higher plants." *Curr Biol* **12**(19): R669–76.

Mathur, J., N. Mathur, *et al.* (2003). "Arabidopsis *CROOKED* encodes for the smallest subunit of the ARP2/3 complex and controls cell shape by region specific fine F-actin formation." *Development* **130**(14): 3137–46.

Mathur, J., P. Spielhofer, *et al.* (1999). "The actin cytoskeleton is required to elaborate and maintain spatial patterning during trichome cell morphogenesis in *Arabidopsis thaliana*." *Development* **126**(24): 5559–68.

Matsubayashi, Y. and Y. Sakagami (1996). "Phytosulfokine, sulfated peptides that induce the proliferation of single mesophyll cells of *Asparagus officinalis* L." *Proc Natl Acad Sci USA* **93**(15): 7623–7.

Matsubayashi, Y. and Y. Sakagami (2006). "Peptide hormones in plants." *Annu Rev Plant Biol* **57**: 649–74.

Mattsson, J., W. Ckurshumova, *et al.* (2003). "Auxin signaling in Arabidopsis leaf vascular development." *Plant Physiol* **131**(3): 1327–39.

Matzke, M. A. and J. A. Birchler (2005). "RNAi-mediated pathways in the nucleus." *Nat Rev Genet* **6**(1): 24–35.

Mayer, K. F. X., H. Schoof, *et al.* (1998). "Role of *WUSCHEL* in regulating stem cell fate in the Arabidopsis shoot meristem." *Cell* **95**(6): 805–15.
McConnell, J. R. and M. K. Barton (1998). "Leaf polarity and meristem formation in Arabidopsis." *Development* **125**(15): 2935–42.
McConnell, J. R., J. Emery, *et al.* (2001). "Role of *PHABULOSA* and *PHAVOLUTA* in determining radial patterning in shoots." *Nature* **411**(6838): 709–13.
McMurray, A. L. and C. H. Miller (1968). "Cucumber sex expression modified by 2-chloroethane-phosphonic acid." *Science* **162**: 1396–7.
McSteen, P. (2006). "Branching out: the ramosa Pathway and the Evolution of Grass Inflorescence Morphology." *Plant Cell* **18**(3): 518–22.
McSteen, P., D. Laudencia-Chingcuanco, *et al.* (2000). "A floret by any other name: control of meristem identity in maize." *Trends in Plant Science* **5**(2): 61–6.
McSteen, P. and O. Leyser (2005). "Shoot branching." *Annu Rev Plant Biol* **56**: 353–74.
Mena, M., B. A. Ambrose, *et al.* (1996). "Diversification of C-function activity in maize flower development." *Science* **274**(5292): 1537–40.
Meyerowitz, E. M. (1989). "Arabidopsis, a useful weed." *Cell* **56**(2): 263–9.
Meyerowitz, E. M. (2002). "Comparative genomics — plants compared to animals: the broadest comparative study of development." *Science* **295**(5559): 1482–5.
Michaels, S. D. and R. M. Amasino (1999). "*FLOWERING LOCUS C* encodes a novel MADS domain protein that acts as a repressor of flowering." *Plant Cell* **11**(5): 949–56.
Michaels, S. D., Y. He, *et al.* (2003). "Attenuation of *FLOWERING LOCUS C* activity as a mechanism for the evolution of summer-annual flowering behavior in Arabidopsis." *Proc Natl Acad Sci USA* **100**(17): 10102–7.
Millar, A. A. and F. Gubler (2005). "The Arabidopsis *GAMYB*-like genes, *MYB33* and *MYB65*, are microRNA-regulated genes that redundantly facilitate anther development." *Plant Cell* **17**: 705–21.
Millar, A. J. (1998). "Molecular intrigue between phototransduction and the circadian clock." *Ann Bot* **81**(5): 581–7.
Miller, C. O., F. Skoog, *et al.* (1955a). "Kinetin, a cell division factor from deoxyribonucleic acid." *J Am Chem Soc* 77: 1392.
Miller, C. O., F. Skoog, *et al.* (1955b). "Structure and synthesis of kinetin." *J Am Chem Soc* **78**: 2662–3.

Minina, E. G. (1938). "On the phenotypical modification of sexual characters in higher plants under the influence of the condition of nutrition and other external factors." *C R Acad Sci USSR (Doklady)* **21**: 298–301.

Mok, D. W. and M. C. Mok (2001). "Cytokinin metabolism and action." *Annu Rev Plant Physiol Plant Mol Biol* **52**: 89–118.

Moon, J., G. Parry, *et al.* (2004). "The ubiquitin–proteasome pathway and plant development." *Plant Cell* **16**(12): 3181–95.

Moore, R. C., S. R. Grant, *et al.* (2005). "Molecular population genetics of redundant floral-regulatory genes in *Arabidopsis thaliana*." *Mol Biol Evolution* **22**(1): 91–103.

Moore, R. C. and M. D. Purugganan (2005). "The evolutionary dynamics of plant duplicate genes." *Curr Opin Plant Biol* **8**(2): 122–8.

Mora-Garcia, S., G. Vert, *et al.* (2004). "Nuclear protein phosphatases with Kelch-repeat domains modulate the response to brassinosteroids in Arabidopsis." *Genes Dev* **18**(4): 448–60.

Mosig, G. and F. Eiserling (1988). Phage T4 structure and metabolism. In: *The Bacteriophages*, Vol. 2. Ed. R. Calender. New York, NY, Plenum Press: pp. 521–606.

Mosquna, A., A. Katz, *et al.* (2004). "Interaction of FIE, a Polycomb protein, with pRb: a possible mechanism regulating endosperm development." *Mol Genet Genomics* **271**(6): 651–7.

Mouradov, A., F. Cremer, *et al.* (2002). "Control of flowering time: interacting pathways as a basis for diversity." *Plant Cell* **14**: S111–30.

Murai, K., M. Miyamae, *et al.* (2003). "*WAP1*, a wheat *APETALA1* homolog, plays a central role in the phase transition from vegetative to reproductive growth." *Plant Cell Physiol* **44**(12): 1255–65.

Nadeau, J. A. and F. D. Sack (2003). "Stomatal development: cross talk puts mouths in place." *Trends Plant Sci* **8**(6): 294–9.

Nakajima, K., G. Sena, *et al.* (2001). "Intercellular movement of the putative transcription factor SHR in root patterning." *Nature* **413**(6853): 307–11.

Napier, R. M., K. M. David, *et al.* (2002). "A short history of auxin-binding proteins." *Plant Mol Biol* **49**(3–4): 339–48.

Navarro, C., N. Efremova, *et al.* (2004). "Molecular and genetic interactions between *STYLOSA* and *GRAMINIFOLIA* in the control of Antirrhinum vegetative and reproductive development." *Development* **131**(15): 3649–59.

Nawy, T., J.-Y. Lee, *et al.* (2005). "Transcriptional profile of Arabidopsis root quiescent center." *Plant Cell* **17**: 1908–25.

Nelson, T. and N. Dengler (1997). "Leaf Vascular Pattern Formation." *Plant Cell* **9**(7): 1121–35.

Nemhauser, J. L. and J. Chory (2004). "Bring it on: new insights into the mechanism of brassinosteroid action." *J Exp Bot* **55**(395): 265–70.

Nemhauser, J. L., T. C. Mockler, *et al.* (2004). "Interdependency of brassinosteroid and auxin signaling in Arabidopsis." *PLoS Biol* **2**(9): 1460–71.

Ng, M. and M. F. Yanofsky (2000). "Three ways to learn the ABCs." *Curr Opin Plant Biol* **3**(1): 47–52.

Ng, M. and M. F. Yanofsky (2001). "Function and evolution of the plant MADS-box gene family." *Nat Rev Genet* **2**(3): 186–95.

Nishimura, C., Y. Ohashi, *et al.* (2004). "Histidine kinase homologs that act as cytokinin receptors possess overlapping functions in the regulation of shoot and root growth in Arabidopsis." *Plant Cell* **16**(6): 1365–77.

North, G. (2005). "Systems and components in biology." *Curr Biol* **15**(11): R395.

Ohashi, Y., A. Oka, *et al.* (2003). "Modulation of phospholipid signaling by *GLABRA2* in root-hair pattern formation." *Science* **300**(5624): 1427–30.

Ohno, S. (1970). *Evolution by Gene Duplication.* Heidelberg, Germany, Springer-Verlag.

Olszewski, N., T. P. Sun, *et al.* (2002). "Gibberellin signaling: biosynthesis, catabolism, and response pathways." *Plant Cell* **14** Suppl: S61–S80.

Ori, N., Y. Eshed, *et al.* (2000). "Mechanisms that control *knox* gene expression in the Arabidopsis shoot." *Development* **127**(24): 5523–32.

Ozkan, H., A. A. Levy, *et al.* (2001). "Allopolyploidy-induced rapid genome evolution in the wheat (*Aegilops–Triticum*) group." *Plant Cell* **13**(8): 1735–47.

Pagnussat, G. C., H.-J. Yu, *et al.* (2005). "Genetic and molecular identification of genes required for female gametophyte development and function in Arabidopsis." *Development* **132**: 603–14.

Pekker, I., J. P. Alvarez, *et al.* (2005). "Auxin response factors mediate Arabidopsis organ asymmetry via modulation of KANADI activity." *Plant Cell* **17**(11): 2899–910.

Pelaz, S., G. S. Ditta, *et al.* (2000). "B and C floral organ identity functions require *SEPALLATA* MADS-box genes." *Nature* **405**(6783): 200–3.

Perez-Rodriguez, M., F. W. Jaffe, *et al.* (2005). "Development of three different cell types is associated with the activity of a specific MYB transcription factor in the ventral petal of *Antirrhinum majus* flowers." *Development* **132**(2): 359–70.

Perl, A., D. Aviv, *et al.* (1990a). "Protoplast-Fusion-Derived CMS Potato Cybrids — Potential Seed-Parents for Hybrid, True-Potato-Seeds." *J Hered* **81**(6): 438–42.

Perl, A., D. Aviv, *et al.* (1990b). "Protoplast-Fusion-Derived Solanum Hybrids — Application and Phylogenetic Limitations." *Theor Appl Genet* **79**(5): 632–40.

Perl-Treves, R. (1999). Male to female conversion along the cucumber shoot: approaches to studying sex genes and floral development in *Cucumis sativus.* In: *Sex Determination in Plants.* Ed. C. C. Ainsworth. Oxford, England, BIOS Scientific Publishers: pp. 189–215.

Perl-Treves, R. and E. Galun (1985). "The Cucumis plastome: physical map, intrageneric variation and phylogenetic relationships." *Theo Appl Genet* **71**: 417–29.

Perl-Treves, R., A. Kahana, *et al.* (1998). "Expression of multiple *AGAMOUS*-like genes in male and female flowers of cucumber (*Cucumis sativus* L)." *Plant Cell Physiol* **39**(7): 701–10.

Perl-Treves, R., D. Zamir, *et al.* (1985). "Phylogeny of Cucumis based on isozyme variability and its comparison with plastome phylogeny." *Theor Appl Genet* **71**: 430–6.

Peters, A. H. F. M. and D. Schubeler (2005). "Methylation of histones: playing memory with DNA." *Curr Opin Cell Biol* **17**: 230–8.

Phizicky, E., P. I. H. Bastiaens, *et al.* (2003). "Protein analysis on a proteomic scale." *Nature* **422**(6928): 208–15.

Pidkowich, M. S., J. E. Klenz, *et al.* (1999). "The making of a flower: control of floral meristem identity in Arabidopsis." *Trends Plant Sci* **4**(2): 64–70.

Pnueli, L., M. Abuabeid, *et al.* (1991). "The MADS box gene family in tomato — temporal expression during floral development, conserved secondary structures and homology with homeotic genes from Antirrhinum and Arabidopsis." *Plant J* **1**(2): 255–66.

Pnueli, L., D. Hareven, *et al.* (1994a). "Isolation of the tomato *AGAMOUS* gene *TAG1* and analysis of its homeotic role in transgenic plants." *Plant Cell* **6**(2): 163–73.

Pnueli, L., D. Hareven, *et al.* (1994b). "The Tm5 MADS box gene mediates organ differentiation in the 3 inner whorls of tomato flowers." *Plant Cell* **6**(2): 175–86.

Prasanth, K. V., S. G. Prasanth, *et al.* (2005). "Regulating gene expression through RNA nuclear retention." *Cell* **123**(2): 249–63.

Preston, J. C. and E. A. Kellogg (2006). "Reconstructing the evolutionary history of paralogous *APETALA1/FRUITFULL*-like genes in grasses (Poaceae)." *Genetics* **174**(1): 421–37.

Raghavan, V. (2003). "Some reflections on double fertilization, from its discovery to the present." *New Phytol* **159**(3): 565–83.

Randolph, L. F. (1936). "The developmental morphology of the caryopsis in maize." *J Agric Res* **53**: 881–916.

Raveh, D., E. Hunbermann, *et al.* (1973). "*In vitro* culture of tobacco protoplasts: use of feeder techniques to support division of cells plated at low densities." *In Vitro* **9**: 216–22.

Razem, F. A., K. Baron, *et al.* (2006a). "Turning on gibberellin and abscisic acid signaling." *Curr Opin Plant Biol* **9**(5): 454–9.

Razem, F. A., A. El-Kereamy, *et al.* (2006b). "The RNA-binding protein FCA is an abscisic acid receptor." *Nature* **439**(7074): 290–4.

Reddy, A. S. N. (2001). "Molecular motors and their functions in plants." *Int Rev Cytol* **204**: 97–178.

Reddy, G. V., M. G. Heisler, *et al.* (2004a). "Real-time lineage analysis reveals oriented cell divisions associated with morphogenesis at the shoot apex of *Arabidopsis thaliana*." *Development* **131**(17): 4225–37.

Reddy, G. V. and E. M. Meyerowitz (2005). "Stem-cell homeostasis and growth dynamics can be uncoupled in the Arabidopsis shoot apex." *Science* **310**(5748): 663–7.

Reddy, V. S., I. S. Day, *et al.* (2004b). "KIC, a novel Ca^{2+} binding protein with one EF-hand motif, interacts with a microtubule motor protein and regulates trichome morphogenesis." *Plant Cell* **16**(1): 185–200.

Reinhardt, D. (2005). "Phyllotaxis — a new chapter in an old tale about beauty and magic numbers." *Curr Opin Plant Biol* **8**(5): 487–93.

Reinhardt, D., M. Frenz, *et al.* (2003). "Microsurgical and laser ablation analysis of interactions between the zones and layers of the tomato shoot apical meristem." *Development* **130**(17): 4073–83.

Reinhardt, D., M. Frenz, *et al.* (2005). "Microsurgical and laser ablation analysis of leaf positioning and dorsoventral patterning in tomato." *Development* **132**(1): 15–26.

Reiser, L., P. Sanchez-Baracaldo, *et al.* (2000). "Knots in the family tree: evolutionary relationships and functions of *knox* homeobox genes." *Plant Mol Biol* **42**(1): 151–66.

Richards, D. E., K. E. King, *et al.* (2001). "How gibberellin regulates plant growth and development: a molecular genetic analysis of gibberellin signaling." *Annu Rev Plant Physiol Plant Mol Biol* **52**: 67–88.

Rijpkema, A. S., S. Royaert, *et al.* (2006). "Analysis of the Petunia *TM6* MADS box gene reveals functional divergence within the *DEF/AP3* lineage." *Plant Cell* **18**(8): 1819–32.

Rodriguez, F. I., J. J. Esch, *et al.* (1999). "A copper cofactor for the ethylene receptor ETR1 from Arabidopsis." *Science* **283**(5404): 996–8.

Rolland-Lagan, A. G., J. A. Bangham, *et al.* (2003). "Growth dynamics underlying petal shape and asymmetry." *Nature* **422**(6928): 161–3.

Ronemus, M. and R. Martienssen (2005). "RNA interference: methylation mystery." *Nature* **433**(7025): 472–3.

Ruiz-Medrano, R., B. Xoconostle-Cazares, *et al.* (2004). "The plasmodesmatal transport pathway for homoeotic proteins, silencing signals and viruses." *Curr Opin Plant Biol* 7(6): 641–50.

Runions, A., M. Fuhrer, *et al.* (2005). "Modeling and visualization of leaf venation patterns." *ACM Trans Graph* **24**(3): 702–11.

Russinova, E., J. W. Borst, *et al.* (2004). "Heterodimerization and endocytosis of Arabidopsis brassinosteroid receptors BRI1 and AtSERK3 (BAK1)." *Plant Cell* **16**(12): 3216–29.

Ryan, E., M. Steer, *et al.* (2001). "Cell biology and genetics of root hair formation in *Arabidopsis thaliana*." *Protoplasma* **215**(1–4): 140–9.

Ryu, K. H., Y. H. Kang, *et al.* (2005). "The WEREWOLF MYB protein directly regulates *CAPRICE* transcription during cell fate specification in the Arabidopsis root epidermis." *Development* **132**(21): 4765–75.

Sablowski, R. W. M. and E. M. Meyerowitz (1998). "Temperature-sensitive splicing in the floral homeotic mutant *APETALA3-1*." *Plant Cell* **10**(9): 1453–63.

Sachs, T. (1989). "The development of vascular networks during leaf development." *Curr Topics Plant Biochem Physiol* **8**: 168–83.

Sachs, T. (1991a). "Cell polarity and tissue patterning in plants." *Development* Suppl **1**: 83–93.

Sachs, T. (1991b). *Pattern Formation in Plant Tissues.* Cambridge, England, Cambridge University Press.

Sakai, H., L. J. Medrano, *et al.* (1995). "Role of *SUPERMAN* in maintaining Arabidopsis floral whorl boundaries." *Nature* **378**(6553): 199–203.

Samach, A. and G. Coupland (2000). "Time measurement and the control of flowering in plants." *Bioessays* **22**(1): 38–47.

Savidge, B., S. D. Rounsley, *et al.* (1995). "Temporal relationship between the transcription of 2 Arabidopsis MADS box genes and the floral organ identity genes." *Plant Cell* 7(6): 721–33.

Sawa, S., T. Ito, *et al.* (1999a). "*FILAMENTOUS FLOWER* controls the formation and development of Arabidopsis inflorescences and floral meristems." *Plant Cell* **11**(1): 69–86.

Sawa, S., K. Watanabe, *et al.* (1999b). "*FILAMENTOUS FLOWER*, a meristem and organ identity gene of Arabidopsis, encodes a protein with a zinc finger and HMG-related domains." *Genes Dev* **13**(9): 1079–88.

Scanlon, M. J. (2003). "The polar auxin transport inhibitor N-1–naphthylphthalamic acid disrupts leaf initiation, KNOX protein regulation, and formation of leaf margins in maize." *Plant Physiol* **133**(2): 597–605.

Scarpella, E., P. Francis, *et al.* (2004). "Stage-specific markers define early steps of procambium development in Arabidopsis leaves and correlate termination of vein formation with mesophyll differentiation." *Development* **131**(14): 3445–55.

Scarpella, E., D. Marcos, *et al.* (2006). "Control of leaf vascular patterning by polar auxin transport." *Genes Dev* **20**(8): 1015–27.

Scheres, B., H. Wolkenfelt, *et al.* (1994). "Embryonic origin of the Arabidopsis primary root and root meristem initials." *Development* **120**(9): 2475–87.

Schiefelbein, J. (2003). "Cell-fate specification in the epidermis: a common patterning mechanism in the root and shoot." *Curr Opin Plant Biol* **6**(1): 74–8.

Schiefelbein, J. W., J. D. Masucci, *et al.* (1997). "Building a root: the control of patterning and morphogenesis during root development." *Plant Cell* **9**(7): 1089–98.

Schmid, M., T. S. Davidson, *et al.* (2005). "A gene expression map of *Arabidopsis thaliana* development." *Nat Genet* **37**: 501–6.

Schmidt, R. J., B. Veit, *et al.* (1993). "Identification and molecular characterization of *Zag1*, the maize homolog of the Arabidopsis floral homeotic gene *Agamous*." *Plant Cell* **5**(7): 729–37.

Schmitz, G. and K. Theres (1999). "Genetic control of branching in Arabidopsis and tomato." *Curr Opin Plant Biol* **2**(1): 51–5.

Schmitz, G. and K. Theres (2005). "Shoot and inflorescence branching." *Curr Opin Plant Biol* **8**(5): 506–11.

Schneeberger, R., M. Tsiantis, *et al.* (1998). "The *rough sheath2* gene negatively regulates homeobox gene expression during maize leaf development." *Development* **125**(15): 2857–65.

Schoof, H., M. Lenhard, *et al.* (2000). "The stem cell population of Arabidopsis shoot meristems is maintained by a regulatory loop between the *CLAVATA* and *WUSCHEL* genes." *Cell* **100**(6): 635–44.

Schultz, L., L. H. J. Kerckhoffs, *et al.* (2001). "Molecular characterization of the brassinosteroid-deficient *lkb* mutant in pea." *Plant Mol Biol* **47**(4): 491–8.

Schwartz, S. H., X. Q. Qin, *et al.* (2004). "The biochemical characterization of two carotenoid cleavage enzymes from Arabidopsis indicates that a carotenoid-derived compound inhibits lateral branching." *J Biol Chem* **279**(45): 46940–5.

Schwarz-Sommer, Z., P. Huijser, *et al.* (1990). "Genetic control of flower development by homeotic genes in *Antirrhinum majus.*" *Science* **250**(4983): 931–6.

Scofield, S. and J. A. Murray (2006a). "The evolving concept of the meristem." *Plant Mol Biol* **60**(6): V-VII.

Scofield, S. and J. A. H. Murray (2006b). "*KNOX* gene function in plant stem cell niches." *Plant Mol Biol* **60**: 929–46.

Sena, G., J. W. Jung, *et al.* (2004). "A broad competence to respond to *SHORT ROOT* revealed by tissue-specific ectopic expression." *Development* **131**(12): 2817–26.

Sergeeva, L. I., J. J. Keurentjes, *et al.* (2006). "Vacuolar invertase regulates elongation of *Arabidopsis thaliana* roots as revealed by QTL and mutant analysis." *Proc Natl Acad Sci USA* **103**(8): 2994–9.

Shaked, H., K. Kashkush, *et al.* (2001). "Sequence elimination and cytosine methylation are rapid and reproducible responses of the genome to wide hybridization and allopolyploidy in wheat." *Plant Cell* **13**(8): 1749–59.

Shantz, E. M. and F. C. Steward (1955). "The identification of compound A from coconut milk as 1,3–diphenylurea." *J Am Chem Soc* **77**: 6351–3.

Shchennikova, A. V., O. A. Shulga, *et al.* (2004). "Identification and characterization of four chrysanthemum MADS-box genes, belonging to the *APETALA1/FRUITFULL* and *SEPALLATA3* subfamilies." *Plant Physiol* **134**(4): 1632–41.

Sheldon, C. C., J. E. Burn, *et al.* (1999). "The FLF MADS box gene: a repressor of flowering in Arabidopsis regulated by vernalization and methylation." *Plant Cell* **11**(3): 445–58.

Shpak, E. D., C. T. Berthiaume, *et al.* (2004). "Synergistic interaction of three ERECTA-family receptor-like kinases controls Arabidopsis organ growth and flower development by promoting cell proliferation." *Development* **131**(7): 1491–501.

Shpak, E. D., M. B. Lakeman, *et al.* (2003). "Dominant-negative receptor uncovers redundancy in the Arabidopsis ERECTA leucine-rich repeat receptor-like kinase signaling pathway that regulates organ shape." *Plant Cell* **15**(5): 1095–110.

Shpak, E. D., J. M. McAbee, *et al.* (2005). "Stomatal patterning and differentiation by synergistic interaction of receptor kinases." *Science* **309**: 290–3.

Sieburth, L. E. (1999). "Auxin is required for leaf vein pattern in Arabidopsis." *Plant Physiol* **121**: 1179–90.

Sieburth, L. E., G. K. Muday, *et al.* (2006). "*SCARFACE* Encodes an ARF-GAP That is Required for Normal Auxin Efflux and Vein Patterning in Arabidopsis." *Plant Cell* **18**(6): 1396–411.

Siegfried, K. R., Y. Eshed, *et al.* (1999). "Members of the *YABBY* gene family specify abaxial cell fate in Arabidopsis." *Development* **126**(18): 4117–28.

Simon, R. and Y. Stahl (2006). "Plant cells CLEave their way to differentiation." *Science* **313**(5788): 773–4.

Simpson, G. G. and C. Dean (2002). "Flowering — Arabidopsis, the rosetta stone of flowering time?" *Science* **296**(5566): 285–9.

Singh, M. B., H. L. Xu, *et al.* (2002). "Developmental expression of polyubiquitin genes and distribution of ubiquitinated proteins in generative and sperm cells." *Sex Plant Reprod* **14**(6): 325–9.

Sinha, N. (1999). "Leaf development in angiosperms." *Annu Rev Plant Physiol Plant Mol Biol* **50**: 419–46.

Smalle, J. and R. D. Vierstra (2004). "The ubiquitin 26S proteasome proteolytic pathway." *Annu Rev Plant Biol* **55**: 555–90.

Smith, H. B. (1999). "Planned parenthood in arabidopsis: *FLOWERING LOCUS C*." *Plant Cell* **11**(5): 763–4.

Smith, L. G. (2003). "Cytoskeletal control of plant cell shape: getting the fine points." *Curr Opin Plant Biol* **6**(1): 63–73.

Smith, L. G., S. Hake, *et al.* (1996). "The *tangled-1* mutation alters cell division orientations throughout maize leaf development without altering leaf shape." *Development* **122**(2): 481–9.

Smyth, D. R. (2005). "Morphogenesis of flowers — our evolving view." *Plant Cell* **17**: 330–41.

Snow, M. and R. Snow (1931). "Experiments on phyllotaxis." *Philos Trans R Soc Lond Ser B* **221**: 1–43.

Soltis, D. E., P. S. Soltis, *et al.* (2002). "Missing links: the genetic architecture of flower and floral diversification." *Trends Plant Sci* 7(1): 22–31.

Soltis, P. S., D. E. Soltis, *et al.* (1999). "Angiosperm phylogeny inferred from multiple genes as a tool for comparative biology." *Nature* **402**(6760): 402–4.

Sommer, H., J. P. Beltran, *et al.* (1990). "*Deficiens*, a homeotic gene involved in the control of flower morphogenesis in *Antirrhinum majus*: the protein shows homology to transcription factors." *EMBO J* **9**(3): 605–13.

Sonthheimer, E. J. (2005). "Assembly and function of RNA silencing complexes." *Nat Rev Mol Cell Biol* **6**: 127–38.

Sorin, C., J. D. Bussell, *et al.* (2005). "Auxin and light control of adventitious rooting in Arabidopsis require *ARGONAUTE1*." *Plant Cell* **17**: 1343–59.

Sorin, C., L. Negroni, *et al.* (2006). "Proteomic analysis of different mutant genotypes of Arabidopsis led to the identification of 11 proteins correlating with adventitious root development." *Plant Physiol* **140**(1): 349–64.

Southerton, S. G., H. Marshall, *et al.* (1998). "Eucalypt MADS-box genes expressed in developing flowers." *Plant Physiol* **118**(2): 365–72.

Spiro, P. A., J. S. Parkinson, *et al.* (1997). "A model of excitation and adaptation in bacterial chemotaxis." *Proc Natl Acad Sci USA* **94**(14): 7263–8.

Spremulli, L. (2000). Protein synthesis, assembly and degradation. In: *Biochemistry and Molecular Biology of Plants.* Eds. B. B. Buchanan, W. Gruissem and R. L. Jones. Rockville, MD, American Society of Plant Physiologists: pp. 412–54.

Sprunck, S., U. Baumann, *et al.* (2005). "The transcript composition of egg cells changes significantly following fertilization in wheat (*Triticum aestivum* L.)." *Plant J* **41**(5): 660–72.

Sridhar, V. V., A. Surendrarao, *et al.* (2004). "Transcriptional repression of target genes by LEUNIG and SEUSS, two interacting regulatory proteins for Arabidopsis flower development." *Proc Natl Acad Sci U S A* **101**(31): 11494–9.

Sridhar, V. V., A. Surendrarao, *et al.* (2006). "*APETALA1* and *SEPALLATA3* interact with *SEUSS* to mediate transcription repression during flower development." *Development* **133**(16): 3159–66.

Staiger, C. J., B. C. Gibbon, *et al.* (1997). "Profilin and actin-depolymerizing factor: modulators of actin organization in plants." *Trends Plant Sci* **2**(7): 275–81.

Steeves, T. A. and I. M. Sussex (1989). *Patterns in Plant Development.* New York, Cambridge University Press.

Stellari, G. M., M. A. Jaramillo, *et al.* (2004). "Evolution of the *APETALA3* and *PISTILLATA* lineages of MADS-box–containing genes in the basal angiosperms." *Mol Biol Evol* **21**(3): 506–19.

Stepanova, A. N. and J. M. Alonso (2005). "Arabidopsis ethylene signaling pathway." *Sci STKE* **2005**(276): cm4.

Stepanova, A. N., J. M. Hoyt, *et al.* (2005). "A link between ethylene and auxin uncovered by the characterization of two root-specific ethylene-insensitive mutants in Arabidopsis." *Plant Cell* **17**(8): 2230–42.

Stevence, P. F. (2005). "Angiosperm phylogeny website." http://www.mobot.org/MOBOT/research/APweb **Version 6**.

Stieger, P. A., D. Reinhardt, *et al.* (2002). "The auxin influx carrier is essential for correct leaf positioning." *Plant J* **32**(4): 509–17.

Stirnberg, P., K. van de Sande, *et al.* (2002). "*MAX1* and *MAX2* control shoot lateral branching in Arabidopsis." *Development* **129**(5): 1131–41.

Struhl, K. (1999). "Fundamentally different logic of gene regulation in eukaryotes and prokaryotes." *Cell* **98**(1): 1–4.

Sugiura, M. and Y. Takeda (2000). Nucleic acids. In: *Biochemistry and Molecular Biology of Plants.* Eds. B. B. Buchanan, W. Gruissem and R. L. Jones. Rockville, MD, American Society of Plant Physiologists: pp. 260–310.

Sundstrom, J. F., N. Nakayama, *et al.* (2006). "Direct regulation of the floral homeotic *APETALA1* gene by APETALA3 and PISTILLATA in Arabidopsis." *Plant J* **46**(4): 593–600.

Sung, S. and R. M. Amasino (2005). "Remembering winter: toward a molecular understanding of vernalization." *Annu Rev Plant Biol* **56**: 491–508.

Sung, S. K., G. H. Yu, *et al.* (1999). "Characterization of *MdMADS2*, a member of the *SQUAMOSA* subfamily of genes, in apple." *Plant Physiol* **120**(4): 969–78.

Sussex, I. M. (1989). "Developmental programming of the shoot meristem." *Cell* **56**(2): 225–9.

Sussex, I. M. and N. M. Kerk (2001). "The evolution of plant architecture." *Curr Opin Plant Biol* **4**(1): 33–7.

Swain, S. M. and D. P. Singh (2005). "Tall tales from sly dwarves: novel functions of gibberellins in plant development." *Trends Plant Sci* **10**: 123–9.

Swarup, R., E. M. Kramer, *et al.* (2005). "Root gravitropism requires lateral root cap and epidermal cells for transport and response to a mobile auxin signal." *Nat Cell Biol* 7(11): 1057–65.

Swiatek, A., M. Lenjou, *et al.* (2002). "Differential effect of jasmonic acid and abscisic acid on cell cycle progression in tobacco BY-2 cells." *Plant Physiol* **128**(1): 201–11.

Swoboda, I., P. L. Bhalla, *et al.* (2001). "Identification of *pronp1*, a tobacco profilin gene activated in tip-growing cells." *Plant Mol Biol* **46**(5): 531–8.

Symons, G. M. and J. B. Reid (2004). "Brassinosteroids do not undergo long-distance transport in pea. Implications for the regulation of endogenous brassinosteroid levels." *Plant Physiol* **135**(4): 2196–206.

Szymanski, D. B. (2005). "Breaking the WAVE complex: the point of Arabidopsis trichomes." *Curr Opin Plant Biol* **8**: 103–12.

Szymanski, D. B., A. M. Lloyd, *et al.* (2000). "Progress in the molecular genetic analysis of trichome initiation and morphogenesis in Arabidopsis." *Trends Plant Sci* **5**(5): 214–9.

Szymanski, D. B., M. D. Marks, *et al.* (1999). "Organized F-actin is essential for normal trichome morphogenesis in Arabidopsis." *Plant Cell* **11**(12): 2331–47.

Takeda, S., Z. Tadele, *et al.* (2004). "*BRU1*, a novel link between responses to DNA damage and epigenetic gene silencing in Arabidopsis." *Genes Dev* **18**(7): 782–93.

Takeda, T., Y. Suwa, *et al.* (2003). "The *OsTB1* gene negatively regulates lateral branching in rice." *Plant J* **33**(3): 513–20.

Talbert, P. B., H. T. Adler, *et al.* (1995). "The *REVOLUTA* gene is necessary for apical meristem development and for limiting cell divisions in the leaves and stems of *Arabidopsis thaliana*." *Development* **121**(9): 2723–35.

Tamaru, H. and E. U. Selker (2001). "A histone H3 methyltransferase controls DNA methylation in *Neurospora crassa*." *Nature* **414**(6861): 277–83.

Tanaka, K., T. Asami, *et al.* (2005). "Brassinosteroid homoeostasis in Arabidopsis is ensured by feedback expressions of multiple genes involved in its metabolism." *Plant Physiol* **138**: 1117–25.

Tanurdzic, M. and J. A. Banks (2004). "Sex-determining mechanisms in land plants." *Plant Cell* **16**: S61–S71.

Tax, F. E. and A. Durbak (2006). "Meristems in the movies: live imaging as a tool for decoding intercellular signaling in shoot apical meristems." *Plant Cell* **18**(6): 1331–7.

Theissen, G. (2001). "Development of floral organ identity: stories from the MADS house." *Curr Opin Plant Biol* **4**(1): 75–85.

Theissen, G., A. Becker, *et al.* (2000). "A short history of MADS-box genes in plants." *Plant Mol Biol* **42**(1): 115–49.

Theissen, G. and H. Saedler (1999). "The golden decade of molecular floral development (1990–1999): a cheerful obituary." *Dev Genet* **25**: 181–93.

Theissen, G. and H. Saedler (2001). "Floral quartets." *Nature* **409**(6819): 469–71.

Thomson, J. M., E. A. Gaucher, *et al.* (2005). "Resurrecting ancestral alcohol dehydrogenases from yeast." *Nat Genet* **37**(6): 630–5.

Tichtinsky, G., V. Vanoosthuyse, *et al.* (2003). "Making inroads into plant receptor kinase signalling pathways." *Trends Plant Sci* **8**(5): 231–7.

Timmermans, M. C. P., A. Hudson, *et al.* (1999). "ROUGH SHEATH2: a Myb protein that represses *knox* homeobox genes in maize lateral organ primordia." *Science* **284**(5411): 151–3.

To, J. P. C., G. Haberer, *et al.* (2004). "Type-A Arabidopsis response regulators are partially redundant negative regulators of cytokinin signaling." *Plant Cell* **16**(3): 658–71.

Tomari, Y. and P. D. Zamore (2005). "Perspective: machines for RNAi. " *Genes Dev* **19**: 517–29.

Torrey, J. G. and E. Galun (1970). "Apolar embryos of Fucus resulting from osmotic and chemical treatments." *Am J Bot* **57**: 111–9.

Traas, J. and I. Bohn-Courseau (2005). "Cell proliferation patterns at the shoot apical meristem." *Curr Opin Plant Biol* **8**(6): 587–92.

Trebitsh, T., J. E. Staub, *et al.* (1997). "Identification of a 1-aminocyclopropane-1-carboxylic acid synthase gene linked to the *female* (*F*) locus that enhances female sex expression in cucumber." *Plant Physiol* **113**(3): 987–95.

Tsai, W. C., C. S. Kuoh, *et al.* (2004). "Four DEF-Like MADS box genes displayed distinct floral morphogenetic roles in Phalaenopsis orchid." *Plant Cell Physiol* **45**(7): 831–44.

Tsiantis, M., R. Schneeberger, *et al.* (1999). "The maize *rough sheath2* gene and leaf development programs in monocot and dicot plants." *Science* **284**(5411): 154–6.

Tucker, E. B. (1982). "Translocation in the Staminal Hairs of *Setcreasea purpurea*. 1. A study of cell ultrastructure and cell-to-cell passage of molecular probes." *Protoplasma* **113**(3): 193–201.

Turing, A. (1952). "The chemical basis of differentiation." *Philos Trans R Soc Lond Ser B* **237**: 37–72.

Turner, J. G., C. Ellis, *et al.* (2002). "The jasmonate signal pathway." *Plant Cell* **14**: S153–64.

Tzeng, T. Y., H. Y. Chen, *et al.* (2002). "Ectopic expression of carpel-specific MADS box genes from lily and lisianthus causes similar homeotic conversion of sepal and petal in Arabidopsis." *Plant Physiol* **130**(4): 1827–36.

Tzeng, T. Y., C. C. Hsiao, *et al.* (2003). "Two lily *SEPALLATA*-like genes cause different effects on floral formation and floral transition in Arabidopsis." *Plant Physiol* **133**(3): 1091–101.

Ueda, M., Y. Koshino-Kimura, *et al.* (2005). "Stepwise understanding of root development." *Curr Opin Plant Biol* **8**: 71–6.

Vale, R. D. (2003). "The molecular motor toolbox for intracellular transport." *Cell* **112**(4): 467–80.

van den Berg, C., V. Willemsen, *et al.* (1995). "Cell fate in the Arabidopsis root meristem determined by directional signaling." *Nature* **378**(6552): 62–5.

van den Berg, C., V. Willemsen, *et al.* (1997). "Short-range control of cell differentiation in the Arabidopsis root meristem." *Nature* **390**(6657): 287–9.

Vandenbussche, M., G. Theissen, *et al.* (2003a). "Structural diversification and neo-functionalization during floral MADS-box gene evolution by C-terminal frameshift mutations." *Nucleic Acids Res* **31**(15): 4401–9.

Vandenbussche, M., J. Zethof, *et al.* (2003b). "Toward the analysis of the petunia MADS box gene family by reverse and forward transposon insertion mutagenesis approaches: B, C, and D floral organ identity functions require *SEPALLATA*-like MADS box genes in petunia." *Plant Cell* **15**(11): 2680–93.

Vandenbussche, M., J. Zethof, *et al.* (2004). "The duplicated B-class heterodimer model: whorl-specific effects and complex genetic interactions in *Petunia hybrida* flower development." *Plant Cell* **16**(3): 741–54.

Vardi, A., P. Spiegel-Roy, *et al.* (1982). "Plant regeneration from Citrus protoplasts: variability in methodological requirement among cultivars and species." *Theor App Genet* **62**: 171–6.

Veit, B., R. J. Schmidt, *et al.* (1993). "Maize floral development: new genes and old mutants." *Plant Cell* **5**(10): 1205–15.

Verbsky, M. L. and E. J. Richards (2001). "Chromatin remodeling in plants." *Curr Opin Plant Biol* **4**(6): 494–500.

Vieten, A., S. Vanneste, *et al.* (2005). "Functional redundancy of PIN proteins is accompanied by auxin-dependent cross-regulation of PIN expression." *Development* **132**(20): 4521–31.

Visa, N. (2005). "Actin in transcription." *EMBO Rep* **6**: 218–9.

Vision, T. J., D. G. Brown, *et al.* (2000). "The origins of genomic duplications in Arabidopsis." *Science* **290**(5499): 2114–7.

Voinnet, O. (2005). "Induction and suppression of RNA silencing: insights from viral infections." *Nat Rev Genet* **6**: 206–20.

Vollbrecht, E., P. S. Springer, *et al.* (2005). "Architecture of floral branch systems in maize and related grasses." *Nature* **436**(7054): 1119–26.

Vollbrecht, E., B. Veit, *et al.* (1991). "The developmental gene *Knotted-1* is a member of a maize homeo box gene family." *Nature* **350**: 241–3.

Wada, T., T. Kurata, *et al.* (2002). "Role of a positive regulator of root hair development, *CAPRICE*, in Arabidopsis root epidermal cell differentiation." *Development* **129**(23): 5409–19.

Waites, R. and A. Hudson (1995). "*Phantastica* — a gene required for dorsoventrality of leaves in *Antirrhinum majus*." *Development* **121**(7): 2143–54.

Waites, R., H. R. N. Selvadurai, *et al.* (1998). "*The PHANTASTICA* gene encodes a MYB transcription factor involved in growth and dorsoventrality of lateral organs in Antirrhinum." *Cell* **93**(5): 779–89.

Wang, J. W., L. J. Wang, *et al.* (2005c). "Control of root cap formation by microRNA-targeted auxin response factors in Arabidopsis." *Plant Cell* **17**(8): 2204–16.

Wang, X., M. B. Goshe, *et al.* (2005a). "Identification and functional analysis of *in vivo* phosphorylation sites of the Arabidopsis BRASSINOSTEROID-INSENSITIVE1 receptor kinase." *Plant Cell* **17**: 1685–703.

Wang, X. L., X. Q. Li, *et al.* (2005b). "Autoregulation and homodimerization are involved in the activation of the plant steroid receptor BRI1." *Dev Cell* **8**(6): 855–65.

Wardlaw, C. W. (1957). "On the organization and reactivity of the shoot apex in vascular plants." *Am J Bot* **44**: 176–85.

Wardlaw, C. W. (1965). *Organization and Evolution in Plants.* London, England, Longmans Green.

Wasteneys, G. O. and M. E. Galway (2003). "Remodelling the cytoskeleton for growth and form: an overview with some new views." *Annu Rev Plant Biol* **54**: 691–722.

Weigel, D. and G. Jurgens (2002). "Stem cells that make stems." *Nature* **415**(6873): 751–4.

Weinscoat, J. (1987). "Out of the garden of Eden." *Nature* **325**: 13.

Weir, I., J. Lu, *et al.* (2004). "*CUPULIFORMIS* establishes lateral organ boundaries in Antirrhinum." *Development* **131**(4): 915–22.

Wessler, S. R. and J. C. Carrington (2005). "The consequences of gene and genome duplication in plants." *Curr Opin Plant Biol* **8**(2): 119–21.

Westergaard, M. (1958). "The mechanism of sex determination in dioecious flowering plants." *Advances in Genet* **9**: 217–81.

Whipple, C. J., P. Ciceri, *et al.* (2004). "Conservation of B-class floral homeotic gene function between maize and Arabidopsis." *Development* **131**(24): 6083–91.

Whipple, C. J. and R. J. Schmidt (2006). "Genetics of grass flower development." *Adv Bot Res* **44**: 385–425.

Wigge, P. A., M. C. Kim, *et al.* (2005). "Integration of spatial and temporal information during floral induction in Arabidopsis." *Science* **309**(5737): 1056–9.

Williams, L. and J. C. Fletcher (2005). "Stem cell regulation in the Arabidopsis shoot apical meristem." *Curr Opin Plant Biol* **8**(6): 582–6.

Williams, R. W., J. M. Wilson, *et al.* (1997). "A possible role for kinase-associated protein phosphatase in the Arabidopsis CLAVATA1 signaling pathway." *Proc Natl Acad Sci USA* **94**(19): 10467–72.

Winter, K. U., C. Weiser, *et al.* (2002). "Evolution of class B floral homeotic proteins: obligate heterodimerization originated from homodimerization." *Mol Biol Evol* **19**(5): 587–96.

Wolfe, K. H., M. Guoy, *et al.* (1989). "Date of the monocot–dicot divergence from chloroplast DNA sequence data" *Proc Nat Acad Sci USA* **86**: 6201–6.

Woll, K., L. A. Borsuk, *et al.* (2005). "Isolation, characterization, and pericycle-specific transcriptome analyses of the novel maize lateral and seminal root initiation mutant *rum1*." *Plant Physiol* **139**(3): 1255–67.

Wolpert, L. (1996). "One hundred years of positional information." *Trends Genet* **12**(9): 359–64.

Wolpert, L. (1998). *Principles of Development.* Oxford, England, Oxford University Press.

Woodward, A. W. and B. Bartel (2005). "Auxin: regulation, action, and interaction." *Ann Bot* **95**(5): 707–35.

Woolfit, M. and K. Wolfe (2005). "The gene duplication that greased society's wheels." *Nat Genet* **37**(6): 566–7.

Wu, X. L., D. Weigel, *et al.* (2002). "Signaling in plants by intercellular RNA and protein movement." *Genes Dev* **16**(2): 151–8.

Xie, Q., G. Frugis, *et al.* (2000). "Arabidopsis NAC1 transduces auxin signal downstream of TIR1 to promote lateral root development." *Genes Dev* **14**(23): 3024–36.

Xie, Q., H. S. Guo, *et al.* (2002). "SINAT5 promotes ubiquitin-related degradation of NAC1 to attenuate auxin signals." *Nature* **419**(6903): 167–70.

Xu, J., H. Hofhuis, *et al.* (2006a). "A molecular framework for plant regeneration." *Science* **311**(5759): 385–8.

Xu, Y., L. L. Teo, *et al.* (2006b). "Floral organ identity genes in the orchid *Dendrobium crumenatum*." *Plant J* **46**: 54–68.

Yamaguchi, T., D. Y. Lee, *et al.* (2006). "Functional diversification of the two C-class MADS box genes *OSMADS3* and *OSMADS58* in *Oryza sativa*." *Plant Cell* **18**(1): 15–28.

Yan, L. L., A. Loukoianov, *et al.* (2004). "The wheat *VRN2* gene is a flowering repressor down-regulated by vernalization." *Science* **303**(5664): 1640–4.

Yanai, O., E. Shani, *et al.* (2005). "Arabidopsis KNOXI proteins activate cytokinin biosynthesis." *Curr Biol* **15**(17): 1566–71.

Yang, Y. W., K. N. Lai, *et al.* (1999a). "Rates of nucleotide substitution in angiosperm mitochondrial DNA sequences and dates of divergence between Brassica and other angiosperm lineages." *J Mol Evol* **48**(5): 597–604.

Yang, Y. W., K. N. Lai, *et al.* (1999b). "Molecular phylogenetic studies of Brassica, rorippa, arabidopsis and allied genera based on the internal transcribed spacer region of 18S–25S rDNA." *Mol Phylogenet Evol* **13**(3): 455–62.

Yanofsky, M. F., H. Ma, *et al.* (1990). "The protein encoded by the Arabidopsis homeotic gene *Agamous* resembles transcription factors." *Nature* **346**(6279): 35–9.

Yanovsky, M. J. and S. A. Kay (2001). "Signaling networks in the plant circadian system." *Curr Opin Plant Biol* **4**(5): 429–35.

Yanovsky, M. J. and S. A. Kay (2003). "Living by the calendar: how plants know when to flower." *Nat Rev Mol Cell Biol* **4**(4): 265–75.

Ye, Z. H. (2002). "Vascular tissue differentiation and pattern formation in plants." *Annu Rev Plant Biol* **53**: 183–202.

Yi, Y. and T. Jack (1998). "An intragenic suppressor of the Arabidopsis floral organ identity mutant *APETALA3-1* functions by suppressing defects in splicing." *Plant Cell* **10**(9): 1465–77.

Yin, Y., D. Vafeados, *et al.* (2005). "A new class of transcription factors mediates brassinosteroid-regulated gene expression in Arabidopsis." *Cell* **120**: 249–59.

Yin, Y. H., Z. Y. Wang, *et al.* (2002). "BES1 accumulates in the nucleus in response to brassinosteroids to regulate gene expression and promote stem elongation." *Cell* **109**(2): 181–91.

Yu, B., Z. Yang, *et al.* (2005). "Methylation as a crucial step in plant microRNA biogenesis." *Science* **307**: 932–5.

Yu, H. and C. J. Goh (2000). "Identification and characterization of three orchid MADS-box genes of the AP1/AGL9 subfamily during floral transition." *Plant Physiology* **123**(4): 1325–36.

Yu, H. and C. J. Goh (2001). "Molecular genetics of reproductive biology in orchids." *Plant Physiol* **127**(4): 1390–3.

Zahn, L. M., H. Kong, *et al.* (2005b). "The evolution of the *SEPALLATA* subfamily of MADS-box genes: a preangiosperm origin with multiple duplications throughout angiosperm history." *Genetics* **169**: 2209–23.

Zahn, L. M., J. Leebens-Mack, *et al.* (2005a). "To B or not to B a flower: the role of *DEFICIENS* and *GLOBOSA* orthologs in the evolution of the angiosperms." *J Hered* **96**(3): 225–40.

Zahn, L. M., J. H. Leebens-Mack, *et al.* (2006). "Conservation and divergence in the *AGAMOUS* subfamily of MADS-box genes: evidence of independent sub- and neofunctionalization events." *Evol Dev* **8**(1): 30–45.

Zambryski, P. and K. Crawford (2000). "Plasmodesmata: gatekeepers for cell-to-cell transport of developmental signals in plants." *Annu Rev Cell Develop Biol* **16**: 393–421.

Zeevaart, J. A. D. (1976). "Physiology of flower formation." *Annu Rev Plant Physiol* **27**: 393–421.

Zhang, B. H., X. P. Pan, *et al.* (2006). "Plant microRNA: a small regulatory molecule with big impact." *Dev Biol* **289**(1): 3–16.

Zhang, L. Q., T. J. Vision, *et al.* (2002). "Patterns of nucleotide substitution among simultaneously duplicated gene pairs in *Arabidopsis thaliana*." *Mol Biol Evol* **19**(9): 1464–73.

Zigmond, S. H. (2004). "Formin-induced nucleation of actin filaments." *Current Opinion in Cell Biology* **16**(1): 99–105.

Zik, M. and V. F. Irish (2003a). "Flower development: initiation, differentiation, and diversification." *Annu Rev Cell Dev Biol* **19**: 119–40.

Zik, M. and V. F. Irish (2003b). "Global identification of target genes regulated by *APETALA3* and *PISTILLATA* floral homeotic gene action." *Plant Cell* **15**(1): 207–22.

Corpus Acknowledgments

The publisher and the author of this book are grateful for the permission given by the following publishers to reproduce their material in this book.

S/No.	Publisher	Figures and Tables
1.	American Association for the Advancement of Science	Figure 69
2.	American Society of Plant Biologists	Figures 6, 9, 12, 17, 21, 22, 25, 33, 35, 41, 46, 51, 53, 58, 59, 60, 61, 65, 71 Table 8
3.	Blackwell Publishing Company	Figure 31 Table 10
4.	Cold Spring Harbor Lab Press	Figure 76
5.	Company of Biologists	Figures 26, 30, 37, 38, 47, 48, 73 Table 3
6.	Elsevier Ltd.	Figures 7, 8, 16, 20, 23, 28, 29, 30, 32, 34, 36, 39, 40, 43, 44, 49, 53, 57, 62, 64, 66, 70, 97 Tables 1, 5, 7, 9
7.	John Wiley & Sons Inc.	Figure 68
8.	Macmillan Publishers Ltd.	Figures 42, 71 Table 2
9.	Oxford University Press	Figures 1, 10, 11, 13, 14, 15, 45, 52, 55, 63, 72, 75

(*Continued*)

(*Continued*)

S/No.	Publisher	Figures and Tables
10.	Proceedings of the National Academy of Science, USA	Figure 24
11.	Springer Verlag	Figures 2, 3, 4, 5, 27, 56, 74 Table 6
12.	Systematic Botany, US National Arboretrum	Figure 19

Note: A few figures of this book contain components from more than one publication; therefore, credit for one figure is given to more than one publisher (e.g. Figure 30).

Index